Grundlehren der mathematischen Wissenschaften 312

A Series of Comprehensive Studies in Mathematics

Springer-Verlag Berlin Heidelberg GmbH

Kai Lai Chung
Zhongxin Zhao

From Brownian Motion to Schrödinger's Equation

With 7 Figures

Springer

Kai Lai Chung
Department of Mathematics
Stanford University
903 Lathrop Drive
Stanford, CA 94305, USA

1st edition 1995
Corrected 2nd Printing 2001

Library of Congress Cataloging-in-Publication Data
Chung, Kai Lai, 1917 –.
From Brownian motion to Schrödinger's Equation / Kai Lai Chung, Zhongxin Zhao.
p. cm. – (Grundlehren der mathematischen Wissenschaften; 312)
Includes bibliographical references and index.
ISBN 978-3-642-63381-2 ISBN 978-3-642-57856-4 (eBook)
DOI 10.1007/978-3-642-57856-4
1. Schrödinger equation. 2. Brownian motion processes. I. Zhao, Zhongxin, 1942 – . II. Title. III. Series. QC174.17.S3C48 1995 530.1'33–dc20 95-1414 CIP

Mathematics Subject Classification (1991): 60J65, 81Q15

ISSN 0072-7830
ISBN 978-3-642-63381-2

http://www.springer.de

Originally published by Springer-Verlag Berlin Heidelberg New York in 1995
Softcover reprint of the hardcover 1st edition 1995

Cover design: MetaDesign plus GmbH, Berlin
Typesetting: TEX output supplied by S. Wilson using a Springer TEX macro-package
Printed on acid-free paper SPIN: 10841678 41/3142/db-5 4 3 2 1 0

And time, which takes survey of all the world,
Must have a stop.

Preface

The title of this book suggests a continuation of my Lectures From Markov Processes to Brownian Motion (Grundlehren 249, 1982), but it is a new departure with a new bent. While the Lectures aspire to a wide range, here the focus is on a few central themes and their accompaniments and developments. It begins with the construction of the Brownian motion process $\{X_t\}$ in $\mathbb{R}^d$ $(d \geq 1)$ as a strong Markov process with continuous paths. This is enriched with a multiplicative functional $\exp\left[\int_0^t q(X_s)\mathrm{d}s\right]$; the result will be called a Feynman–Kac process. The random structure is deepened when a domain D is fixed and the paths are stopped at its exit, yielding the stopping time τ_D and the stopping place $X(\tau_D)$. A theory of q-harmonicity then emerges as an extension of the classic harmonic theory, in which probability has played a belated but spectacular role. A continuous weak-sense solution of Schrödinger's equation $(\frac{\Delta}{2} + q)\varphi = 0$ will be called a q-harmonic function. With this coinage we are *in procinto* to take on the great problems of quantum potential theory: the representation of a q-harmonic function by its boundary values, the unique solvability of the Dirichlet boundary value problem, Poisson's equation, Harnack's inequality, ..., and all the lush Green landscape now tinted with the ubiquitous q. The following reader's guide should be used in conjunction with the section headings in the Contents.

Chapters 1 and 2 contain a substantial review of the classic theory $(q \equiv 0)$ adapted to later recourse. Chapter 3 introduces the class J of q that is particularly amenable to the probabilistic treatment, in both its old and new forms, followed by a study of the Feynman–Kac semigroup. Chapters 4 and 5 constitute the core methodology in which the *gauge* reigns. This is the function of x defined by $E^x\left(\tau_D < \infty; \exp\left[\int_0^{\tau_D} q(X_s)\mathrm{d}s\right]\right)$, and is the seemingly innocuous, purely probabilistic object that confronted me at the outset. "When is it finite" was my first question, some sixteen years ago. Several answers will be given in this book, but the chief result is a dichotomy: either the gauge is infinite everywhere in D, or it is bounded there, indeed in its closure. This *gauge theorem* holds true when D has finite Lebesgue measure, and is given three proofs: first for the special case of a bounded q, then for any q in J via detailed computations with a small ball, and finally in a general framework requiring no balls and allowing the paths to be discontinuous. For any domain D and any q in J, the pair (D, q) will be called *gaugeable* when the gauge is bounded in D. This is the main assumption under which all the problems mentioned above will be solved by explicit expressions closely related to the gauge itself. The next step is motivated by Doob's theory of conditional process. This is expounded in Chapter 5 and leads to the conditional gauge theorem in Chapter 7, which is an extension of the gauge theorem for a bounded Lipschitz domain. The specialization of the domain is needed to reduce one kind of conditioning to the intuitive conditioning with respect to the variable $X(\tau_D)$. Much labor is spent in the calculations, in particular for an inequality connecting five or six Green functions. One cannot help wondering if there is not a more perceptible approach to these matters? The conditional gauge has various applications some of which are given in Chapter 8. In that chapter we also study the variation of the gaugeability of (D, q) with D or with q, either directly or by way of the principal eigenvalue of the associated Schrödinger spectrum. There is a discussion at the end of Section 8.4 including an open problem which shows the delicate balance between gaugeability and boundary regularity of a domain. Chapter 9 contains my earliest results in the area, refurbished with a few vignettes from ordinary differential equations, including a probabilistic treatment of the initial value problem.

The actual preparation of the manuscript for the book began in 1986. After a broad outline was laid down, successive chapters were drafted, then revised and revised again, some parts undergoing no less than the legendary seven transmutations. Traces of this have been left in some places, as they do no harm and may even aid the novice. Personally, I have often taken as much pleasure (and time) in polishing up other folks' results as my own. Zhongxin Zhao came to Stanford University from Beijing to study with me in 1982. He wrote the first drafts of Chapters 6 and 7 which incorporate his work on the conditional gauge. He was also charged with the drafting of various other sections, specially the prerequisites from operator and spectrum theory, and Lipschitzian geometry. In the course of continual revisionism we found not a few new wrinkles as well as old stumbles, some of which are mentioned in the Notes at the end of Chapters. These notes are written entirely by myself, and any opinions expressed or sentiments betrayed therein are mine alone. Here and there I have inserted some true anecdotes, "for my own diversion" as Chinese writers like to say, but also to remind the young reader that mathematicians are human. The References were compiled by Zhao and are largely confined to items cited in the body of the book.

We are grateful to several persons who read parts of the manuscript: Neil Falkner for Chapter 3, Bruce Erickson for Chapter 9, Eugene Fabes for Section 6.2, Wolfgang Stummer for an early draft of initial sections. I am indebted to Heinz Bauer for stimulating comments, to Barry Simon for interesting correspondence, to Pat Fitzsimmons for a little bibliographic verification, and to Rick Durrett for inviting me to Cornell this July and listening to ten hours of lectures on the central portion of the book, during which I made more amendments in the galley. Julie Riddleberger typed the entire manuscript and made all the revisions expertly and tirelessly. Mrs Priscilla Feigen helped with the dispatch of an uncountable number of faxes during the proof-reading period. Toward the end, I made a compact with Zhao to catch each other's misses and to donate a sum to environmental causes for each unbalanced count. Some wildlife stands to gain no matter who loses. All will end well.

Stanford, California, U.S.A.
August 21, 1994

Kai Lai Chung

Table of Contents

1. Preparatory Material

In this chapter we collect together a number of introductory topics for later use. Our aim is to give a self-contained efficient presentation rather than a full discussion of the material. A general treatise such as Chung (1982a) should be consulted for a more deliberate treatment.

1.1 Basic Notation

Let $\mathbb{R}$ denote the set of real numbers. For each natural number $d \geq 1$, let $\mathbb{R}^d$ denote the d-dimensional Euclidian space $[x = (x_1, \cdots, x_d) : x_i \in \mathbb{R}, 1 \leq i \leq d]$. We write

$$|x| = \left[\sum_{i=1}^{d} x_i^2\right]^{1/2},$$

so that the distance between x and y is $|x - y|$. The distance between two sets A and B will be denoted by $\rho(A, B)$; when A is the point x, this becomes $\rho(x, B)$.

$\overline{\mathbb{R}^d} = \mathbb{R}^d \cup \{\infty\}$ is the one-point compactification of $\mathbb{R}^d$; in particular, $\overline{\mathbb{R}^1}$ is the set of extended real numbers. We denote $\mathbb{R}_+ = [0, \infty)$ and $\overline{\mathbb{R}_+} = [0, \infty]$. $\mathcal{B}^d$ is the Borel tribe (Borel σ-field) of $\mathbb{R}^d$ and $\overline{\mathcal{B}^d}$ is that of $\overline{\mathbb{R}^d}$. Sometimes we shall write simply $\mathcal{B}$ for $\mathcal{B}^d$.

The Lebesgue measure in $\mathbb{R}^d$ will be denoted by m, where the dimension d is understood from the context and $m(dx)$ is often abbreviated to dx. The associated $(d - 1)$-dimensional Lebesgue measure will be denoted by σ. For $x \in \mathbb{R}^d$, $r > 0$, we denote the open ball with center x and radius r by

$$B(x, r) = \{y \in \mathbb{R}^d : |y - x| < r\},$$

and its bounding sphere by

$$S(x, r) = \{y \in \mathbb{R}^d : |y - x| = r\}.$$

For $d \geq 1$ we have:

$$v_d(r) \equiv m(B(x,r)) = \frac{\pi^{d/2}}{\Gamma(d/2+1)} r^d;$$

$$\sigma_{d-1}(r) \equiv \sigma(S(x,r)) = \frac{2\pi^{d/2}}{\Gamma(d/2)} r^{d-1}.$$

Note that for $d = 1$, σ is the counting measure, i.e. $\sigma(\{x\}) = 1$ for each $x \in \mathbb{R}^1$. When d is understood we shall write $v(r)$ for $v_d(r)$ and $\sigma(r)$ for $\sigma_{d-1}(r)$.

For any subset A of $\mathbb{R}^d$, we denote its *complement* by $A^c = \mathbb{R}^d - A$, its *closure* by $\overline{A}$, and its *boundary* by $\partial A = \overline{A} \cap \overline{A^c}$. These three derived sets are all subsets of $\mathbb{R}^d$, not of $\overline{\mathbb{R}^d}$. This crucial decision has been made to avoid the pitfall of treating '∞' as though it were an ordinary point. Thus, when A is unbounded, although ∞ is in its closure with respect to $\overline{\mathbb{R}^d}$, it is not in $\overline{A}$ or ∂A according to our definition.

We shall use the following notation, where $A \in \mathcal{B}^d$.

$\mathcal{B}(A)$:= the class of extended-real-valued Borel measurable functions on A.

$C(A)$:= the class of real-valued continuous functions on A.

$L^p(A)$ $(1 \le p < \infty)$:= the class of f in $\mathcal{B}(A)$ such that

$$\|f\|_p^p \equiv \int_A |f(x)|^p \mathrm{d}x < \infty.$$

$L^\infty(A)$:= the class of f in $\mathcal{B}(A)$ such that

$$\|f\|_\infty = \operatorname{ess\,sup}_{x \in A} |f(x)| = \inf\{M : m(x \in A : |f(x)| > M) = 0\} < \infty.$$

Each $L^p(A)$, $1 \le p \le \infty$, is a Banach space with the norm $\| \ \|_p$. We shall frequently omit the subscript '∞' in $\| \ \|_\infty$ when the context is clear.

For an open set A, we also set:

$$L^1_{\text{loc}}(A) \quad := \quad \text{the class of } f \text{ in } \mathcal{B}(A) \text{ such that } f \in L^1(C) \text{ for every compact subset } C \text{ of } A.$$

Such a function is said to be *locally integrable* in A.

Observe that we require Borel (not Lebesgue) measurability in L^p, $1 \le p \le \infty$. This is mandated by the use of probability methods in the book; the newcomer to probability theory should ask himself why. We use the subscripts '+,' 'b,' and 'c' to denote the subclasses of 'positive', 'bounded' and 'compact-supported' functions, respectively; the last only in an open set A. In this book, 'positive' means '≥ 0' and 'strictly positive' means '> 0'; 'increasing' for a function f means 'if $x < y$, then $f(x) \le f(y)$'; similarly for 'negative' and 'decreasing'. If $x_n \in \mathbb{R}$, $x_n \uparrow x$ ($x_n \uparrow\uparrow x$) means $x_n \le x_{n+1}$ ($x_n < x_{n+1}$) for all n and $\lim_n x_n = x$. If A and B are sets in $\mathbb{R}^d$, $A \subset\subset B$ means $\overline{A} \subset B$; $A_n \uparrow B$ ($A_n \uparrow\uparrow B$) means $A_n \subset A_{n+1}$ ($A_n \subset\subset A_{n+1}$) for all n and $\cup_n A_n = B$; similarly for $\downarrow$ and $\downarrow\downarrow$.

For any set A, its indicator is the function defined by

$$1_A(x) = \begin{cases} 1 & \text{if } x \in A; \\ 0 & \text{if } x \notin A. \end{cases}$$

For $a \in \mathbb{R}$, $b \in \mathbb{R}$, we write $a \vee b = \max(a,b)$ and $a \wedge b = \min(a,b)$.

1.2 Markov Process and Strong Markov Property

We begin by defining a Markov process on $\mathbb{R}^d$ as state space. Let $\Omega = M(\mathbb{R}_+, \mathbb{R}^d)$ be the collection of all mappings from $\mathbb{R}_+$ to $\mathbb{R}^d$. Thus, each element ω of Ω is a function $\omega(\cdot)$ on $\mathbb{R}_+$, called a 'sample function' or 'path'. For each $t \in \mathbb{R}_+$, we set

$$X_t(\omega) = X(t, \omega) = \omega(t),$$

and write X_t for the function $X_t(\cdot)$ on Ω. The tribe generated by $\{X_s, 0 \le s \le t\}$ is denoted by $\mathcal{F}_t$ and that generated by $\{X_s, 0 \le s < \infty\}$ is denoted by $\mathcal{F}_\infty$, or more simply $\mathcal{F}$. Thus, for each $t \in \mathbb{R}_+$, we have $X_t \in \mathcal{F}_t \subset \mathcal{F}$. We define a family of mappings (called 'shifts') $\{\theta_t, t \in \mathbb{R}_+\}$ on Ω as follows:

$$(\theta_t \omega)(s) = \omega(t+s), \quad s \in \mathbb{R}_+.$$

It is trivial to see that each θ_t maps Ω into Ω. The same is true if Ω is the subspace of all right continuous functions on $\mathbb{R}_+$ or that of all continuous functions on $\mathbb{R}_+$. For any $B \in \mathcal{F}$ and $t \in \mathbb{R}_+$, we have

$$\theta_t^{-1} B = \{\omega \in \Omega : \theta_t \omega \in B\} \in \mathcal{F}.$$

Let $\{P^x, x \in \mathbb{R}^d\}$ be a family of probability measures on $(\Omega, \mathcal{F})$. Then each X_t is a random variable in the probability space $(\Omega, \mathcal{F}, P^x)$ taking values in $\mathbb{R}^d$. The mathematical expectation E^x associated with P^x is defined as follows: for each $Y \in \mathcal{F}$:

$$E^x(Y) = \int_\Omega Y(\omega) P^x(\mathrm{d}\omega)$$

provided that the integral exists. For a set $A \in \mathcal{F}$ we shall write:

$$E^x(A; Y) = E^x(1_A Y).$$

Now suppose that the following conditions are satisfied:

(i) $\forall x \in \mathbb{R}^d, P^x(X_0 = x) = 1$;

(ii) $\forall B \in \mathcal{F},$ the function $x \to P^x(B)$ belongs to $\mathcal{B}(\mathbb{R}^d)$;

(iii) $\forall t > 0$, $x \in \mathbb{R}^d$, $A \in \mathcal{F}_t$ and $B \in \mathcal{F}$:

$$P^x(A \cap \theta_t^{-1} B) = E^x(A; P^{X_t}(B)).$$

Then $\{X_t\}$ is called a *Markov process* on $(\Omega, \mathcal{F}_t, P^x)$, where '$t \in \mathbb{R}_+$, $x \in \mathbb{R}^d$' is understood. Observe that it is a consequence of (ii) that $P^{X_t}(B) \in \mathcal{F}_t$. The property (iii) is called the *Markov property*.

For any probability measure μ on $\mathcal{B}^d$, we define the probability measure P^μ on $\mathcal{F}$ as follows:

$$P^\mu(B) = \int_{\mathbb{R}^d} P^y(B) \mu(\mathrm{d}y).$$

When μ is the point-mass ϵ_x at x, P^μ reduces to P^x. We refer to μ as the initial distribution of $\{X_t\}$. A set B in $\mathcal{F}$ such that $P^x(B) = 0$ for all $x \in \mathbb{R}^d$ is called a *null set*. A property that is true for all ω in Ω except a null set is said to hold *almost surely*, abbreviated to 'a.s.'; alternatively, we say 'almost all' sample functions have this property. More specifically, 'P^x-a.s.' means 'for all ω except a set of P^x-measure zero'.

Next we shall construct a Markov process from a given transition function. A function $(t, x, B) \to P(t; x, B)$ on $\mathbb{R}_+ \times \mathbb{R}^d \times \mathcal{B}^d$ satisfying the following conditions is called a transition function.

(a) $\forall t \in \mathbb{R}_+, x \in \mathbb{R}^d : P(t; x, \cdot)$ is a probability measure on $\mathcal{B}^d$;
$P(0; x, \cdot) = \epsilon_x(\cdot)$;

(b) $\forall t \in \mathbb{R}_+, B \in \mathcal{B}^d : P(t; \cdot, B) \in \mathcal{B}(\mathbb{R}^d)$;

(c) $\forall t \in \mathbb{R}_+, s \in \mathbb{R}_+, x \in \mathbb{R}^d, B \in \mathcal{B}^d$:

$$P(t + s; x, B) = \int_{\mathbb{R}^d} P(t; x, \mathrm{d}y) P(s; y, B). \tag{1}$$

Associated with the transition function P, we define for each $t \in \mathbb{R}_+$ the operator P_t as follows. For each $f \in \mathcal{B}(\mathbb{R}^d)$,

$$P_t f(x) = \int_{\mathbb{R}^d} P(t; x, \mathrm{d}y) f(y), \quad x \in \mathbb{R}^d, \tag{2}$$

whenever the integral exists (possibly equal to $\pm\infty$). It follows from (b) that P_t maps $\mathcal{B}_+(\mathbb{R}^d)$ into $\mathcal{B}_+(\mathbb{R}^d)$ and $L^\infty(\mathbb{R}^d)$ into $L^\infty(\mathbb{R}^d)$. It follows from (c) that the family $\{P_t : t \in \mathbb{R}_+\}$ forms a semigroup in $L^\infty(\mathbb{R}^d)$, i.e. for all $t \in \mathbb{R}_+$, $s \in \mathbb{R}_+$, and $f \in L^\infty(\mathbb{R}^d)$, we have

$$P_0 f = f; \quad P_{t+s} f = P_t(P_s f) \in L^\infty(\mathbb{R}^d).$$

This is called the *transition semigroup* of the Markov process $\{X_t\}$ and will be denoted by $\{P_t\}$.

A Markov process can be constructed from a given transition function as described below.

Theorem 1.1 *Let P be a transition function on $\mathbb{R}^d$. Then there exists a Markov process $\{X_t, t \in \mathbb{R}_+\}$ on $(\Omega, \mathcal{F}_t, P^x)$, where $\Omega = M(\mathbb{R}_+, \mathbb{R}^d)$, such that $\forall t \in \mathbb{R}_+$, $x \in \mathbb{R}^d$, $B \in \mathcal{B}^d$:*

$$P^x(X_t \in B) = P(t; x, B). \tag{3}$$

Proof We must construct the family of probabilities $\{P^x\}$. For a fixed $x \in \mathbb{R}^d$, we define the family of finite-dimensional distributions, for $n \geq 1$, $0 \leq t_1 < \cdots < t_n$, and $B_k \in \mathcal{B}^d$, $1 \leq k \leq n$:

$$\begin{aligned} &F^x_{t_1,\cdots,t_n}(B_1, \cdots, B_n) \qquad (4)\\ &= \int_{B_1} \cdots \int_{B_n} P(t_1; x, \mathrm{d}y_1) P(t_2 - t_1; y_1, \mathrm{d}y_2) \cdots P(t_n - t_{n-1}; y_{n-1}, \mathrm{d}y_n). \end{aligned}$$

It is easy to check from the conditions (a)–(c) for the transition function that the family $F^x_{t_1,\cdots,t_n}(B_1,\cdots,B_n)$ satisfies the consistency condition, for $1 \le i \le n$:

$$F^x_{t_1,\cdots,t_n}(B_1,\cdots,B_{i-1},\mathbb{R}^d,B_{i+1},\cdots,B_n) \\ = F^x_{t_1,\cdots,t_{i-1},t_{i+1},\cdots,t_n}(B_1,\cdots,B_{i-1},B_{i+1},\cdots,B_n).$$

Hence, by Kolmogorov's extension theorem (see Doob (1953)), there exists a probability measure P^x on $\Omega = M(\mathbb{R}_+,\mathbb{R}^d)$ such that

$$P^x(X_{t_1} \in B_1,\cdots,X_{t_n} \in B_n) = F^x_{t_1,\cdots,t_n}(B_1,\cdots,B_n). \tag{5}$$

We shall now verify the properties (i)–(iii) for a Markov process. Property (i) is true by the definition of $P(0;x,\cdot)$. It is sufficient to verify property (ii) when the set B is of the form below:

$$A = \{X_{t_1} \in B_1,\cdots,X_{t_k} \in B_k\},\quad 0 \le t_1 < \cdots < t_k.$$

This follows by repeated use of the fact that P_t maps $L^\infty(\mathbb{R}^d)$ into $L^\infty(\mathbb{R}^d)$, as noted above. Finally, for $t > 0$, let A be as indicated above but with $t_k \le t$; $G_i \in \mathcal{B}^d$, $1 \le i \le m$ and

$$C = \{X_{s_1} \in G_1,\cdots,X_{s_m} \in G_m\},\quad 0 \le s_1 < \cdots < s_m.$$

Then we have

$$\begin{aligned} &P^x(A \cap \theta_t^{-1}C) \\ &= P^x(X_{t_1} \in B_1,\cdots,X_{t_k} \in B_k, X_{t+s_1} \in G_1,\cdots,X_{t+s_m} \in G_m) \\ &= \int_{B_1}\cdots\int_{B_k}\int_{G_1}\cdots\int_{G_m} P(t_1;x,\mathrm{d}y_1)\cdots P(t_k-t_{k-1};y_{k-1},\mathrm{d}y_k) \\ &\quad \cdot P(t-t_k+s_1;y_k,\mathrm{d}u_1)P(s_2-s_1;u_1,\mathrm{d}u_2)\cdots P(s_m-s_{m-1};u_{m-1},\mathrm{d}u_m), \end{aligned}$$

and

$$\begin{aligned} &E^x(A;P^{X_t}(C)) \\ &= \int_{B_1}\cdots\int_{B_k}\int_{\mathbb{R}^d} P(t_1;x,\mathrm{d}y_1)\cdots P(t_k-t_{k-1};y_{k-1},\mathrm{d}y_k) \\ &\quad \cdot P(t-t_k;y_k,\mathrm{d}u)P^u(C) \\ &= \int_{B_1}\cdots\int_{B_k}\int_{\mathbb{R}^d}\int_{G_1}\cdots\int_{G_m} P(t_1;x,\mathrm{d}y_1)\cdots P(t_k-t_{k-1};y_{k-1},\mathrm{d}y_k) \\ &\quad \cdot P(t-t_k;y_k,\mathrm{d}u)P(s_1;u,\mathrm{d}u_1)P(s_2-s_1;u_1,\mathrm{d}u_2)\cdots \\ &\quad \cdot P(s_m-s_{m-1};u_{m-1},\mathrm{d}u_m). \end{aligned}$$

Since by (1)

$$P(t-t_k+s_1;y_k,\mathrm{d}u_1) = \int_{\mathbb{R}^d} P(t-t_k;y_k,\mathrm{d}u)P(s_1;u,\mathrm{d}u_1),$$

it follows that

$$P^x(A \cap \theta_t^{-1}C) = E^x(A; P^{X(t)}(C)) \tag{6}$$

for all A and C of the forms indicated above. Hence, by a monotone class argument, (6) holds for all $A \in \mathcal{F}_t$ and $C \in \mathcal{F}$, and property (iii) of a Markov process is verified. □

Corollary to Theorem 1.1 *If $\Phi \in \mathcal{F}$ and $E^x(\Phi)$ exists for all $x \in \mathbb{R}^d$, then the function $x \to E^x(\Phi)$ belongs to $\mathcal{B}(\mathbb{R}^d)$.*

We introduce an important class of functions for an open set A as follows:

$C_0(A)$ = the class of f in $C_b(A)$ such that for any $z \in \partial A$ and also for $z = \infty$ when A is unbounded, we have $\lim_{\substack{x \to z \\ x \in A}} f(x) = 0.$

For each f in $C_0(A)$ there is a unique function $\overline{f}$ in $C_b(\overline{A} \cup \{\infty\})$ such that $\overline{f} = f$ in A, and $\overline{f} = 0$ in $(\partial A) \cup \{\infty\}$. If A is bounded, we may omit $\{\infty\}$ in the above (but we may also keep it without harm!). The set $C_0(A)$ is a Banach space with $\| \;\|_\infty$ as its norm, abbreviated to $\| \;\|$.

The semigroup $\{P_t\}$ is said to have the *Feller property* iff for each $t \geq 0 : P_t$ maps $C_0(\mathbb{R}^d)$ into $C_0(\mathbb{R}^d)$; and for each $f \in C_0(\mathbb{R}^d)$, we have

$$\lim_{t \downarrow 0} \|P_t f - f\| = 0. \tag{7}$$

The semigroup $\{P_t\}$ is said to have the *strong Feller property* iff for each $t > 0$, P_t maps $L^\infty(\mathbb{R}^d)$ into $C_b(\mathbb{R}^d)$. Note that the strong Feller property does not imply the Feller property! Later we shall use both terms when $\mathbb{R}^d$ is replaced by an open set D.

A Markov process whose transition semigroup $\{P_t\}$ has the Feller property is called a Feller process. This is an important class of processes, one of the most fundamental properties of which is that almost all its sample functions are right continuous in $\mathbb{R}_+$ (see Chung (1982a, Chapter 2)). More precisely, in the construction given in Theorem 1.1, the sample space $M(\mathbb{R}_+, \mathbb{R}^d)$ may be replaced by the smaller collection of all mappings from $\mathbb{R}_+$ to $\mathbb{R}^d$ that are right continuous in $\mathbb{R}_+$, so that for each $\omega \in \Omega, X(\cdot, \omega)$ is a right continuous function in $\mathbb{R}_+$. We do not need to invoke this rather sophisticated result since we shall only be dealing with a particular case (Brownian motion) for which more specific properties will be proved in Section 1.3. However, we would add the fundamental observation that for a process with right continuous sample functions, the function $(t, \omega) \to X(t, \omega)$ is $\mathcal{B}^1 \times \mathcal{F}$ measurable (see Chung (1982a, page 19)).

We now introduce the notion of an *optional time*. This is the single tool that separates probabilistic methods from others, without which the theory of Markov processes would lose much of its strength and depth.

A function τ from Ω to $\overline{\mathbb{R}_+} = [0, \infty]$ is called an *optional time* iff

$$\forall t \in \mathbb{R}_+ : \{\omega \in \Omega : \tau(\omega) < t\} \in \mathcal{F}_t.$$

This is equivalent to

$$\forall t \in \mathbb{R}_+ : \{\omega \in \Omega : \tau(\omega) \leq t\} \in \mathcal{F}_{t+},$$

where

$$\mathcal{F}_{t+} = \bigcap_{s>t} \mathcal{F}_s.$$

From here on, a set like $\{\omega \in \Omega : \tau(\omega) < t\}$ will be written as $\{\tau < t\}$. For each optional τ, we define

$$\mathcal{F}_{\tau+} = \{A \in \mathcal{F}_\infty : \forall t \geq 0 : A \cap \{\tau < t\} \in \mathcal{F}_t\}.$$

It can be shown that $\mathcal{F}_{\tau+}$ is a tribe included in $\mathcal{F}_\infty = \mathcal{F}$. Simple properties of τ and $\mathcal{F}_{\tau+}$ which are easy to verify will be taken for granted below; see Chung (1982a). A prime example of τ is given at the beginning of Section 1.5.

The shift operator will be extended as follows:

$$(\theta_\tau \omega)(t) = \omega(\tau(\omega) + t), \text{ on } \{\tau < \infty\},$$

or equivalently,

$$X_t(\theta_\tau(\omega)) = X(\tau(\omega) + t, \omega), \quad t \in \mathbb{R}_+,$$

which is abbreviated to

$$X_t(\theta_\tau) = X(\tau + t) = X_{\tau+t}.$$

However, the novice must be careful about such concise notation. For instance, observe that on $\{\tau < \infty\}$, $X_\tau(\omega) = X(\tau(\omega), \omega)$ so that

$$X_\tau(\theta_t \omega) = X(\tau(\theta_t \omega), \theta_t \omega) = X(t + \tau(\theta_t \omega), \omega), \tag{8}$$

which is not the same as $X_t(\theta_\tau \omega)$!

If τ is optional, then for each $t \geq 0$, $\tau + t$ is optional and $X_{\tau+t} \in \mathcal{F}_{(\tau+t)+}$.

We are now ready for the following fundamental result, which is known as the *strong Markov property* and will be invoked repeatedly later.

Theorem 1.2 *Suppose that the transition semigroup $\{P_t\}$ of the Markov process $\{X_t\}$ has the Feller property and that almost all sample functions are right continuous. Then for each optional time τ, $A \in \mathcal{F}_{\tau+}$ and $\Phi \in \mathcal{F}$, where Φ is either positive or bounded, we have for each $x \in \mathbb{R}^d$:*

$$E^x[A \cap (\tau < \infty); \Phi(\theta_\tau)] = E^x[A \cap (\tau < \infty); E^{X(\tau)}(\Phi)]. \tag{9}$$

Proof It is sufficient to prove (9) when Φ is of the form below, where $0 \leq t_1 < \cdots < t_m$ and $f_i \in C_0(\mathbb{R}^d)$, $1 \leq i \leq m$:

$$\Phi = f_1(X_{t_1}) \cdots f_m(X_{t_m}). \tag{10}$$

If so, by standard arguments based on a monotone class theorem, (9) will hold as stated; for details, see Chung (1982a, Section 1.1). Using the Markov and Feller properties, it is easy to prove by induction on m that for the above Φ the function $x \to E^x\{\Phi\}$ belongs to $C_0(\mathbb{R}^d)$. We now define for each $n \geq 1$:

$$\tau_n = \begin{cases} k2^{-n} & \text{if } (k-1)2^{-n} \leq \tau < k2^{-n}, \quad k = 1, 2, \cdots \\ \infty & \text{if } \tau = \infty. \end{cases}$$

Then $\tau_n \downarrow \tau$ as $n \to \infty$. Since τ is optional and $A \in \mathcal{F}_{t+}$, it follows that $A \cap \{(k-1)2^{-n} \leq \tau < k2^{-n}\} \in \mathcal{F}_{k2^{-n}}$; hence we have by the Markov property:

$$\begin{aligned} & E^x \left[A \cap (\tau < \infty); E^{X(\tau_n)}(\Phi)\right] \\ &= \sum_{k=1}^{\infty} E^x[A \cap ((k-1)2^{-n} \leq \tau < k2^{-n}); E^{X(k2^{-n})}(\Phi)] \\ &= \sum_{k=1}^{\infty} E^x[A \cap ((k-1)2^{-n} \leq \tau < k2^{-n}); \Phi(\theta_{k2^{-n}})] \\ &= E^x \left[A \cap (\tau < \infty); f_1(X_{t_1+\tau_n}) \cdots f_m(X_{t_m+\tau_n})\right]. \end{aligned}$$

Letting $n \to \infty$ in the above, we obtain (9) by the almost sure right continuity of $t \to x_t$. □

A convenient form of (9) is given in terms of conditional expectation, as follows.

Corollary to Theorem 1.2 *For all $x \in \mathbb{R}^d$, we have P^x-a.s. on $\{\tau < \infty\}$:*

$$E^x \left[\Phi(\theta_\tau) | \mathcal{F}_{\tau+}\right] = E^{X(\tau)}[\Phi].$$

When τ is a constant $t \geq 0$, then it is easy to verify that $\mathcal{F}_{\tau+}$ reduces to the previously defined $\mathcal{F}_{t+}$, and (9) implies that the property (iii) in the definition of a Markov process is true under the present assumption for $A \in \mathcal{F}_{t+}$. This extension from $\mathcal{F}_t$ to $\mathcal{F}_{t+}$ is important, although it may not be evident in the context of our developments. However, the following result, known as Blumenthal's zero-or-one law, is an immediate consequence.

Proposition 1.3 *If (9) is true, then for any $A \in \mathcal{F}_{0+}$ and any $x \in \mathbb{R}^d$, we have $P^x(A) = 0$ or 1.*

Proof Apply (9) with $\tau = 0$ to the given set A with $\Phi = 1_A$. We obtain

$$\begin{aligned} E^x[A; 1_A(\theta_0)] &= E^x[A; E^{X(0)}(1_A)] \\ &= E^x[A; E^x(1_A)] = P^x(A)^2. \end{aligned}$$

But the first term above is just $P^x(A \cap A) = P^x(A)$; hence the conclusion follows. □

1.3 Construction of Brownian Motion

For $t > 0, x \in \mathbb{R}^d$, $y \in \mathbb{R}^d$, we set

$$p(t; x, y) = \frac{1}{(2\pi t)^{d/2}} \exp\left(-\frac{|x-y|^2}{2t}\right). \tag{11}$$

The function $x \to p(t; 0, x)$ is the d-dimensional normal density with the zero vector as mean and $t \cdot I^d$ as the variance matrix, where I^d denotes the $(d \times d)$-identity matrix. The corresponding probability distribution is the d-dimensional normal (or Laplace–Gauss) distribution. A transition function is derived from (11) by setting

$$P(t; x.B) = \int_B p(t; x, y) \mathrm{d}y. \tag{12}$$

This transition function may be used to construct a Markov process, according to Theorem 1.1. This particular process is the basis for our probabilistic study. It is called the *Brownian motion (process)* in $\mathbb{R}^d$. The most significant feature of this process is that all its sample functions are continuous in $\mathbb{R}_+$. In other words, if we use Theorem 1.1 to construct a Markov process on the sample space $M(\mathbb{R}_+, \mathbb{R}^d)$, then it can be proved that, by virtue of the special properties of the transition function given in (12), almost all sample functions are continuous. Such an approach is described in Chung (1982a, page 77). Here, we present a more direct construction, due to Lévy (1948), in which the sample space is taken at the outset to be $C(\mathbb{R}_+, \mathbb{R}^d)$, the collection of all continuous functions from $\mathbb{R}_+$ to $\mathbb{R}^d$.

Theorem 1.4 *There exists a Markov process* $\{X_t, t \in \mathbb{R}_+\}$ *on* $(\Omega, \mathcal{F}_t, P^x)$, *where* $\Omega = C(\mathbb{R}_+, \mathbb{R}^d)$ *and (3) holds with the P defined by (11) and (12).*

Proof We first observe that for this new $\Omega, \omega \in \Omega$ still implies that $\theta_t \omega \in \Omega$ for each $t \in \mathbb{R}_+$; hence, property (iii) of a Markov process makes sense.

Let V be the subset of $\mathbb{R}_+$ of the form $k2^{-n}$, where n and k range over all positive integers; elements of V will be called binary numbers. Using our method, we first construct a process $\{X_t\}$ indexed by $t \in V$, with the given $P(t; \cdot, \cdot)$ similarly restricted to $t \in V$. Kolmogorov's theorem applies equally well to such an index set and yields for each $x \in \mathbb{R}^d$ a probability measure P_V^x on $(\Omega_V, \mathcal{F}_V)$, where $\Omega_V = M(V, \mathbb{R}^d)$ and $\mathcal{F}_V$ is the tribe generated by $\{X_t, t \in V\}$. Now for each integer $N \geq 1$, let

$$\begin{aligned} \Lambda_N &= \{\omega \in \Omega_V : \omega(\cdot) \text{ is uniformly continuous on } V \cap [0, N]\}, \\ \Lambda &= \bigcap_{N=1}^{\infty} \Lambda_N. \end{aligned}$$

We claim that $P_V^x(\Lambda) = 1$. To show this it is sufficient to prove that $P_V^x(\Lambda_1) = 1$. For $n \geq 1$, set

$$S_n = (\omega \in \Omega_V : |\omega(k2^{-n}) - \omega((k-1)2^{-n})| < 2^{-n/8}, k = 1, \cdots, 2^n).$$

The assertion will follow from

$$\Lambda_1 \supset \liminf_n S_n \tag{13}$$

and

$$P_V^x(\limsup_n S_n^c) = 0. \tag{14}$$

For any $\omega \in \liminf_n S_n$, there exists $N = N(\omega)$ such that $\omega \in \cap_{n \geq N} S_n$. Fix such an ω and N. Then, for any u and v both in $[0,1] \cap V$, $0 < v - u < 2^{-N}$, we take an $s = q2^{-p}$ in (u, v) with the smallest possible integer p and expand $v - s$ and $s - u$ as binary numbers:

$$\begin{aligned} v - \frac{q}{2^p} &= \frac{1}{2^{n_1}} + \cdots + \frac{1}{2^{n_k}}, \quad p \leq n_1 < n_2 < \cdots < n_k; \\ \frac{q}{2^p} - u &= \frac{1}{2^{m_1}} + \cdots + \frac{1}{2^{m_r}}, \quad p \leq m_1 < m_2 < \cdots < m_r. \end{aligned}$$

Considering the increasing numbers:

$$\begin{aligned} u &= \frac{q}{2^p} - \sum_{i=1}^{r} \frac{1}{2^{m_i}}, \ \frac{q}{2^p} - \sum_{i=1}^{r-1} \frac{1}{2^{m_i}}, \cdots, \frac{q}{2^p} - \frac{1}{2^{m_1}}, \ \frac{q}{2^p}, \\ & \quad \frac{q}{2^p} + \frac{1}{2^{n_1}}, \cdots, \frac{q}{2^p} + \sum_{i=1}^{k} \frac{1}{2^{n_i}} = v, \end{aligned}$$

we see that there is a partition $u = s_0 < s_1 < \cdots < s_l = v$ such that $[s_i, s_{i+1}]$ is of the form $[k2^{-n}, (k+1)2^{-n}]$, $i = 0, \cdots, l-1$, where any given integer n may occur at most twice in the denominator 2^n. This is a consequence of our choice of p.

Let $1/2^w = \max_{0 \leq i < l} |s_{i+1} - s_i|$. Thus, $w \geq N$ and

$$\begin{aligned} |\omega(v) - \omega(u)| &\leq \sum_{i=0}^{l-1} |\omega(s_{i+1}) - \omega(s_i)| \leq 2 \sum_{n=w}^{\infty} 2^{-n/8} \leq \frac{2}{1 - 2^{-1/8}} 2^{-w/8} \\ &\leq \frac{2}{1 - 2^{-1/8}} |v - u|^{1/8}. \end{aligned}$$

This shows that $\omega \in \Lambda_1$, whence (13) holds. Next we have

$$\begin{aligned} \sum_{n=1}^{\infty} P_V^x(S_n^c) &\leq \sum_{n=1}^{\infty} \sum_{k=1}^{2^n} P_V^x \left[|\omega(k2^{-n}) - \omega((k-1)2^{-n})| \geq 2^{-n/8} \right] \\ &\leq \sum_{n=1}^{\infty} \sum_{k=1}^{2^n} 2^{n/2} E_V^x \left[|\omega(k2^{-n}) - \omega((k-1)2^{-n})|^4 \right] \\ &\leq \sum_{n=1}^{\infty} 2^{(3n/2)} \left[3d2^{-2n} + d(d-1)2^{-2n} \right] \\ &= \sum_{n=1}^{\infty} \frac{d(d+2)}{2^{n/2}} < \infty. \end{aligned}$$

In the third inequality above, we used the fourth moment of the normal distribution with covariance matrix $2^{-n} \cdot I^d$. Thus, (14) follows from the Borel–Cantelli lemma.

Each ω in Λ may be uniquely extended to $C(\mathbb{R}_+, \mathbb{R}^d)$ by uniform continuity. Thus, there exists a 1–1 mapping T from $C(\mathbb{R}_+, \mathbb{R}^d)$ onto Λ:

$$T : [\omega(t), t \in \mathbb{R}_+] \to [\omega(t), t \in V].$$

If $A \in \mathcal{F}$, then $TA \in \mathcal{F}_V$. Now we define P^x on $(\Omega, \mathcal{F})$ by

$$P^x(A) = P_V^x(TA), \quad A \in \mathcal{F}.$$

Then, by construction, we have for each $t \in V, x \in \mathbb{R}^d, B \in \mathcal{B}^d$:

$$P^x(X_t \in B) = P(t; x, B). \tag{15}$$

But $X(\cdot, \omega) = \omega(\cdot) \in C(\mathbb{R}_+, \mathbb{R}^d)$. Hence, for any $t \in \mathbb{R}_+$, if $t_n \in V$, $t_n \to t$, then X_{t_n} converges to X_t, P^x-almost surely. Therefore, for any $f \in C_c(\mathbb{R}^d)$, we have, using the notation of (2):

$$E^x\{f(X_t)\} = \lim_{n\to\infty} E^x\{f(X_{t_n})\} = \lim_{n\to\infty} P_{t_n} f(x) = P_t f(x). \tag{16}$$

The last relation follows from the continuity of $t \to p(t; \cdot, \cdot)$ in $(0, \infty)$ and dominated convergence in

$$P_t f(x) = \int_{\mathbb{R}^d} p(t; x, y) f(y) \mathrm{d}y.$$

As in the proof of Theorem 1.2, the fact that (16) is true for all $f \in C_c$ implies that it is true for $f = 1_B$, which is (15). The proof of Theorem 1.4 is complete. □

Corollary to Theorem 1.4 *For any* $B \in \mathcal{B}^{dn}$, *and* $0 = t_0 < t_1 < \cdots < t_n$, *we have*

$$P^x[(X_{t_1}, \cdots, X_{t_n}) \in B] = \int_B \prod_{i=1}^n p(t_i - t_{i-1}; y_{i-1}, y_i) \mathrm{d}y_1 \cdots \mathrm{d}y_n, \tag{17}$$

where $y_0 = x$.

This gives the density of the joint distribution of the process. Since $p(t; x, y) = p(t; 0, y-x)$, it follows from the product form in (17) that $\{X_{t_i} - X_{t_{i-1}}, 1 \le i \le n\}$ are independent random variables with the normal distributions $P(t_i - t_{i-1}, 0, \cdot)$. In fact, the last statement is another characterization of the Brownian motion; see Chung (1982a, Section 4.1).

We proceed to show that the transition semigroup $\{P_t\}$ for the Brownian motion has both the strong Feller property and the Feller property. The former is included in the next proposition.

Proposition 1.5 *For each* $t > 0$, $f \in L^\infty(\mathbb{R}^d)$ *or* $f \in L^1(\mathbb{R}^d)$, *we have* $P_t f \in C(\mathbb{R}^d)$.

Proof Let $f \in L^\infty(\mathbb{R}^d)$. If $x_n \to x$, then

$$|P_t f(x_n) - P_t f(x)| \le \|f\| \int_{\mathbb{R}^d} |p(t; x_n, y) - p(t; x, y)| dy.$$

The above integral converges to zero as $n \to \infty$ because the part outside $B(0, r)$ is arbitrarily small for large r, while the part over $B(0, r)$ converges to zero by bounded convergence. Hence $P_t f$ is continuous at x. When $f \in L^1(\mathbb{R}^d)$ a simpler proof may be obtained using dominated convergence. □

Proposition 1.6 *For any $t > 0$ and any bounded $\mathcal{F}$-measurable ϕ, the function*

$$x \to E^x[\phi(\theta_t)]$$

is continuous on $\mathbb{R}^d$.

Proof Let $f(x) = E^x(\phi)$, then $f \in L^\infty(\mathbb{R}^d)$ by the Corollary to Theorem 1.1. Since

$$E^x[\phi(\theta_t\omega)] = E^x\left[E^{X_t}(\phi)\right] = E^x[f(X_t)],$$

the result follows from Proposition 1.5. □

Proposition 1.7 *$\{P_t\}$ has the Feller property.*

Proof We first verify that for each $t > 0$, P_t maps $C_0(\mathbb{R}^d)$ into $C_0(\mathbb{R}^d)$. In view of Proposition 1.5, we need only prove that for any $t > 0$ and $f \in C_0(\mathbb{R}^d)$,

$$\lim_{x \to \infty} P_t f(x) = 0. \tag{18}$$

For any $\varepsilon > 0$, there exists $r > 0$ such that $|f(y)| < \varepsilon$ if $|y| \ge r$. Hence we have

$$|P_t f(x)| \le \varepsilon + \|f\| \int_{B(0,r)} p(t; x, y) dy.$$

The second term on the right side goes to zero when $x \to \infty$, proving (18).

Next, let $f \in C_0(\mathbb{R}^d)$; then f is uniformly continuous on $\mathbb{R}^d$. Hence for any $\varepsilon > 0$, there exists $\delta > 0$ such that $|f(x) - f(y)| < \varepsilon$ when $|y - x| < \delta$. Thus, we have

$$\begin{aligned} \|P_t f - f\| &\le \sup_{x \in \mathbb{R}^d} \int_{\mathbb{R}^d} p(t; x, y)|f(y) - f(x)| dy \\ &\le \varepsilon + 2\|f\| \int_{|u| \ge \delta} p(t; 0, u) du. \end{aligned}$$

The last-written integral goes to zero when $t \downarrow 0$. Hence the second condition for the Feller property is true. □

1.4 Harmonic Function and Poisson Kernel

For a domain D in $\mathbb{R}^d$, we let $C^n(D)$ $(n \geq 1)$ denote the collection of all n times continuously differentiable functions on D.

If $f \in C^2(D)$, its Laplacian Δf is defined by

$$\Delta f(x) = \sum_{i=1}^{d} \frac{\partial^2 f}{\partial x_i^2}(x), \quad x = (x_1, \cdots, x_d) \in D.$$

Definition *A function h is said to be* harmonic *in D iff $h \in C^2(D)$ and satisfies Laplace's equation*

$$\Delta h = 0 \tag{19}$$

in D.

Obviously all linear functions are harmonic in $\mathbb{R}^d$ $(d \geq 1)$, and all harmonic functions are linear in $\mathbb{R}^1$.

An important class of harmonic functions in $\mathbb{R}^d$ $(d \geq 2)$ is given by the following proposition.

Proposition 1.8 *The class of all harmonic functions in $\mathbb{R}^d \backslash \{0\}$ depending only on $|x|$ consists of functions of the form*

$$c_1 \ln |x| + c_2 \text{ in } \mathbb{R}^2 \text{ and } c_1 |x|^{2-d} + c_2 \text{ in } \mathbb{R}^d,\ d \geq 3, \tag{20}$$

where c_1 and c_2 are constants.

Proof Writing r for $|x|$, we see that Laplace's equation for $r \to h(r)$ reduces to

$$\frac{\mathrm{d}^2 h}{\mathrm{d}r^2} + \frac{d-1}{r} \frac{\mathrm{d}h}{\mathrm{d}r} = 0.$$

Solving this differential equation, we obtain the general solutions given in (20). □

The following characterization of a harmonic function is fundamental. We recall from Section 1.1 that $S(x, r) = \partial B(x, r)$.

Theorem 1.9 *A finite, real-valued function h in a domain D of $\mathbb{R}^d$ $(d \geq 1)$ is harmonic in D if and only if* (a) *$h \in L^1_{\text{loc}}(D)$ and* (b) *for any $x \in D$ and $B(x, r) \subset\subset D$ the integral below exists and we have*

$$h(x) = \frac{1}{\sigma(S(x, r))} \int_{S(x,r)} h(y) \sigma(\mathrm{d}y). \tag{21}$$

In this case, we also have

$$h(x) = \frac{1}{m(B(x, r))} \int_{B(x,r)} h(y) m(\mathrm{d}y). \tag{22}$$

Moreover, $h \in C^\infty(D)$.

Proof We first prove that under the assumption (a), (21) implies (22). This follows from the formula for 'spherical integration':

$$\int_0^r \int_{S(x,s)} h(y)\sigma(\mathrm{d}y)\mathrm{d}s = \int_{B(x,r)} h(y)m(\mathrm{d}y);$$

but the reader is advised to prove it using (a)! Next, we define for any $\delta > 0$ a function ϕ on $\mathbb{R}_+$ as follows:

$$\phi(r) = \begin{cases} C\exp\left(\frac{1}{r^2-\delta^2}\right) & \text{if } 0 \le r < \delta; \\ 0 & \text{if } \delta \le r < \infty; \end{cases}$$

where the constant C is so chosen that

$$\int_0^\infty \phi(r)\sigma(r)\mathrm{d}r = 1,$$

with $\sigma(r) = \sigma(S(x,r))$. It is easy to verify that $\phi \in C^\infty((0,\infty))$.

Now let h satisfy (a) and (b). We then have by (21), if $\rho(x, \partial D) > \delta$ so that (21) holds for $0 < r < \delta$:

$$\begin{aligned} h(x) &= \int_0^\infty \left[\frac{1}{\sigma(r)}\int_{S(x,r)} h(y)\sigma(\mathrm{d}y)\right]\phi(r)\sigma(r)\mathrm{d}r \\ &= \int_{\mathbb{R}^d} h(y)\phi(|y-x|)m(\mathrm{d}y). \end{aligned}$$

Here again, the integrability of h over $B(x,r)$, together with the local boundedness of ϕ is used to justify the transformation of the integrals by Fubini's theorem. The infinite differentiability of h then follows from that of ϕ. Thus, we have proved that (a) and (b) imply that $h \in C^\infty(D)$.

For any $h \in C^2(D)$, if $\rho(x, \partial D) > \delta$, we write for $0 < r \le \delta$:

$$I(r) \equiv \frac{1}{\sigma(r)}\int_{S(x,r)} h(y)\sigma(\mathrm{d}y) = \frac{1}{\sigma(1)}\int_{S(0,1)} h(x+rz)\sigma(\mathrm{d}z)$$

and differentiate the right-hand side of the above with respect to r. An application of Gauss's 'divergence formula' for a smooth function (C^2 is sufficient) in a ball yields the derivative:

$$\begin{aligned} I'(r) &= \frac{1}{\sigma(1)}\int_{S(0,1)} \frac{\partial h}{\partial r}(x+rz)\sigma(\mathrm{d}z) \\ &= -\frac{1}{\sigma(r)}\int_{S(x,r)} \frac{\partial h}{\partial n}(y)\sigma(\mathrm{d}y) \\ &= \frac{1}{\sigma(r)}\int_{B(x,r)} \Delta h(y)m(\mathrm{d}y), \end{aligned}$$

where $\frac{\partial}{\partial n}$ is the inner normal derivative.

If h satisfies (a) and (b), then for $0 < r < \delta$, $I(r)$ is constant, and so $I'(r) = 0$. Since Δh is continuous, $\lim_{r \downarrow 0} I'(r) = \Delta h(x)$ by the last expression above. Hence $\Delta h = 0$, in other words, h is harmonic in D.

Conversely, if h is harmonic in D, then $h \in C^2(D)$ and the above calculation is valid and yields $I'(r) = 0$ for $0 < r \leq \delta$. Since $\lim_{r \downarrow 0} I(r) = h(x)$, we obtain (21). Thus, h satisfies (a) and (b). □

Corollary to Theorem 1.9 *A finite, real-valued function h in a domain D is harmonic in D if and only if (22) is true for every $B(x, r) \subset\subset D$.*

Proof The 'only if' part is contained in the theorem. To prove the 'if' part, observe first that implicit in the present hypothesis is that h is integrable over each ball strictly contained in D, hence $h \in L^1_{\text{loc}}(D)$. Now for any x and x' in D, we have by applying (22) to both points:

$$|h(x) - h(x')| \leq \frac{1}{v(r)} \int_C |h(y)| m(\mathrm{d}y),$$

where $C = B(x, r) \Delta B(x', r)$ and $v(r) = m(B(x, r))$. As $|x - x'| \to 0$, $m(C) \to 0$, and so h is continuous in D. This being so, we can rewrite the right-hand side of (22) as follows:

$$v(r)h(x) = \int_0^r \int_{S(x,s)} h(y)\sigma(\mathrm{d}y)\mathrm{d}s.$$

The continuity of h implies that the above surface integral is continuous in s; hence, by differentiating with respect to r, we obtain

$$v'(r)h(x) = \int_{S(x,r)} h(y)\sigma(\mathrm{d}y).$$

This is (21), because $v'(r) = \sigma(r)$. Therefore condition (b) of the theorem holds as well as (a), and h is harmonic in D by the theorem. □

Remark The properties given in (21) and (22) will be referred to as 'sphere averaging' and 'ball averaging', respectively.

The next result is known as the 'maximum principle'.

Proposition 1.10 *If a harmonic function in a domain D takes its supremum or infimum at some point in D, then it is constant on D.*

Proof Suppose there exists $x_0 \in D$ such that $h(x_0) = \sup_{x \in D} h(x)$. If $h(x) = h(x_0)$, then (22) shows that $h(y) = h(x_0)$ for m-a.e. y in $B(x, r)$. Hence by continuity, $h(y) = h(x_0)$ for all y in $B(x, r)$. Thus, the set $D_1 = [x \in D : h(x) = h(x_0)]$ is a nonempty open set containing x_0. But $D \backslash D_1 = [x \in D : h(x) < h(x_0)]$ is also an open set because h is continuous. Since D is connected, we have $D = D_1$. The proof in the case of the infimum is similar, alternatively we may consider $-h$ instead of h. □

Let $f \in \mathcal{B}(\partial D)$. A function ϕ defined on D is said to have boundary value f on ∂D iff for each $z \in \partial D, \phi(x) \to f(z)$ as $x \to z$, $x \in D$. We then write

$$\phi\,|_{\partial D} = f\,.$$

Definition *Let D be a domain and f a bounded and continuous function on ∂D. A solution to the (classical)* Dirichlet problem *for (D, f) is a harmonic function in D having boundary value f on ∂D; in other words,*

$$\begin{cases} \Delta\phi & = 0 \quad \text{in } D \\ \phi|_{\partial D} & = f. \end{cases}$$

Thus, ϕ can be continuously extended to $\overline{D}$ with $\phi = f$ on ∂D. Remember that ∞ is not a point in $\overline{D}$!

The following result is an easy consequence of Proposition 1.10.

Corollary 1.11 *A function that is harmonic in a bounded domain with boundary value zero must be identically zero in the domain.*

It follows that the solution to the Dirichlet problem in a bounded domain, if it exists, is unique. The existence under a regularity condition will be proved in the next section by probabilistic methods.

We first introduce a particular case.

Definition *Let D be a bounded domain and σ a Borel measure on ∂D. A positive and continuous function K on $D \times \partial D$ is called a* (generalized) Poisson kernel *for D iff the solution to the Dirichlet problem for any $f \in C(\partial D)$ can be expressed by*

$$\int_{\partial D} K(x,z)f(z)\sigma(\mathrm{d}z). \tag{23}$$

When D is a C^∞ (or more generally C^1) domain, we will take σ to be the usual area measure.

It follows from Corollary 1.11 and the continuity of $K(x,\cdot)$ that the Poisson kernel, if it exists, is unique.

We have the following criterion.

Theorem 1.12 *Let D and σ be as in the above. Suppose K is a positive and continuous function on $D \times \partial D$ and satisfies the following conditions*

(i) $\forall z \in \partial D$, *$K(\cdot, z)$ is harmonic in D;*

(ii) $\forall x \in D$, $\int_{\partial D} K(x,z)\sigma(\mathrm{d}z) = 1$;

(iii) $\forall w \in \partial D$, $\delta > 0$ $\lim_{\substack{x\to w \\ x \in D}} \int_{\partial D \cap B(w,\delta)^c} K(x,z)\sigma(\mathrm{d}z) = 0$.

Then K is a Poisson kernel for D.

Proof For any $f \in C(\partial D)$, we define the function h in D by (23). We have to prove that h is the solution to the Dirichlet problem for (D, f).

The assertion that h is harmonic follows from the sphere-averaging characterization of Theorem 1.9, condition (i) and Fubini's theorem.

For any $w \in \partial D$, $\varepsilon > 0$, there exists $\delta > 0$ such that if $z \in \partial D \cap B(w, \delta)$, then $|f(z) - f(w)| < \frac{\varepsilon}{2}$. Hence, by condition (iii) there exists $r > 0$ such that if $x \in D$ and $|x - w| < r$, then

$$\int_{\partial D \cap B(w,\delta)^c} K(x, z)\sigma(\mathrm{d}z) \leq \varepsilon/4M,$$

where $M = \sup_{z \in \partial D} |f(z)|$. Thus, we have, by condition (ii),

$$\begin{aligned} |h(x) - f(w)| &= \left| \int_{\partial D} K(x, z)[f(z) - f(w)]\sigma(\mathrm{d}z) \right| \\ &\leq \frac{\varepsilon}{2} + 2M \int_{\partial D \cap B(w,\delta)^c} K(x, z)\sigma(\mathrm{d}z) \leq \varepsilon. \end{aligned}$$

This shows that

$$\lim_{\substack{x \to w \\ x \in D}} h(x) = f(w),$$

so that $h|_{\partial D} = f$. Hence h is a solution of (D, f). □

Theorem 1.13 *For any ball $B(a, r)$ in $\mathbb{R}^d$ $(d \geq 1)$, the function*

$$K(x, z) = \frac{\Gamma(d/2)}{2\pi^{d/2} r} \cdot \frac{r^2 - |x - a|^2}{|x - z|^d}, \quad x \in B(a, r),\ z \in S(a, r) \tag{24}$$

is the Poisson kernel for $B(a, r)$.

Proof It is sufficient to check conditions (i)–(iii) in Theorem 1.12 for the given $K(x, z)$. Without loss of generality, we may suppose that $a = 0$. For any fixed $z \in S(0, r)$, set

$$g(x) = \begin{cases} -2|x - z| & \text{if } d = 1; \\ -2 \ln |x - z| & \text{if } d = 2; \\ 2(d - 2)^{-1} |x - z|^{-(d-2)} & \text{if } d \geq 3. \end{cases}$$

By Proposition 1.8, g is harmonic in $B(0, r)$. Hence so are its partial derivatives, and therefore also the following linear combination:

$$\sum_{i=1}^{d} \left[z_i \frac{\partial}{\partial x_i} g(x) \right] - \frac{1}{|x - z|^{d-2}} = \frac{r^2 - |x|^2}{|x - z|^d} \quad (d \geq 1).$$

Thus K satisfies condition (i).

Next consider the integral $\int_{S(0,r)} (r^2 - |x|^2)|x - z|^{-d} \sigma(\mathrm{d}z)$. By Theorem 1.9 and Fubini's theorem, this is also harmonic for $x \in B(0, r)$. Inspection shows that it is a function of $|x|$ alone, hence it is of the form (20) in Proposition 1.8. Since its value at $x = 0$ is equal to $r^{2-d}\sigma(S(0, r))$, we must have $c_1 = 0$ and

$c_2 = r^{2-d}\sigma(S(0,r))$ in (20). The value of $\sigma(S(0,r))$ is given in Section 1.1. This verifies condition (ii); condition (iii) is obvious from (24). □

Corollary to Theorem 1.13 *If $h \in C(\overline{B(a,r)})$ and h is harmonic in $B(a,r)$ then for $x \in B(a,r)$:*

$$h(x) = \frac{\Gamma(d/2)}{2\pi^{d/2}r} \int_{S(a,r)} \frac{r^2 - |x-a|^2}{|x-z|^d} h(z)\sigma(\mathrm{d}z). \tag{25}$$

Theorem 1.14 (Harnack's inequality) *Let D be a domain in $\mathbb{R}^d$ $(d \geq 1)$ and A a compact subset of D. Let $H_+(D)$ denote the set of all functions h that are positive and harmonic in D. Then there exists a constant C depending only on D and A such that for all $h \in H_+(D)$, $x \in A$, $y \in A$, we have*

$$h(x) \leq Ch(y).$$

Proof Let $B(a,r) \subset\subset D$. Then (25) holds for any $h \in H_+(D)$. If $x \in B(a, \frac{r}{2})$, then clearly for $z \in S(a,r)$,

$$\begin{aligned}
\frac{2^{d-2}}{3^{d-1}} \frac{r^2}{|a-z|^d} &= \frac{2^{d-2}}{3^{d-1}} \frac{1}{r^{d-2}} \\
&= \frac{3r^2/4}{(3r/2)^d} \\
&\leq \frac{r^2 - |x-a|^2}{|x-z|^d} \leq \frac{2^d}{r^{d-2}} = 2^d \frac{r^2}{|a-z|^d}.
\end{aligned}$$

Hence for each $a \in D$, there is a ball $B_a \subset\subset D$ such that

$$\forall h \in H_+(D),\ \forall x \in B_a : \quad \frac{2^{d-2}}{3^{d-1}} h(a) \leq h(x) \leq 2^d h(a). \tag{26}$$

Now for each $a \in D$, let D_a be the set of all $x \in D$ for which there exists a constant C_x such that

$$\forall h \in H_+(D) : h(x) \leq C_x h(a). \tag{27}$$

It follows from (26) that D_a is an open set containing a and that $D \backslash D_a$ is also open. Since D is connected, we must have $D_a = D$. Thus, for each $a \in D$, (27) is true for all $x \in D$, but the constant C_x may depend on a as well as x.

Suppose the assertion of the theorem is false. Then there exist a compact subset A of D, $a_n \in A$, $x_n \in A$, and $h_n \in H_+(D)$ such that

$$h_n(x_n) \geq nh_n(a_n).$$

Without loss of generality, we may suppose that $a_n \to a \in A$ and $x_n \to x \in A$ and that $a_n \in B_a$, $x_n \in B_x$. Then we have by (26):

$$h_n(x) \geq 2^{-d}h_n(x_n) \geq 2^{-d}nh_n(a_n) \geq \frac{1}{4}3^{1-d}nh_n(a).$$

For these h_n, inequality (27) is false for any C. This contradiction proves the theorem. □

1.5 Exit Time and Place

Suppose that $\{X_t\}$ is a d-dimensional Markov process. For any open or closed set A in $\mathbb{R}^d$, let

$$T_A(\omega) = \inf\{t > 0 : X_t(\omega) \in A\}. \tag{28}$$

(The infimum of an empty set is defined to be ∞). This is called the (first) *hitting time* of A. The hitting time of A^c is called the *exit time* from A and denoted by τ_A. There are measure-theoretic difficulties in dealing with T_A for a general set A (see Chung (1982a)); however, we shall deal with special cases below.

An optional time τ is said to be strictly optional if $\{\tau \leq t\} \in \mathcal{F}_t$ for all $t > 0$. It is easy to see that this implies optionality.

Proposition 1.15 *If all sample functions of $\{X_t\}$ are right continuous, then for any open set A, T_A is an optional time. If all sample functions of $\{X_t\}$ are continuous, then for any closed set A, T_A is strictly optional.*

Proof First suppose that A is open. The right continuity of $\{X_t\}$ implies for each $t > 0$:

$$\{T_A < t\} = \bigcup_{r \in \mathbb{Q} \cap (0,t)} \{X_r \in A\},$$

where $\mathbb{Q}$ is the set of rational numbers. The right member of the above belongs to $\mathcal{F}_t$; hence T_A is optional.

Now suppose that A is closed so that A^c is open. Let B_n, $n \geq 1$, be open sets with $B_n \uparrow\uparrow A^c$. For each $t > 0$, we have

$$\begin{aligned} \{T_A > t\} &= \{\forall s \in (0,t] : X_s \in A^c\} \\ &= \bigcap_{k=1}^{\infty} \bigcup_{n=1}^{\infty} \left\{\forall s \in \left[t(k+1)^{-1}, t\right] : X_s \in \overline{B}_n\right\}. \end{aligned}$$

In the second equation above, we use the fact that the range of the continuous sample function $X(\cdot)$ from $[t(k+1)^{-1}, t]$ $(k \geq 1)$ is a compact set in $\mathbb{R}^d$. By (left as well as right) continuity, we may replace each member on the right side of the above by

$$\left\{\forall s \in \mathbb{Q} \cap \left[t(k+1)^{-1}, t\right] : X_s \in \overline{B}_n\right\},$$

which belongs to $\mathcal{F}_t$. Therefore, we have

$$\{T_A \leq t\} = \{T_A > t\}^c \in \mathcal{F}_t,$$

proving the strict optionality of T_A. □

In the rest of this section, we let $\{X_t : t \in \mathbb{R}_+\}$ be the Brownian motion in $\mathbb{R}^d$, and D be open or closed as in Proposition 1.15. Actually the results are true for any Borel set D, and the only reason that we cannot deal with the general case is the optionality of τ_D used in the proof below!

Proposition 1.16 *Let $0 < \theta < 1$. For all $x \in \mathbb{R}^d$ and $t > 0$, we have*

$$P^x\{\tau_D > t\} \leq \theta^{(t/u)-1},$$

where

$$u = \frac{1}{2\pi}\left[\frac{m(D)}{\theta}\right]^{2/d}.$$

Proof We may suppose $m(D) < \infty$. For $x \in D$ and any $u > 0$:

$$\begin{aligned} P^x\{\tau_D > u\} &\leq P^x\{X_u \in D\} \\ &= \int_D p(t; x, y)\mathrm{d}y \leq \frac{m(D)}{(2\pi u)^{d/2}}. \end{aligned}$$

We denote the last member by θ, then choose u so that $\theta < 1$. Using either the form of the Markov property with respect to $\{\mathcal{F}_t\}$ (see the remark after the Corollary to Theorem 1.2) or the strict optionality of τ_D (Proposition 1.15), we have for all $n \geq 1$:

$$\begin{aligned} P^x\{\tau_D > (n+1)u\} &= E^x\{\tau_D > nu; P^{X(nu)}[\tau_D > u]\} \\ &\leq P^x\{\tau_D > nu\} \cdot \theta, \end{aligned}$$

and consequently by induction,

$$P^x\{\tau_D > nu\} \leq \theta^n. \tag{29}$$

Therefore, for any $t > 0$ we have

$$P^x\{\tau_D > t\} \leq \theta^{[t/u]} \leq \theta^{(t/u)-1},$$

where $[t/u]$ denotes the greatest integer $\leq t/u$. □

Theorem 1.17 *We have*

$$\sup_{x \in \mathbb{R}^d} E^x\{\tau_D\} \leq A_d m(D)^{2/d}, \tag{30}$$

where

$$A_d = \frac{d+2}{2\pi d}\left(\frac{d+2}{2}\right)^{2/d}.$$

Proof By an elementary inequality, we have

$$E^x\left\{\frac{\tau_D}{u}\right\} \leq \sum_{n=0}^{\infty} P^x\left\{\frac{\tau_D}{u} > n\right\};$$

hence by (29),

$$E^x\left\{\frac{\tau_D}{u}\right\} \leq \frac{1}{1-\theta}.$$

From the relationship between u and θ, we obtain

$$E^x\{\tau_D\} \leq \frac{u}{1-Cu^{-d/2}} \text{ where } C = \frac{m(D)}{(2\pi)^{d/2}}.$$

Denoting the right member of the above inequality by $\phi(u)$, we use calculus to determine its minimum value for $u > C^{2/d}$. This is attained at $u_0 = (C(d+2)/2)^{2/d}$ with $\phi(u_0) = (d+2)u_0/d$. The θ_0 corresponding to u_0 is $2/(d+2)$. This proves the assertion. □

Remark It is known that for fixed $m(D)$, the optimal constant A_d in (30) is attained when D is a ball; see Aizenman and Simon (1982). While our estimate is less sharp, our method applies to any Markov process with transition density satisfying a simple inequality; see Chung (1992).

A trivial consequence of (30) is that if $m(D) < \infty$, then $\tau_D < \infty$ almost surely. Whenever the latter holds, the qualification '$\tau_D < \infty$' in various formulae will be omitted.

Proposition 1.18 *Set* $\lambda_0 = \pi d e^{-1} m(D)^{-2/d}$. *Then, for* $0 < \lambda < \lambda_0$, *we have:*

$$\sup_{x\in\mathbb{R}^d} E^x\{e^{\lambda\tau_D}\} \leq 1 + \frac{e^{d/2}\lambda}{\lambda_0 - \lambda}.$$

For each real λ, *as* $m(D)$ *converges to 0 the supremum above converges to 1.*

Proof

$$\begin{aligned} E^x\{e^{\lambda\tau_D}\} &= -\int_0^\infty e^{\lambda t} dP^x\{\tau_D > t\} \\ &= 1 + \lambda\int_0^\infty P^x\{\tau_D > t\}e^{\lambda t}dt \leq 1 + \frac{\lambda}{\theta}\int_0^\infty \theta^{t/u}e^{\lambda t}dt, \end{aligned}$$

where θ and u are as in the proof of Proposition 1.16. A simple computation shows that if we set

$$\lambda(\theta) = -2\pi \ln\theta \left[\frac{\theta}{m(D)}\right]^{2/d},$$

then we have

$$E^x\{e^{\lambda\tau_D}\} \leq 1 + \frac{\lambda}{\theta(\lambda(\theta)-\lambda)}$$

provided $0 < \lambda < \lambda(\theta)$. The function $\lambda(\theta)$, $0 < \theta < 1$, attains its maximum at $\theta_0 = e^{-d/2}$, with $\lambda(\theta_0) = \pi d e^{-1} m(D)^{-2/d}$. Using this value θ_0 for θ in the above we obtain the asserted inequality.

Now for a fixed $\lambda > 0$, as $\lambda_0 \to \infty$, the left member of the inequality converges to 1. Next, by the Cauchy–Schwarz inequality, we have for all x:

$$1 \le E^x\{e^{-\lambda\tau_D}\}E^x\{e^{\lambda\tau_D}\}.$$

Hence the convergence result for $\lambda < 0$ follows from that for $\lambda > 0$. □

There is another way of deriving an inequality like Proposition 1.18 from one like Theorem 1.17, based on a general method explained in Lemma 3.7.

Corollary to Proposition 1.18 *There exist constants $a > 0$ and $b > 0$ such that*

$$\sup_{x\in\mathbb{R}^d} P^x(t < \tau_D) \le a e^{-bt}.$$

The next result is true for a general Markov process having the strong Feller property.

Proposition 1.19 *Let D be a nonempty open set in $\mathbb{R}^d$. For each $t > 0$, the function $x \to P^x(t < \tau_D)$ is upper semi-continuous in $\mathbb{R}^d$.*

Proof As a consequence of the definition of τ_D, for any $s > 0$ on $\{s < \tau_D\}$ we have:

$$\tau_D = s + \tau_D \circ \theta_s. \tag{31}$$

This is true even for an arbitrary set D. Hence, for a fixed $t > 0$ and $0 < s < t$, the set $(t - s < \tau_D \circ \theta_s)$ decreases as $s \downarrow 0$ to

$$(t < \tau_D) = \bigcap\nolimits_{0<s<t} (t - s < \tau_D \circ \theta_s).$$

Therefore,

$$\begin{aligned} P^x(t < \tau_D) &= \lim_{s\downarrow\downarrow 0} \downarrow P^x(t - s < \tau_D \circ \theta_s) \\ &= \lim_{s\downarrow\downarrow 0} \downarrow P^x[\theta_s^{-1}(t - s < \tau_D)]. \end{aligned}$$

By Proposition 1.6, for $0 < s < t$, the function

$$x \to P^x[\theta_s^{-1}(t - s < \tau_D)]$$

is continuous in $\mathbb{R}^d$. Hence the proposition follows. □

Remark For the Brownian motion, we shall prove later (see Theorem 2.2) that the function $P^x(t < \tau_D)$ is actually continuous in D. The following supplement is convenient.

Proposition 1.20 *For any $t > 0$ and $x \in \mathbb{R}^d$, we have*

$$P^x(\tau_D = t) = 0.$$

Proof Let μ be a probability measure on $\mathbb{R}^d$ with a strictly positive density function such as the normal density and consider the distribution of τ_D under P^μ. This has at most a countable set C of discontinuities so that

$$P^\mu(\tau_D = u) = 0 \text{ if } u \notin C.$$

Hence there exists a set $B \subset \mathbb{R}^d$ with $m(B) = 0$ such that

$$P^y(\tau_D = u) = 0 \text{ if } y \notin B,\ u \notin C. \tag{32}$$

Now for a given $t > 0$, let $u \in (0, t)$, $u \notin C$. Then for any x:

$$P^x(\tau_D = t) \leq E^x[P^{X(t-u)}(\tau_D = u)] = 0$$

by (32) because $P^x[X(t-u) \in B] = 0$. □

We turn our attention to the exit place, $X(\tau_D)$, and begin with an intuitively obvious result.

Proposition 1.21 *For any ball $B = B(a, r) \subset \mathbb{R}^d$, and any $A \in \mathcal{B}(S(a, r))$, we have*

$$P^a(X(\tau_B) \in A) = \frac{\sigma(A)}{\sigma(S(a, r))}.$$

Proof Without loss of generality, we may suppose that $a = 0$. Let T denote any rotation in $\mathbb{R}^d$ and set $\tilde{X}_t = T^{-1}(X_t)$. It is easy to see that $\{\tilde{X}_t\}$ and $\{X_t\}$ have the same distribution under P^0. Thus,

$$P^0(X(\tau_B) \in TA) = P^0(\tilde{X}(\tau_B) \in A) = P^0(X(\tau_B) \in A).$$

This shows that the measure on $S(0, r)$, $P^0(X(\tau_B) \in \cdot)$, is invariant under each rotation T. It is well known that the unique probability measure having this property is the uniform distribution on $S(0, r)$. □

1.6 Dirichlet Boundary Value Problem

The classical Dirichlet boundary value problem is not always solvable! The next definition is essential for its solution.

Definition *A point z is called a* regular boundary point *of D iff $z \in \partial D$ and $P^z(\tau_D = 0) = 1$. The set of regular boundary points of D is denoted by $(\partial D)_r$ and D is said to be* regular *iff $\partial D = (\partial D)_r$.*

It is a deep result in potential theory that $\partial D \backslash (\partial D)_r$ is a *polar set*, in particular, $m(\partial D \backslash (\partial D)_r) = 0$; see Chung (1982a, page 186). If $z \in \partial D \backslash (\partial D)_r$, then by Proposition 1.3, we have $P^z(\tau_D = 0) = 0$, i.e. $P^z(\tau_D > 0) = 1$.

A sufficient condition for regularity, known as the cone condition, is given in the following proposition.

Proposition 1.22 *Let $z \in \partial D$. If there exists a cone A with vertex z such that $A \cap B(z, r) \subset D^c$ for some $r > 0$, then z is regular.*

Proof Set

$$C = \frac{\sigma[A \cap S(z,r)]}{\sigma[S(z,r)]} > 0$$

and

$$B_n = B(z, \frac{r}{n}), \quad A_n = A \cap S(z, \frac{r}{n}), \; n \geq 1.$$

We have under P^z, $(\tau_D = 0) \supset \limsup_n (X(\tau_{B_n}) \in A_n)$. Hence

$$\begin{aligned} P^z(\tau_D = 0) &\geq P^z(\limsup_n (X(\tau_{B_n}) \in A_n)) \\ &\geq \limsup_n P^z(X(\tau_{B_n}) \in A_n) = C. \end{aligned}$$

Since $(\tau_D = 0) \in \mathcal{F}_{0+}$, it follows from Proposition 1.3 that $P^z(\tau_D = 0) = 1$. □

Remark Actually, a sharper argument shows that the above cone may be replaced by a flattened one of lower dimension; see Chung (1982a).

We are now ready to solve the classical Dirichlet boundary problem using the methods of probability theory.

Theorem 1.23 *For any domain D and any $f \in L^\infty(\partial D)$, the function $H_D f$ defined in $\mathbb{R}^d$ by*

$$H_D f(x) = E^x[\tau_D < \infty; f(X(\tau_D))] \tag{33}$$

is harmonic in D. If, in addition, $z \in (\partial D)_r$ and f is continuous at z, then

$$\lim_{\substack{x \to z \\ x \in \overline{D}}} H_D f(x) = f(z). \tag{34}$$

Proof This is the first time the strong Markov property is applied, so we shall explain the symbolic formalities in detail. Hereafter, such steps will be left to the reader. Let $x \in D$, and $B = B(x,r) \subset\subset D$. In (9) we take:

$$\tau = \tau_B, \quad A = \{\tau_B < \tau_D\}, \quad \Phi = 1_{\{\tau_D < \infty\}} f(X(\tau_D)).$$

By the remark after Theorem 1.17, $\tau_B < \infty$ almost surely. Since $\tau_B < \tau_D$ by geometry, (31) implies $\tau_D = \tau_B + \tau_D \circ \theta_{\tau_B}$, while (8) implies $X(\tau_D) \circ \theta_{\tau_B} = X(\tau_D)$; whence $\Phi(\theta_\tau) = \Phi$. Therefore, (9) reduces in this case to

$$E^x\{\tau_D < \infty; f(X(\tau_D))\} = E^x\{E^{X(\tau_B)}[\tau_D < \infty; f(X(\tau_D))]\}.$$

Using Proposition 1.21, we may write this as

$$H_D f(x) = \frac{1}{\sigma[S(x,r)]} \int_{S(x,r)} H_D f(y) \sigma(\mathrm{d}y), \quad x \in D,$$

which shows the sphere-averaging property of $H_D f$. Hence $H_D f$ is harmonic in D by Theorem 1.9.

Now suppose that $z \in (\partial D)_r$ and f is continuous at z. Given any $\varepsilon > 0$, there exists $\delta > 0$ such that for all $w \in \partial D \cap B(z, \delta)$, we have

$$|f(w) - f(z)| < \frac{\varepsilon}{2}.$$

We set $M = \|f\|$. Since $P^x\left[\tau_{B(x,\delta/2)} > 0\right] = 1$ by path-continuity, there exists $s > 0$ such that for all x:

$$P^x\left[\tau_{B(x,\delta/2)} \leq s\right] < \frac{\varepsilon}{8M}.$$

Note that this probability does not depend on x. For this fixed s we have by Proposition 1.19:

$$\limsup_{x \to z} P^x(\tau_D > s) \leq P^z(\tau_D > s) = 0, \tag{35}$$

because z is regular. Hence there exists $\delta' > 0$ such that if $|x - z| < \delta'$, then

$$P^x(\tau_D > s) < \frac{\varepsilon}{8M}.$$

Now for any two random variables Y and Z and any $s \in \mathbb{R}^1$, we have $(Y \leq Z) \subset (Y \leq s) \cup (Z > s)$. Hence it follows from the above that

$$P^x\left[\tau_{B(x,\delta/2)} \leq \tau_D\right] \leq P^x\left[\tau_{B(x,\delta/2)} \leq s\right] + P^x\left[\tau_D > s\right] < \frac{\varepsilon}{4M}.$$

If $|x - z| < \frac{\delta}{2}$, then under P^x we have $\tau_{B(x,\delta/2)} \leq \tau_{B(z,\delta)}$ and so

$$P^x\left[\tau_{B(z,\delta)} \leq \tau_D\right] < \frac{\varepsilon}{4M}.$$

Therefore, we have for $x \in \overline{D}$ with $|x - z| < \delta' \wedge (\delta/2)$:

$$\begin{aligned} &E^x[\tau_D < \infty; |f(X(\tau_D)) - f(z)|] \\ &\leq P^x\left[\tau_D < \tau_{B(z,\delta)}\right] \frac{\varepsilon}{2} + P^x\left[\tau_{B(z,\delta)} \leq \tau_D\right] 2M \\ &\leq \frac{\varepsilon}{2} + \frac{\varepsilon}{4M} \cdot 2M = \varepsilon. \end{aligned}$$

Since ε is arbitrary, this implies that

$$\lim_{\substack{x \to z \\ x \in \overline{D}}} E^x\left[\tau_D < \infty; f(X(\tau_D))\right] = \lim_{\substack{x \to z \\ x \in \overline{D}}} P^x(\tau_D < \infty) f(z) = f(z).$$

The last equality follows from (35). Thus (34) is proved. □

Theorem 1.23 contains a solution of the classical Dirichlet boundary value problems as defined in Section 1.4. At the same time, it gives a probabilistic representation of a harmonic function in a domain. Both these fundamental results will be extended in Chapter 4. Here we confine ourselves to a bounded domain, as follows.

Theorem 1.24 (a) *Suppose D is a bounded regular domain. Then the function $H_D f$ defined in (33) is the unique solution to the Dirichlet problem (D, f).*

(b) *Suppose D is a bounded domain and h is harmonic in D and continuous in $\overline{D}$. Then we have the representation*

$$h(x) = E^x\{h(X(\tau_D))\}, \quad x \in D. \tag{36}$$

Remark If D is also regular in part (b), then (36) holds for $x \in \overline{D}$.

Proof Let us be more precise here. For part (a), (33) and (34) show that $H_D f$ is a solution of (D, f). Suppose ϕ is another solution of (D, f). Then $H_D f - \phi$ is a solution to the Dirichlet problem $(D, 0)$, By Corollary 1.11, we must have $H_D f - \phi \equiv 0$ in D.

For part (b), let D_n be regular domains such that $D_n \uparrow\uparrow D$. Such sub-domains exist by elementary topology; in fact, we can take each D_n to satisfy the cone condition in Proposition 1.22 (see Appendix to Chapter 1). Then by part (a), we have for each n:

$$h(x) = E^x\{h(X(\tau_{D_n}))\}, \ x \in D_n.$$

Note that $\tau_{D_n} < \tau_D < \infty$ almost surely. Since $h \in C(\overline{D})$, h is bounded in $\overline{D}$ as well as continuous. As $n \to \infty$, $\tau_{D_n} \uparrow\uparrow \tau_D$ and $h(X(\tau_{D_n})) \to h(X(\tau_D))$ by the continuity of paths and that of h. Therefore (36) follows by bounded convergence. □

In certain formulations of the boundary value problem it is customary to define a *harmonic measure* as follows. For each $x \in D$ and $B \in \mathcal{B}(\partial D)$, let

$$H(x, B) = P^x\{\tau_D < \infty; X(\tau_D) \in B\}; \tag{37}$$

so that (33) may be written as

$$H_D f(x) = \int_{\partial D} f(z) H(x, \mathrm{d}z), \ x \in D.$$

Then $H(\cdot, B)$ is harmonic in D by Theorem 1.23, and $H(x, \cdot)$ is obviously a measure on ∂D with total mass $P^x\{\tau_D < \infty\} \leq 1$. When D is bounded and regular and the Poisson kernel for D defined in Section 1.4 exists it follows from Theorem 1.24(a) and (23) that

$$H(x, \mathrm{d}z) = K(x, z)\sigma(\mathrm{d}z). \tag{38}$$

For example, when D is a ball, K is given by (24). While analysts may feel more comfortable with such a formula, it should be clear from our treatment here that the completely general formula (37) is scarcely worth an explicit definition. Indeed, $H(x, \cdot)$ is simply the distribution of $X(\tau_D)$ under P^x. In Chapter 5, the deeper idea of conditioning with respect to $X(\tau_D)$ will be explored.

Appendix to Chapter 1

Theorem A.1 *Let D be an arbitrary domain in $\mathbb{R}^d$, $d \geq 1$. Then for any compact set A in D, there exists a bounded regular domain U such that $A \subset U \subset\subset D$.*

Proof Let $x_0 \in D$ and $B^N = B(x_0, N)$, $N = 1, 2, \cdots$, $D^N = D \cap B(x_0, N)$, and let C^N be the connected component of D^N to which x_0 belongs. Since D is connected, for any x in D, $x \neq x_0$, there is a continuous map f from [0,1] into D such that $f(0) = x_0$ and $f(1) = x$. The image of f is a compact subset of D, hence it is contained in D^N for some N, and $x \in C^N$, by definition. This shows that $\cup_{N=1}^{\infty} C^N = D$. Thus, for any compact subset A of D, there exists $N_0 \geq 1$ such that $A \subset C^{N_0}$. Consequently, it is sufficient to prove the theorem for a bounded D.

Let $x_0 \in D$ and $\rho(x_0, \partial D) = \varepsilon_0$. From here on $0 < \varepsilon < \varepsilon_0$. We define

$$D_\varepsilon = \{x \in D : \rho(x, \partial D) > \varepsilon\}.$$

Then $x_0 \in D_\varepsilon$. Let us first prove that D_ε is regular. If $y \in \partial D_\varepsilon$, then $\rho(y, \partial D) = \varepsilon$ and there exists $z \in \partial D$ such that $|y - z| = \varepsilon$. Thus $y \in \overline{B}(z, \varepsilon)$, but $\overline{B}(z, \varepsilon)$ does not intersect D_ε. Hence any point on ∂D_ε belongs to a closed ball lying outside D_ε; therefore, by obvious geometry it also belongs to a truncated cone lying outside D_ε. This implies that D_ε is regular by Proposition 1.22.

However, D_ε need not be connected. Let E_ε denote the component of D_ε containing x_0. Then $\partial E_\varepsilon \subset \partial D_\varepsilon$ and the (probabilistic) definition of regularity implies that E_ε is regular since D_ε is. We now prove that $\cup_{\varepsilon>0} E_\varepsilon = D$. Let $x \in D$, then since D is connected there is a continuous map f from $[0, 1]$ into D such that $f(0) = x_0$ and $f(1) = x$. We set

$$\delta = \frac{1}{2} \inf_{0 \leq t \leq 1} \rho(f(t), \partial D).$$

Then $0 < \delta \leq \varepsilon_0/2$, and for all $t \in [0, 1]$ we have $\rho(f(t), \partial D) \geq 2\delta$. Therefore, the image of f is contained in D_δ and so x_0 and x belong to the same component of D_δ; in other words, $x \in E_\delta$. Thus $\cup_{\varepsilon>0} E_\varepsilon = D$.

Now for any compact set $A \subset D$, there exists $\varepsilon > 0$ such that $A \subset E_\varepsilon$. Thus the theorem follows by taking $U = E_\varepsilon$. □

We are indebted to Neil Falkner for most of the preceding elegant proof, not easily found in books.

From Theorem A.1 we deduce the following approximation results which are used in the present book.

Theorem A.2 (i) *For an arbitrary domain D, there exists a sequence of regular bounded domains $\{D_n\}$ such that $D_n \uparrow\uparrow D$.*

(ii) *For an arbitrary bounded domain D, there exists a sequence of regular bounded domains $\{D_n\}$ such that $D_n \downarrow\downarrow \overline{D}$.*

Proof (i) Apply Theorem A.1 with the given D and $A = \{x_0\}$, where x_0 is any point in D. Denote the corresponding U by D_1. Now for any $n \geq 1$, suppose we have already constructed a bounded regular $D_n \subset\subset D$. We set

$$\delta_n = \frac{1}{n} \wedge \frac{1}{2}\rho(D_n, \partial D),$$

and

$$A_n = \overline{B}(x_0, n) \cap \{x \in D : \rho(x, \partial D) \geq \delta_n\}.$$

Apply Theorem A.1 with $A = A_n$ and denote the corresponding U by D_{n+1}. It is easy to see that the sequence $\{D_n\}$ satisfies the requirements.

(ii) For a bounded domain D, let D_1 be a ball containing $\overline{D}$. For $n \geq 1$, suppose we have already constructed a regular domain D_n such that $D_n \supset\supset D$. We set

$$E_n = \{x \in \mathbb{R}^d : \rho(x, \overline{D}) < \frac{1}{2}\rho(\partial D_n, \overline{D})\}.$$

Then E_n is a domain since $E_n = \cup_{x \in \overline{D}} B(x, \frac{1}{2}\rho(\partial D_n, \overline{D}))$. Apply Theorem A.1 with $D = E_n$ and $A = \overline{D}$ and denote the corresponding U by D_{n+1}. The sequence $\{D_n\}$ satisfies the requirements. □

Theorem A.2 is sufficient for applications in this book except for a few instances such as Proposition 2.12. There we need the approximating domains D_n to belong to C^2 in order to apply a theorem by Widman and the divergence theorem. In fact, we can even make them all belong to C^∞ using well-known techniques in analysis (see Edmunds and Evans (1987, Chapter 5, Theorem 4.20)). But we must stop here.

Notes on Chapter 1

The Brownian motion process is also known as Wiener space or measure. Its mathematical foundation was laid in Wiener (1923), in which it was proved that all sample functions may be taken to be continuous. That proof uses Daniell's extension theorem, but another proof was given in the book by Paley and Wiener (1934) using Fourier transform theory. Paul Lévy in his autobiography (Lévy 1970) regretted that he had missed this fundamental discovery. Theorem 1.4 follows Lévy's proof given in Lévy (1948, Chapter 7): it is undoubtedly the most direct way. Lévy gave another more ingenious construction in Lévy (1970) using the interpolation of normal random variables. It is curious to note that another proof by Loève given in the appendix there quite missed the point and was corrected later in the second edition with the addition of a note about 'separability'. Nowadays one can also derive the result from a general criterion beginning with a right-continuous Markov process as considered in Theorem 1.2 see e.g. Chung (1982a, page 77).

In this book we have tried to tone down the various measurability questions which tend to crowd the exposition at the outset. Thus, we try to make do with the natural filtration $\{\mathcal{F}_t\}$ without augmentation. The reader may consult Chung (1982a, Chapter 2) for certain ramifications of the general notions briefly and sparingly discussed in Sections 1.2 and 1.3.

Wiener constructed his process and solved his generalized Dirichlet problem, but he did not apply the former to the latter. It was Kakutani who made the connections. In Kakutani (1944) he treated the problem in terms of the harmonic measure, 'in the sense of R. Nevanlinna', namely when $f = 1_A$ with $A \subset \partial D$, in Theorem 1.23. In Kakutani (1945) he considered the problem of a positive linear mapping from $C(\partial D)$ to $H(\overline{D})$ (the space of functions continuous in $\overline{D}$ and harmonic in D) and obtained the solution as the limit of a sequence of 'sweeping-out' (balayage) operations of the form $f \to H_{B_n} f$ where the B_n are balls with closures in D. Whereas $H_{B_n} f$ can be expressed analytically by the Poisson formula, only the purely probabilistic incarnation exists for $H_D f$ given in (33). Let us quote a sentence from Kakutani (1944) to illustrate the historical perspective. Referring to the argument for (34) with f as in the above, he said, '[it] is also intuitively clear; but it is not so easy to prove it in a rigorous way'.

2. Killed Brownian Motion

2.1 Feller Properties

Let $\{X_t\}$ be the d-dimensional Brownian motion and D a domain in $\mathbb{R}^d$ $(d \geq 1)$. Adjoin an extra point ∂ to D and set

$$X_t^D = \begin{cases} X_t & \text{on } (t < \tau_D) \\ \partial & \text{on } (t \geq \tau_D). \end{cases}$$

This is called the Brownian motion killed outside D. Its state space is $D_\partial = D \cup \{\partial\}$ and its transition function is given by

$$P_t^D(x, A) = P^D(t; x, A) = P^x(t < \tau_D; X_t \in A), \quad t > 0, \; x \in D, \; A \in \mathcal{B}(D).$$

Thus, for $t > 0$, $x \in D$, and $f \in L^\infty(D)$:

$$P_t^D f(x) = \int_D P^D(t; x, \mathrm{d}y) f(y) = E^x[t < \tau_D; f(X_t)]. \tag{1}$$

If we use the convention that $f(\partial) = 0$ for any f, then the last term above is just $E^x\{f(X_t^D)\}$.

In Chapter 1 we proved that the Brownian motion $\{X_t\}$ has the Feller property (Proposition 1.7) and the strong Feller property (Proposition 1.5). In this section we shall prove these properties for $\{X_t^D\}$. We need a simple lemma:

Lemma 2.1 *For any compact set $K \subset D$, we have*

$$\lim_{s \downarrow 0} \sup_{x \in K} P^x\{\tau_D \leq s\} = 0.$$

Proof Set

$$r = \rho(K, \partial D) > 0.$$

Then for any $x \in K$, we have

$$P^x\{\tau_D \leq s\} \leq P^x\{\tau_{B(x,r)} \leq s\}, \quad s > 0.$$

For any $x \in \mathbb{R}^d$,

$$P^x\{\tau_{B(x,r)} \leq s\} = P^0\{\tau_{B(0,r)} \leq s\}, \quad s > 0.$$

By the continuity of the paths of $\{X_t\}$ we have

$$\lim_{s\downarrow 0} P^0\{\tau_{B(0,r)} \le s\} = 0.$$

This proves the lemma. □

We now give the main result in this section.

Theorem 2.2 *For any domain $D \subset \mathbb{R}^d$ we have*

$$P_t^D f \in C_b(D), \textit{ for any } t > 0, f \in L^\infty(D).$$

Moreover, if D is regular, then $P_t^D f \in C_0(D)$ for any $f \in C_0(D)$. In the latter case, $\{X_t^D\}$ on D has both the Feller and the strong Feller property.

Proof For $f \in L^\infty(D)$, fixed $t > 0$ and $0 < s < t$, we have

$$P_t^D f(x) = E^x\{s < \tau_D; E^{X_s}[t - s < \tau_D; f(X_{t-s})]\}. \tag{2}$$

Set

$$\phi_s(x) = E^x[t - s < \tau_D; f(X_{t-s})].$$

By the Corollary to Theorem 1.1, $\phi_s \in L^\infty(D)$. By the strong Feller property (Proposition 1.5), $P_s\phi_s \in C_b(\mathbb{R}^d)$. Clearly, by (2), we have

$$|P_s\phi_s(x) - P_t^D f(x)| \le P^x(\tau_D \le s)\|f\|_\infty.$$

By Lemma 2.1, this converges uniformly to zero as $s \downarrow 0$ on any compact subset of D. Hence $P_t^D f$ is continuous in D, and so $P_t^D f \in C_b(D)$.

If D is regular, then for all $z \in \partial D$ we have by (1.35):

$$\limsup_{x\to z} |P_t^D f(x)| \le \|f\|_\infty \limsup_{x\to z} P^x(t < \tau_D) = 0.$$

If D is unbounded, then we must also check that

$$\lim_{x\to\infty} P_t^D f(x) = 0, \text{ for any } t > 0 \text{ and } f \in C_0(D).$$

This is trivial because $|f| \in C_0(D)$ and

$$\lim_{x\to\infty} P_t^D|f|(x) \le \lim_{x\to\infty} P_t|f|(x) = 0.$$

Thus $P_t^D f \in C_0(D)$ and we have proved that P_t^D maps $C_0(D)$ into $C_0(D)$. Next we have

$$P_t^D f(x) - f(x) = \int_D P_t^D(x, \mathrm{d}y)[f(y) - f(x)] - f(x)[1 - P_t^D(x, D)]$$

and consequently

$$|P_t^D f(x) - f(x)| \le \int_D P_t^D(x, \mathrm{d}y)|f(y) - f(x)| + |f(x)|P^x\{t \ge \tau_D\}.$$

The integral on the right side of the above converges to zero uniformly for all x in $\mathbb{R}^d$, as shown in the proof of Proposition 1.7. Now for any compact subset K of D we have

$$\sup_{x\in D} |f(x)|P^x\{t \geq \tau_D\} \leq \sup_{x\in D\setminus K} |f(x)| + \|f\| \sup_{x\in K} P^x\{t \geq \tau_D\}.$$

We can choose K so that the first 'sup' on the right side in the above is as small as we please, because $f \in C_0(D)$; then the second 'sup' converges to zero as $t \downarrow 0$ by Lemma 2.1. It follows that

$$\lim_{t\downarrow 0} \|P_t^D f - f\|_\infty = 0,$$

which is the second condition of the Feller property. □

As a consequence of Theorem 2.2, we have

Proposition 2.3 *Let D be a regular domain in $\mathbb{R}^d$. Then $\{P_t^D\}$ forms a strongly continuous operator semigroup in $C_0(D)$.*

2.2 Transition Density

In this section, we shall construct the density function for P_t^D, $t > 0$ and establish its fundamental properties.

For $t > 0$, x, $y \in \mathbb{R}^d$, set

$$r^D(t;x,y) = E^x[\tau_D < t; p(t-\tau_D; X(\tau_D), y)] \tag{3}$$

and

$$p^D(t;x,y) = p(t;x,y) - r^D(t;x,y), \tag{4}$$

where $p(t;x,y)$ is given in (1.11).

The following result is due to G.A. Hunt (Hunt 1956).

Theorem 2.4 *Let D be a domain in $\mathbb{R}^d$. Then for any $t > 0$,*

$$P^x(t < \tau_D; X_t \in A) = \int_A p^D(t;x,y)\mathrm{d}y, \quad x \in \mathbb{R}^d, A \in \mathcal{B}(\mathbb{R}^d). \tag{5}$$

The function $p^D(t;\cdot,\cdot)$ is symmetric on $\mathbb{R}^d \times \mathbb{R}^d$, continuous on $(\mathbb{R}^d\setminus\partial D) \times (\mathbb{R}^d\setminus\partial D)$, and strictly positive on $D \times D$.

For any $t > s > 0$, x, $y \in \mathbb{R}^d$, we have

$$p^D(t;x,y) = \int_{\mathbb{R}^d} p^D(s;x,z)p^D(t-s;z,y)\mathrm{d}z. \tag{6}$$

For any $t > 0$, $y \in D$ and $z \in (\partial D)_r$, we have

$$\lim_{\substack{x\to z\\ x\in D}} p^D(t;x,y) = 0. \tag{7}$$

Proof For any $t > 0$, $A \in \mathcal{B}(\mathbb{R}^d)$ and $x \in \mathbb{R}^d$, we have by Proposition 1.20,

$$\begin{aligned} P^x(t < \tau_D; X_t \in A) &= P^x(X_t \in A) - P^x(\tau_D \leq t; X_t \in A) \\ &= \int_A p(t; x, y)\mathrm{d}y - P^x(\tau_D < t; X_t \in A). \end{aligned} \tag{8}$$

For any $0 < u < t$, $n \geq 1$, and $1 \leq k \leq 2^n$, set

$$T_n = \begin{cases} ku2^{-n} & \text{if } (k-1)u2^{-n} \leq \tau_D < ku2^{-n} \\ \infty & \text{if } \tau_D \geq u. \end{cases}$$

Then

$$\begin{aligned} &P^x(\tau_D < u; X_t \in A) \\ &= \sum_{k=1}^{2^n} P^x\left(\frac{(k-1)u}{2^n} \leq \tau_D < \frac{ku}{2^n}; X_t \in A\right) \\ &= \sum_{k=1}^{2^n} E^x\left[\frac{(k-1)u}{2^n} \leq \tau_D < \frac{ku}{2^n}, P^{X(ku2^{-n})}(X(t - ku2^{-n}) \in A)\right] \\ &= E^x[\tau_D < u; \int_A p(t - T_n; X(T_n), y)\mathrm{d}y]. \end{aligned} \tag{9}$$

On $(\tau_D < u)$, $t - T_n \geq t - u > 0$ $(n \geq 1)$ and $T_n \to \tau_D$. By letting first $n \to \infty$, then $u \uparrow t$ in (9), we obtain

$$\begin{aligned} P^x(\tau_D < t; X_t \in A) &= \int_A E^x[\tau_D < t; p(t - \tau_D; X(\tau_D), y)]\mathrm{d}y \\ &= \int_A r^D(t; x, y)\mathrm{d}y. \end{aligned}$$

Thus (5) follows from (8)

Next we prove that $r^D(t; \cdot, \cdot)$ is continuous and symmetric on $D \times D$. For any u and v both in D, take open neighbourhoods $U \subset\subset D$ and $V \subset\subset D$ of u and v, respectively. Set

$$\delta = \min[\rho(U, \partial D), \rho(V, \partial D)] > 0.$$

It is easy to see that for any $\alpha > 0$, $p(t; x, y)$ is bounded and uniformly continuous on the set $\{(t; x, y) : t > 0, |x - y| \geq \alpha, \ x, y \in \mathbb{R}^d\}$. Set

$$M_\alpha = \sup_{\substack{t>0 \\ |x-y|\geq\alpha}} p(t; x, y).$$

For any $0 < s < t$, $x \in U$, $y \in V$, we have

$$\begin{aligned} r^D(t; x, y) &= E^x[\tau_D \leq s; p(t - \tau_D; X(\tau_D), y)] \\ &\quad + E^x[s < \tau_D; E^{X(s)}[\tau_D < t - s; \ p(t - s - \tau_D; X(\tau_D), y)]]. \end{aligned}$$

Thus

$$|r^D(t;x,y) - E^x[E^{X(s)}[\tau_D < t-s;\ p(t-s-\tau_D; X(\tau_D), y)]]|$$
$$\leq 2M_\delta P^x(\tau_D \leq s) \downarrow 0 \qquad (10)$$

as $s \downarrow 0$, uniformly on $U \times V$.

Set

$$\phi_s(x,y) = E^x[E^{X(s)}[\tau_D < t-s;\ p(t-s-\tau_D; X(\tau_D), y)]].$$

For fixed $y \in V$, $\phi_s(\cdot, y)$ is continuous on U by the strong Feller property (Proposition 1.5), while the family $\{\phi_s(x,\cdot), x \in U\}$ is equi-continuous on V by the uniform continuity of $p(t;x,y)$ on $(t > 0, |x-y| \geq \delta)$. It follows that $\phi_s(\cdot,\cdot)$ is jointly continuous on $U \times V$, as is $r^D(t;,\cdot,\cdot)$ by (10).

Next, take a sequence of domains $(D_n : n \geq 1)$ such that $D_n \uparrow\uparrow D$. Then, for any A and B in $\mathcal{B}(D)$, we have, using the symmetry of $p(s;\cdot,\cdot)$ for every $s > 0$,

$$\int_D \int_D p^D(t;x,y) 1_A(x) 1_B(y) \mathrm{d}x\mathrm{d}y = \int_A P^x(t < \tau_D; X_t \in B)\mathrm{d}x$$
$$= \lim_n \lim_m \int_A P^x\left[X\left(\frac{kt}{2^m}\right) \in \overline{D}_n, k = 1, 2, \cdots, 2^m - 1; X_t \in B\right] \mathrm{d}x$$
$$= \lim_n \lim_m \int_{\underbrace{\overline{D}_n \times \cdots \times \overline{D}_n}_{2^m+1 \text{ factors}}} 1_A(x_0) \prod_{i=0}^{2^m-1} p\left(\frac{t}{2^m}; x_i, x_{i+1}\right) 1_B(x_{2^m}) \mathrm{d}x_0 \mathrm{d}x_1 \cdots \mathrm{d}x_{2^m}$$
$$= \lim_n \lim_m \int_{\underbrace{\overline{D}_n \times \cdots \times \overline{D}_n}_{2^m+1 \text{ factors}}} 1_B(x_{2^m}) \prod_{i=0}^{2^m-1} p\left(\frac{t}{2^m}; x_{i+1}, x_i\right) 1_A(x_0) \mathrm{d}x_{2^m} \cdots \mathrm{d}x_0$$
$$= \int_D \int_D p^D(t;x,y) 1_B(x) 1_A(y) \mathrm{d}x\mathrm{d}y.$$

This shows that

$$p^D(t;x,y) = p^D(t;y,x), \qquad (m \times m)\text{-a.e. on } D \times D.$$

The joint continuity of $p^D(t;\cdot,\cdot)$ on $D \times D$ then implies that the same is true for all $(x,y) \in D \times D$.

For any $t > s > 0$, we have by (5) and the Markov property,

$$\int_A p^D(t;x,y)\mathrm{d}y = E^x[s < \tau_D; P^{X(s)}[t-s < \tau_D; X_{t-s} \in A]]$$
$$= \int_A \left[\int_D p^D(s;x,z) p^D(t-s;z,y) \mathrm{d}z\right] \mathrm{d}y,$$

for any $A \in \mathcal{B}(D)$. Hence (6) follows by continuity.

To prove that p^D is strictly positive, we first prove that for any $c > 0$, if $0 < t \leq \frac{c^2}{d}$ and $|x-y| < c \leq \rho(x, \partial D) \wedge \rho(y, \partial D)$, then $p^D(t;x,y) > 0$. It is easy to verify that $(2\pi t)^{-d/2}\exp(-c^2/2t)$ is increasing in t for $0 < t \le c^2/d$. Hence, for the above t, x, y, we have

$$r^D(t;x,y) \le (2\pi t)^{-d/2}\exp(-c^2/2t),$$

and consequently

$$p^D(t;x,y) \ge (2\pi t)^{-d/2}[\exp(-|x-y|^2/2t) - \exp(-c^2/2t)] > 0.$$

Now for any $t > 0$ and $x, y \in D$, we connect x and y by a curve Γ in D such that

$$r = \rho(\Gamma, \partial D) > 0.$$

Then we can choose a sufficiently large integer n such that $\frac{t}{n} \le \frac{r^2}{4d}$ and such that there exist points $a_0, a_1, \cdots, a_{n+1}$ on Γ with $a_0 = x$, $a_{n+1} = y$ and $a_i \in B(a_{i-1}, \frac{r}{6})$, $i = 1, 2, \cdots, n+1$. Then for any $x_i \in B(a_i, \frac{r}{6})$, we have $|x_i - x_{i+1}| \le |x_i - a_i| + |a_i - a_{i+1}| + |a_{i+1} - x_{i+1}| < \frac{r}{2}$ and $\rho(x_i, \partial D) \ge r - \frac{r}{6} > \frac{r}{2}$. Therefore, by (5), we have

$$\begin{aligned} p^D(t;x,y) &= \int_D \cdots \int_D p^D(\frac{t}{n};x,x_1)p^D(\frac{t}{n};x_1,x_2)\cdots p^D(\frac{t}{n};x_n,y)\mathrm{d}x_1\cdots\mathrm{d}x_n \\ &\ge \int_{B(a_1,\frac{r}{6})} \cdots \int_{B(a_n,\frac{r}{6})} p^D(\frac{t}{n};x,x_1)\cdots p^D(\frac{t}{n};x_n,y)\mathrm{d}x_1\cdots\mathrm{d}x_n > 0. \end{aligned}$$

Now suppose that $z \in (\partial D)_r$. For any $t > 0$, $y \in D$, we have by Theorem 1.23,

$$\lim_{x\to z} E^x[\tau_D < \infty; p(t; X(\tau_D), y)] = p(t;z,y). \tag{11}$$

Set $\lambda = \rho(y, \partial D)$. For any $\varepsilon > 0$, there exists $\delta > 0$ such that for any $t > 0$, $s > 0$, $|t-s| \le \delta$ and $|x-y| \ge \lambda$, we have

$$|p(t;x,y) - p(s;x,y)| < \varepsilon.$$

Thus

$$\begin{aligned} &|r^D(t;x,y) - E^x[\tau_D < \infty; p(t;X(\tau_D),y)]| \\ &\quad \le E^x[\tau_D < t; |p(t-\tau_D; X(\tau_D), y) - p(t; X(\tau_D), y)|] + M_\lambda P^x(\tau_D \ge t) \\ &\quad \le \varepsilon + 2M_\lambda P^x(\tau_D \ge \delta) + M_\lambda P^x(\tau_D \ge t). \end{aligned}$$

Since for any $u > 0$, $\lim_{x\to z} P^x(\tau_D \ge u) = 0$, it follows from (11) that

$$\lim_{x\to z} r^D(t;x,y) = p(t;z,y).$$

This is the last assertion of the theorem. □

2.3 Green Potential and Function

For a general Markov process $\{X_t\}$ with the transition function $P(t; x, y)$ as in Section 1.1, its *potential kernel* $U(\cdot, \cdot)$ is defined as follows:

$$U(x, B) = \int_0^\infty P(t; x, B)\mathrm{d}t = E^x \left\{ \int_0^\infty 1_B(X_t)\mathrm{d}t \right\},$$

provided that for each x and B, $P(\cdot; x, B)$ is a Borel measurable function on $\mathbb{R}_+$. This is the case when $\{X_t\}$ is the killed Brownian motion. The equality of the two expressions above is due to the Fubini–Tonelli theorem, whether the result is finite or infinite. When $D = \mathbb{R}^d$ and the transition density $p(t; x, y)$ exists and is Borel measurable in t, we can define the *potential density* by

$$u(x, y) = \int_0^\infty p(t; x, y)\mathrm{d}t,$$

provided this integral is finite. For any $f \in \mathcal{B}(\mathbb{R}^d)$, the function Uf defined by

$$Uf(x) = \int_0^\infty P_t f(x)\mathrm{d}t = E^x \left\{ \int_0^\infty f(X_t)\mathrm{d}t \right\} \tag{12}$$

is called the potential of f, provided it is well defined, i.e. when Uf^+ or Uf^- is finite. When the potential density exists, we therefore have

$$Uf(x) = \int_{\mathbb{R}^d} u(x, y)f(y)\mathrm{d}y.$$

Now we apply these general notions to the free and killed Brownian motion processes in $\mathbb{R}^d$. Deferring to a time-honored notation, we write G_D for both the U and u above; thus

$$G_D f(x) = \int_0^\infty P_t^D f(x)\mathrm{d}t = E^x \left\{ \int_0^{\tau_D} f(X_t)\mathrm{d}t \right\} \tag{13}$$

and

$$G_D(x, y) = \int_0^\infty p^D(t; x, y)\mathrm{d}t, \quad (x, y) \in D \times D. \tag{14}$$

The operator G_D is the *Green potential* operator, and the function G_D is the *Green function*. (There is an unresolved dispute as to whether the syntax would be improved if we were to use the expression "Green's function" instead).

Let us pause to confirm that the third member in (13) is a Borel measurable function of x, provided it is well defined. We may suppose that $f \in \mathcal{B}_+$. First of all, by a remark in Section 1.2, the function $(t, \omega) \to f(X(t, \omega))$ is $\mathcal{B}^1 \times \mathcal{F}$ measurable. Hence so is the Lebesgue integral $\int_0^T f(X(t, \omega))\mathrm{d}t$ for any constant T. Since $\tau_D \in \mathcal{F}$ by Proposition 1.15, it follows that $\int_0^{\tau_D} f(X_t)\mathrm{d}t \in \mathcal{F}$. Hence the third member in (13) belongs to $\mathcal{B}^d$. Having spelt out this argument here, we may pass over similar questions of measurability elsewhere.

Observe that $G_D f$ is defined by (13) for x and f in $\mathbb{R}^d$ and not merely in D, while $G_D(x,y)$ is defined by (14) for $(x,y) \in \mathbb{R}^d \times \mathbb{R}^d$, because $p^D(t;x,y)$ is. However, it is often expedient to consider the Green function $G_D(\cdot,\cdot)$ on $D \times D$ only. This is in conformity with our definitions at the beginning of Section 2.1 and spares us some pains involving the crossing of the boundary of D.

We begin with the case $D = \mathbb{R}^d$. When $d \geq 3$, a direct elementary calculation using (1.11) yields the vital formula:

$$\int_0^\infty p(t;x,y)\mathrm{d}t = \frac{C_d}{|x-y|^{d-2}}, \quad C_d = \frac{\Gamma(\frac{d}{2}-1)}{2\pi^{d/2}}. \tag{15}$$

For $d = 1$ or 2, the corresponding result is ∞, and a compensated potential density is obtained as follows (see Chung (1982a)). For $d = 2$, we set $x_0 = (0,0)$ and $x_\delta = (0,\delta)$, so that

$$p(t;x_0,x_\delta) = \frac{1}{2\pi t}\mathrm{e}^{-\frac{\delta^2}{2t}}.$$

Then we have

$$\int_0^\infty [p(t;x,y) - p(t;x_0,x_\delta)]\mathrm{d}t = \frac{1}{\pi}\ln\frac{\delta}{|x-y|}. \tag{16}$$

For $d = 1$, we have

$$\int_0^\infty [p(t;x,y) - p(t;0,0)]\mathrm{d}t = -|x-y|.$$

We define the function g in $\mathbb{R}^d$ as follows:

$$g(x) = \begin{cases} C_d|x|^{-(d-2)} & \text{if } d \geq 3; \\ \pi^{-1}\ln|x|^{-1} & \text{if } d = 2; \\ -|x| & \text{if } d = 1. \end{cases} \tag{17}$$

For $(x,y) \in \mathbb{R}^d \times \mathbb{R}^d$, $d \geq 1$, we set

$$g(x,y) = g(x-y) = g(|x-y|).$$

A domain D will be called *Greenian* iff $G_D(x,y) < \infty$ for $(x,y) \in D \times D$, $x \neq y$. It follows from (14) and (15) that any domain in $\mathbb{R}^d$, $d \geq 3$ is Greenian. In $\mathbb{R}^1$, any domain except $\mathbb{R}^1$ itself is Greenian. In fact, the Green function of an arbitrary bounded or unbounded interval except $(-\infty,\infty)$ will be given explicitly at the end of this section. The situation for $\mathbb{R}^2$ is more complicated (see Hunt (1956)) and will be treated only under the special assumption below, which is referred to later as 'Green-boundedness' for all $d \geq 1$:

$$\|G_D 1\|_\infty = \sup_x E^x\{\tau_D\} < \infty, \tag{18}$$

where it makes no difference whether the 'sup' is taken over D or $\mathbb{R}^d$. This assumption is adequate for our later applications. By Theorem 1.17, a domain D

with $m(D) < \infty$ is Green-bounded. An example of a Green-bounded domain with $m(D) = \infty$ is the infinite strip in $\mathbb{R}^2$: $\{(x, y) \in \mathbb{R}^2 : |x| < a\}$ ($a > 0$). We shall return to this example in Section 4.2.

We begin with a basic relationship between the Green functions g and G_D, known as the 'balayage formula'.

Theorem 2.5 *Let D be any domain in $\mathbb{R}^d$, $d \geq 3$, or any Green-bounded domain in $\mathbb{R}^2$. Then for $(x, y) \in D \times D$ we have:*

$$G_D(x, y) = g(x, y) - E^x\{\tau_D < \infty; g(X(\tau_D), y)\}. \tag{19}$$

The last term in (19) is a harmonic function of x in D.

Proof In what follows, we shall write G for G_D. From the definition of r^D, by the Fubini–Tonelli theorem, and using an obvious substitution for t, we have:

$$\begin{aligned}\int_0^\infty r^D(t; x, y)\mathrm{d}t &= \int_0^\infty E^x\{\tau_D < t; p(t - \tau_D; X(\tau_D), y)\}\mathrm{d}t \\ &= E^x\{\tau_D < \infty; \int_0^\infty p(t; X(\tau_D), y)\mathrm{d}t\},\end{aligned}$$

which reduces to the expectation given in (19). This result holds whether the above quantity is finite or infinite.

For $d \geq 3$, we have

$$E^x\{\tau_D < \infty; g(X(\tau_D), y)\} \leq g(\delta) = \frac{C_d}{\delta^{d-2}} < \infty,$$

where $\delta = \rho(y, \partial D)$. Hence, by (14) and (15), we obtain

$$G(x, y) = \int_0^\infty p(t; x, y)\mathrm{d}t - \int_0^\infty r^D(t; x, y)\mathrm{d}t,$$

which is (19). It follows that $0 \leq G(x, y) < \infty$ if $x \neq y$, and $G(x, x) = \infty$ for all $(x, y) \in D \times D$.

For $d = 2$, we take a detour using Laplace transforms. For each $\delta > 0$, we have by monotone convergence:

$$\begin{aligned}G(x, y) &= \lim_{s\downarrow 0} \int_0^\infty \mathrm{e}^{-st}[p(t; x, y) - r^D(t; x, y)]\mathrm{d}t \\ &= \lim_{s\downarrow 0} \int_0^\infty \mathrm{e}^{-st}[p(t; x, y) - p(t; x_0, x_\delta)]\mathrm{d}t \\ &\quad - \lim_{s\downarrow 0} \int_0^\infty \mathrm{e}^{-st}[r^D(t; x, y) - p(t; x_0, x_\delta)]\mathrm{d}t.\end{aligned}$$

The first limit in the above is equal to $\frac{1}{\pi} \ln \frac{\delta}{|x-y|} = g(\frac{|x-y|}{\delta})$ by (16). After a substitution for t, the integral in the second limit may be written as

$$E^x\left\{\int_0^\infty \mathrm{e}^{-s(t+\tau_D)}[p(t;X(\tau_D),y)-p(t;x_0,x_\delta)]\mathrm{d}t\right\}$$
$$+E^x\left\{\int_0^\infty \mathrm{e}^{-st}(\mathrm{e}^{-s\tau_D}-1)p(t;x_0,x_\delta)\mathrm{d}t\right\}. \tag{20}$$

Now choose $\delta = \rho(y,\partial D)$, so that $|X(\tau_D)-y| \geq \delta$ almost surely. Then the integrand in the first integral in (20) is negative (≤ 0) and therefore, by monotone convergence, the limit of the first expectation in (20) as $s \downarrow 0$ is equal to $E^x\{g(|X(\tau_D)-y|/\delta)\}$ by (16), noting that $\tau_D < \infty$ (a.s.) for a Green-bounded D. For the second expectation in (20), we use the inequality $1-\mathrm{e}^{-s\tau_D} \leq s\tau_D$ to obtain the upper bound

$$E^x\{\tau_D\}s\int_0^\infty \mathrm{e}^{-st}p(t;x_0,x_\delta)\mathrm{d}t.$$

As $s \downarrow 0$, this converges to zero by calculus. Collecting these results, we have proved that for each $y \in D$ and $\delta = \rho(y,\partial D)$:

$$G(x,y) = g\left(\frac{|x-y|}{\delta}\right) - E^x\left\{g\left(\frac{|X(\tau_D)-y|}{\delta}\right)\right\},$$

which is (19) after cancellation of δ.

Observe that

$$E^x\{\ln|X(\tau_D)-y|\} \geq \ln\delta,$$

so that the preceding equation stands regardless whether the two members are finite or infinite. The question of finiteness is a little more delicate than might have been expected. In fact using $\ln|x| < |x|$, we have:

$$E^x\{\ln|X(\tau_D)-y|\} \leq E^x\{|X(\tau_D)-x|\} + |x-y|. \tag{21}$$

Now it is a well-known property of the Brownian motion process in $\mathbb{R}^d$ that $\{|X(t)-X(0)|^2 - \mathrm{d}t,\ t \geq 0\}$ forms a martingale; so that, for all $t \geq 0$, we have:

$$E^x\{|X(t\wedge\tau_D)-x|^2\} = \mathrm{d}E^x\{t\wedge\tau_D\}.$$

Letting $t \to \infty$, we obtain by Fatou's lemma:

$$E^x\{|X(\tau_D)-x|^2\} \leq \mathrm{d}E^x\{\tau_D\}. \tag{22}$$

(Actually, equality holds in the above, by a longer argument). It follows that the left member in (21) is finite for all x and y in D.

Define the function h on $D \times D$ as follows:

$$h(x,y) = E^x\{g(X(\tau_D),y)\}.$$

It follows from the above that

$$\ln\rho(y,\partial D) \leq -\pi h(x,y) \leq (2E^x\{\tau_D\})^{1/2} + |x-y|.$$

Hence $h(\cdot, y) \in L^1_{\text{loc}}(D)$ by (18). The argument in the proof of Theorem 1.23 shows that $h(\cdot, y)$ has the sphere-averaging property in D. Hence $h(\cdot, y)$ is harmonic in D by Theorem 1.9. □

Corollary to Theorem 2.5 *For each $y \in D$, $G_D(\cdot, y)$ is harmonic in $D \backslash \{y\}$.*

This follows from (19) and the harmonicity of $g(\cdot, y)$ in $\mathbb{R}^d \backslash \{y\}$ by Proposition 1.8.

Remark For $d = 1$, (19) holds if D is bounded. This can be proved by the method used for $d = 2$, but direct verification is simpler. For $D = (0, \infty)$, (19) is false.

Theorem 2.6 *Let D be as in Theorem 2.5 and $G = G_D$. Then we have for $(x, y) \in D \times D$:*

(i) *$G(x, y) = G(y, x)$; $0 < G(x, y) < \infty$ if $x \neq y$; $G(x, x) = \infty$.*

(ii) *For $d \geq 3, G(x, y) \leq g(x, y)$; for $d = 2$,*

$$G(x, y) \leq g^+(x, y) + C, \tag{23}$$

where C is a constant depending only on $\|G_D 1\|_\infty$.

(iii) *$G(\cdot, \cdot)$ is extended continuous in $D \times D$.*

(iv) *For each $z \in (\partial D)_r$, we have*

$$\lim_{x \to z} G(x, y) = 0.$$

(v) *If D is unbounded, we have $\lim_{D \ni y \to \infty} G(x, y) = 0$.*

Proof (i) The strict positivity and symmetry are consequences of these properties for $p^D(t; x, y)$ given in Theorem 2.4. The rest follows from (19).

(ii) For $d \geq 3$, $G \leq g$ is an immediate consequence of (19). To prove (23) for $d = 2$, we begin with the following inequalities, which are valid for $x \neq y$, $x \neq z$:

$$\begin{aligned} \ln \frac{|z-y|}{|x-y|} &\leq \ln\left(1 + \frac{|z-x|}{|x-y|}\right) \leq \ln 2 + (\ln|z-x| - \ln|x-y|)^+ \\ &\leq \ln 2 + (|z-x| - \ln|x-y|)^+ \\ &\leq \ln 2 + |z-x| + \ln^+ \frac{1}{|x-y|}, \end{aligned}$$

where we have used the inequality $\ln|x| < |x|$, as in (21). Hence, if $x \neq y$, we have

$$\pi G(x, y) = E^x \left\{ \ln \frac{|X(\tau_D) - y|}{|x-y|} \right\} \leq \ln 2 + E^x\{|X(\tau_D) - x|\} + \ln^+ \frac{1}{|x-y|}.$$

Substituting this into (19) and using (22) as before, we obtain

$$G(x, y) \leq g^+(x, y) + \frac{\ln 2}{\pi} + \frac{1}{\pi}(2G_D 1(x))^{1/2}.$$

Since D is assumed to be Green-bounded, (23) now follows.

(iii) For $d \geq 3$, we give a proof based on Theorem 2.4. First let $x_0 \in D$, $y_0 \in D$, $x_0 \neq y_0$. Set $|x_0 - y_0| = 2\delta > 0$, and let $x_n \to x_0$, $y_n \to y_0$; then $|x_n - y_n| \geq \delta$ for all $n \geq n_0$. For each $t_0 > 0$, we have

$$\int_{t_0}^{\infty} p^D(t;x,y)\mathrm{d}t \leq \int_{t_0}^{\infty} \frac{\mathrm{d}t}{(2\pi t)^{d/2}} \tag{24}$$

for all $(x,y) \in D \times D$. On the other hand, for all $n \geq n_0$,

$$p^D(t;x_n,y_n) \leq \frac{1}{(2\pi t)^{d/2}} \mathrm{e}^{-\frac{\delta^2}{2t}},$$

and the right member above is integrable in $(0, t_0)$. Hence, by the continuity of $p^D(t;\cdot,\cdot)$ for all $t > 0$ and dominated convergence, we deduce that

$$\int_0^{t_0} p^D(t;x_n,y_n)\mathrm{d}t \to \int_0^{t_0} p^D(t;x_0,y_0)\mathrm{d}t. \tag{25}$$

Since the left member of (24) converges to zero uniformly in (x,y) as $t_0 \to \infty$, it follows that (25) remains true when t_0 is replaced by ∞. Thus by (14), $G(x_n,y_n) \to G(x_0,y_0)$; in other words, $G(\cdot,\cdot)$ is continuous at (x_0,y_0). Now, according to Fatou's lemma, for each $x_0 \in D$, and $x_n \to x_0$, $y_n \to x_0$ we have:

$$\underline{\lim}_{n\to\infty} \int_0^{\infty} p^D(t;x_n,y_n)\mathrm{d}t \geq \int_0^{\infty} p^D(t;x_0,x_0) = G(x_0,x_0).$$

The last term is ∞ by Theorem 2.5. Thus $G(x_n,y_n) \to G(x_0,x_0)$, and so $G(\cdot,\cdot)$ is extended continuous at all (x_0,x_0).

For $d = 2$, we use a different method based on (19). Consider the function f on $D \times D$:

$$f(x,y) = E^x\{\ln|X(\tau_D) - y|\}.$$

We have proved above that f is finite in $D \times D$. Let y, y', and z be three different points of $\mathbb{R}^2$. Then, if $|z-y| \geq |z-y'|$, we have

$$\ln\frac{|z-y|}{|z-y'|} \leq \ln\left(1 + \frac{|y-y'|}{|z-y'|}\right) \leq \frac{|y-y'|}{|z-y'|}.$$

Interchanging y and y', we obtain

$$\left|\ln\left|\frac{z-y}{z-y'}\right|\right| \leq |y-y'|\left(\frac{1}{|z-y|} \vee \frac{1}{|z-y'|}\right).$$

Consequently, we have if $y \in D$, $y' \in D$:

$$\begin{aligned} |f(x,y) - f(x,y')| &= \left|E^x\left\{\ln\frac{|X(\tau_D)-y|}{|X(\tau_D)-y'|}\right\}\right| \\ &\leq |y-y'|\left(\frac{1}{\rho(y,\partial D)} \vee \frac{1}{\rho(y',\partial D)}\right). \end{aligned}$$

Thus, for $x \in D$, $f(x, \cdot)$ is not only continuous in D but equi-continuous in the manner shown. Because both G and g are symmetric in (x, y), so is f. Hence for each $y \in D$, $f(\cdot, y)$ is continuous in $x \in D$. (Actually, it is even harmonic by Theorem 2.5). These properties imply the continuity of f in $D \times D$, by elementary analysis. Hence the same is true for G_D, by (19).

This method can also be used in the case $d \geq 3$.

(iv) In the case that $d \geq 3$, since $g(\cdot, y) \in C_b(\partial D)$, this follows from (19) and Theorem 1.23 with $f = g(\cdot, y)$.

In the case that $d = 2$, since $g(w, y) = \frac{1}{\pi} \ln \frac{1}{|w-y|}$ is not bounded, we must prove the result directly using the argument in the proof of Theorem 1.23. Denote the number in (18) by M; then we have by (22),

$$\sup_x E^x(|X(\tau_D) - x|^2) \leq 2M.$$

Let $z \in (\partial D)_r$. For any given $\varepsilon > 0$, there exists $\delta > 0$ such that if $|w - z| < \delta$, $w \neq y$, $z \neq y$:

$$\left| \ln \frac{1}{|w-y|} - \ln \frac{1}{|z-y|} \right| < \frac{\varepsilon}{4}.$$

As in the proof of Theorem 1.23, for given $y \neq z$, there exists $\delta' > 0$ depending on y such that if $|x - z| < \frac{\delta}{2} \wedge \delta'$, then

$$P^x[\tau_{B(z,\delta)} \leq \tau_D] < \frac{|z-y|^2 \varepsilon^2}{8M}.$$

Now for any x in D with $|x - z| < \frac{\delta}{2} \wedge \delta' \wedge \frac{|z-y|}{4} \varepsilon$, we have, as in the last steps of the proof of Theorem 1.23:

$$\begin{aligned} &E^x \left[\left| \ln \frac{1}{|X(\tau_D) - y|} - \ln \frac{1}{|z-y|} \right| \right] \\ &\leq \frac{\varepsilon}{4} + E^x \left[\tau_{B(z,\delta)} \leq \tau_D; \left| \ln \frac{|X(\tau_D) - y|}{|z-y|} \right| \right]. \end{aligned}$$

The last expectation in the above can be estimated as follows:

$$\begin{aligned} &E^x \left[\tau_{B(z,\delta)} \leq \tau_D; \ln \left(1 + \frac{|X(\tau_D) - z|}{|z-y|} \right) \right] \\ &\leq E^x \left[\tau_{B(z,\delta)} \leq \tau_D; \frac{|x-z| + |X(\tau_D) - x|}{|z-y|} \right] \\ &\leq \frac{\varepsilon}{4} + \frac{1}{|z-y|} E^x \left[\tau_{B(z,\delta)} \leq \tau_D; |X(\tau_D) - x| \right] \\ &\leq \frac{\varepsilon}{4} + \frac{1}{|z-y|} \left[P^x(\tau_{B(z,\delta)} \leq \tau_D) \right]^{1/2} \left[E^x |X(\tau_D) - x|^2 \right]^{1/2} \\ &\leq \frac{\varepsilon}{4} + \frac{1}{|z-y|} \left(\frac{|z-y|^2 \varepsilon^2}{8M} \right)^{1/2} (2M)^{1/2} = \frac{3}{4} \varepsilon. \end{aligned}$$

It follows that

$$\lim_{x \to z} E^x \left\{ \ln \frac{1}{|X(\tau_D) - y|} \right\} = \ln \frac{1}{|z-y|}.$$

(v) For $d \geq 3$, this is trivial by (ii). For $d = 2$, we have by (19),

$$G(x,y) = E^x \left\{ \ln \frac{|X(\tau_D) - y|}{|x-y|} \right\}.$$

The above integrand converges to zero as $y \to \infty$ and is dominated by

$$\ln \left(1 + \frac{|X(\tau_D) - x|}{|x-y|} \right) \leq \frac{|X(\tau_D) - x|}{|x-y|} \leq |X(\tau_D) - x|$$

for $|x-y| \geq 1$. The last term in the above is integrable under E^x by (22). Hence (v) follows by the dominated convergence theorem. □

Corollary to Theorem 2.6 *If D is bounded and regular, then for each $y \in D$, the function*

$$g(\cdot, y) - G_D(\cdot, y)$$

is the unique harmonic function in D having the boundary value $g(\cdot, y)$.

This follows from (iv) and Corollary 1.11.

Thus, this corollary enables us to define the Green function for a bounded and regular domain. Historically, this is how George Green constructed the function named after him, based on electromagnetic considerations. Here is a more striking characterization of $G(\cdot, y)$. It is the unique positive harmonic function in $D \backslash \{y\}$ converging to zero on ∂D, up to a positive multiple. The classical proof of this result depends on Bôcher's theorem (see Wermer (1974)).

As an interesting example we shall give the explicit analytical expression for the Green function for a ball.

For a ball $B = B(a,r)$ in $\mathbb{R}^d$ $(d \geq 2)$, the Kelvin inversion transformation with respect to B is defined as follows:

$$y^* = a + \frac{r^2}{|y-a|^2}(y-a), \quad y \in \mathbb{R}^d \backslash \{a\}.$$

For any $x \neq a$, $y \neq a$, we have

$$|y^* - x^*|^2 = r^4 \left| \frac{y-a}{|y-a|^2} - \frac{x-a}{|x-a|^2} \right|^2 = \frac{r^4 |y-x|^2}{|y-a|^2 |x-a|^2},$$

so that

$$|y^* - x^*| = \frac{r^2 |y-x|}{|y-a||x-a|}. \tag{26}$$

Let $d \geq 3$. For $y \in B\backslash\{a\}$, we define

$$G(x,y) = \frac{C_d}{|x-y|^{d-2}} - \frac{C_d r^{d-2}}{|y-a|^{d-2}|x-y^*|^{d-2}}, \quad x \in B, \tag{27}$$

where C_d is given in (15).

Since $x \to |x-y^*|^{2-d}$ is harmonic in B, $g(\cdot,y) - G(\cdot,y)$ is harmonic in B. Next, if $z \in \partial B$ then $z^* = z$, and $|y^* - z| = \frac{r|y-z|}{|y-a|}$ by (26), so that $G(z,y) = 0$. Hence by continuity, G satisfies the conditions of the Corollary to Theorem 2.6. Therefore G is the Green function for B. In particular, it is symmetric in $D \times D$.

For $d = 2$, we define, for any $y \in B\backslash\{a\}$,

$$G(x,y) = \frac{1}{\pi}\ln\frac{1}{|x-y|} - \frac{1}{\pi}\ln\frac{r}{|y-a||x-y^*|}, \quad x \in B. \tag{28}$$

It is easy to verify the same conditions for G as in the case $d \geq 3$. Hence G is the Green function for B.

For $d = 1$, $B(a,r) = (a-r, a+r)$. It is easy to check that the function

$$G(x,y) = \begin{cases} r^{-1}(x-a+r)(a+r-y), & \text{if } a-r < x < y < a+r \\ r^{-1}(a+r-x)(y-a+r), & \text{if } a-r < y \leq x < a+r \end{cases}$$

is the Green function for B. After a change of coordinates, the Green function for (a,b) is given by:

$$G(x,y) = \begin{cases} 2(b-a)^{-1}(x-a)(b-y), & \text{if } a < x < y < b; \\ 2(b-a)^{-1}(b-x)(y-a), & \text{if } a < y \leq x < b. \end{cases} \tag{29}$$

Let f be the indicator of a bounded Borel set in (a,∞). Then we have by (29), for $x \in (a,b)$,

$$G_{(a,b)}f(x) = \frac{2(b-x)}{b-a}\int_a^x (y-a)f(y)dy + \frac{2(x-a)}{b-a}\int_x^b (b-y)f(y)dy.$$

Letting $b \to \infty$, we obtain by bounded convergence

$$2\int_a^x (y-a)f(y)dy + 2(x-a)\int_x^\infty f(y)dy = 2\int_a^\infty [(x-a)\wedge(y-a)]f(y)dy.$$

On the other hand, it is clear from the second expression in (13) that

$$G_{(a,b)}f \uparrow G_{(a,\infty)}f.$$

It follows from these considerations that

$$G_{(a,\infty)}(x,y) = 2(x-a) \wedge 2(y-a). \tag{30}$$

This formula may also be derived using the explicit form for $p^{(0,\infty)}(t;x,y)$ given in Chapter 9.

2.4 Compactness and Spectrum

For any domain D and $1 \leq p < \infty$, $L^p(D)$ will be called an 'appropriate space' for D. If in addition D is regular, then $C_0(D)$ will also be called an appropriate space for D. This terminology is used only for the convenience of later reference.

An operator T from a Banach space S to another S' is called *compact* iff any bounded sequence $\{s_n\}$ in S contains a subsequence $\{s_{n_k}\}$ such that $T(s_{n_k})$ converges in S'. When $S = S'$, we say that T is a compact operator on S.

Theorem 2.7 *Let D be a domain, Then for each appropriate space S for D, $\{P_t^D : t \geq 0\}$ forms a strongly continuous semigroup in S. If, in addition, $m(D) < \infty$, then for each $t > 0$, P_t^D is a bounded operator from S_1 to S_2 for any two appropriate spaces S_1 and S_2 for D. Furthermore, P_t^D is a compact operator and has the same eigenvalues $\{\exp(\lambda_k t) : k = 1, 2, \cdots\}$ with $\lambda_k < 0$ in all the appropriate spaces for D.*

Proof Let $t > 0$, $1 \leq p < \infty$. For $f \in L^p(D)$, $f \geq 0$, we have by Hölder's inequality applied with the sub-probability measure $P_t^D(x, dy)$:

$$(P_t^D f(x))^p = \left(\int_D f(y) p^D(t; x, y) dy\right)^p \leq \int_D f(y)^p p^D(t; x, y) dy.$$

Hence by Fubini's theorem and the symmetry of $p^D(t; \cdot, \cdot)$:

$$\int_D (P_t^D f(x))^p dx \leq \int_D \left(\int_D p^D(t; x, y) dx\right) f(y)^p dy \leq \int_D f(y)^p dy.$$

It follows that for each $t > 0$:

$$\|P_t^D\|_p \leq 1, \quad 1 \leq p < \infty. \tag{31}$$

By the Feller property (Theorem 2.2), we have for any $f \in C_0(D)$:

$$\lim_{t \downarrow 0} \|P_t^D f - f\|_\infty = 0. \tag{32}$$

Since $C_0(D)$ is dense in each appropriate space S for D, whether it is regular or not, we have by (31) and (32), for any $f \in S$,

$$\lim_{t \downarrow 0} \|P_t^D f - f\|_S = 0, \tag{33}$$

where the norm $\|\cdot\|_S$ is that for the space S. This proves that $\{P_t^D : t \geq 0\}$ is a strongly continuous semigroup in each appropriate space S for D.

Now suppose that $m(D) < \infty$. We have for all $p \in [1, \infty]$, $L^\infty(D) \subset L^p(D)$ and for any $f \in L^\infty(D)$:

$$\|f\|_p \leq \|f\|_\infty [m(D)]^{1/p}.$$

Hence the identity embedding I from $L^\infty(D)$ to $L^p(D)$ is a bounded operator. For $t > 0$, $f \in L^p(D)$, we have for each $x \in D$:

$$|P_t^D f(x)| \le (2\pi t)^{d/2}[m(D)]^{1/p'} \|f\|_p,$$

where $p' \in [1, \infty]$ and $\frac{1}{p} + \frac{1}{p'} = 1$. Therefore $P_t^D f \in L^\infty(D)$ and P_t^D is a bounded operator from $L^p(D)$ to $L^\infty(D)$. Now for any $p \in [1, \infty]$ and $r \in [1, \infty]$, we can represent P_t^D as follows:

$$L^p(D) \xrightarrow{P_t^D} L^\infty(D) \xrightarrow{I} L^r(D);$$

hence, P_t^D is a bounded operator from $L^p(D)$to $L^r(D)$.

If D is also a regular domain, then by Theorem 2.4, we have for each $y \in D$:

$$\lim_{x \to \partial D \cup \{\infty\}} p^D(t; x, y) = 0. \tag{34}$$

Hence for any $f \in L^p(D)$ $(1 \le p \le \infty)$, we have $P_t^D f \in C_0(D)$ by the dominated convergence theorem. As a bounded operator from $L^p(D)$ to $L^\infty(D)$ with a range contained in $C_0(D)$, P_t^D is also a bounded operator from $L^p(D)$ to $C_0(D)$. Since $C_0(D) \subset L^\infty(D)$, P_t^D is also a bounded operator from $C_0(D)$ to $L^p(D)$. Thus we have proved that for any two appropriate spaces S_1 and S_2, P_t^D is a bounded operator from S_1 to S_2.

For any fixed $t > 0$, P_t^D is a compact operator in $L^p(D)$ for $1 < p < \infty$ because

$$\int_D p^D(t;, x, y)^{p'} dy \le (2\pi t)^{-p'd/2} m(D) < \infty,$$

where $\frac{1}{p} + \frac{1}{p'} = 1$, by a known criterion which requires a finite measure space (see Dunford and Schwartz (1958, Volume I, page 518)).

For the appropriate space $S = L^1(D)$ or $C_0(D)$, we can decompose the operator P_t^D in S as follows:

$$S \xrightarrow{P_{t/3}^D} L^2(D) \xrightarrow{P_{t/3}^D} L^2(D) \xrightarrow{P_{t/3}^D} S.$$

Hence P_t^D is a compact operator in S, being the composition of a compact operator $P_{t/3}^D$ in $L^2(D)$ with bounded operators.

For each $t > 0$, by the spectral property of compact operators (see Dunford and Schwartz (1958)), P_t^D has a countable set of eigenvalues with the only possible accumulation point being 0. We shall prove that 0 is not an eigenvalue of P_t^D in any of these appropriate spaces; i.e. if there exists $f \in S$ such that $P_t^D f = 0$, then $f = 0$. Suppose first that $f \in L^2(D)$. Since $p^D(t; \cdot, \cdot)$ is symmetric, P_t^D is a symmetric operator in $L^2(D)$, then

$$\|P_{t/2}^D f\|_2^2 = (P_t^D f, f) = 0$$

and inductively, $P_{t/2^n}^D f = 0$ for all $n \ge 1$. Hence we have $f = 0$ by (33). Next, if $f \in L^p(D) \subseteq L^1(D)$, $1 \le p < \infty$, then for $0 < \varepsilon < t$, $P_\varepsilon^D f \in L^2(D)$ by (31), and $P_{t-\varepsilon}^D(P_\varepsilon^D f) = 0$. Applying the above result with t replaced by $t - \varepsilon$ and f by

$P_\varepsilon^D f$, we obtain $P_\varepsilon^D f = 0$. This being true for all $\varepsilon < t$, we conclude that $f = 0$ by (33).

Now let S_1 and S_2 be any two appropriate spaces for D. If $\beta \neq 0$ is an eigenvalue of P_t^D in S_1 and $f \in S_1$ is its corresponding eigenfunction, then

$$f = \frac{1}{\beta} P_t^D f \in S_2;$$

hence β is also an eigenvalue of P_t^D in S_2. This shows that for each $t > 0$, P_t^D has the same eigenvalues in all the appropriate spaces and all eigenfunctions are contained in each appropriate space, hence in $L^2(D)$. Since for each $t > 0$, P_t^D is symmetric and bounded, it is self-adjoint. We have by the spectral resolution theorem for a semigroup of self-adjoint operators in L^2 (see Yosida (1980, page 313)):

$$P_t^D = \int_{-\infty}^{\alpha_0} e^{\lambda t} dE_\lambda, \quad t > 0, \tag{35}$$

where $\{E_\lambda : -\infty < \lambda \leq \alpha_0\}$ is the spectral family of projection operators for A_2; and

$$A_2 = \int_{-\infty}^{\alpha_0} \lambda dE_\lambda$$

is the infinitesimal generator of $\{P_t^D : t \geq 0\}$ in $L^2(D)$.

It follows from the compactness of P_t^D $(t > 0)$ and the representation (35) that A_2 has the eigenvalues $\{\lambda_n\} : \lambda_1 > \lambda_2 > \cdots > \lambda_n > \cdots$ and for each $t > 0$, P_t^D has the eigenvalues $\{e^{\lambda_n t}\}$. Now let f be an eigenfunction corresponding to λ_1. Then for $t > 0$, by Proposition 1.16,

$$e^{\lambda_1 t} \|f\|_\infty = \|P_t^D f\|_\infty \leq \|f\|_\infty \|P_t^D 1\|_\infty = \|f\|_\infty \sup_{x \in D} P^x(t < \tau_D) \to 0$$

as $t \to \infty$. This proves that $\lambda_1 < 0$. □

2.5 Laplacian as Generator

In this section we review the basic properties of the Laplacian as an inverse to the Green operator. The operator Δ is defined in Section 1.4 in the strict, or, classical, sense. It is necessary to generalize it in the sense of generalized functions (Schwartz 'distributions'). Let

$$C_c^\infty(D) = \{f \in C^\infty(D) : f \text{ has compact support in } D\}.$$

The functions in $C_c^\infty(D)$ will be called 'test functions'. We recall the definition of $L_{\text{loc}}^1(D)$ from Section 1.1. Since a (locally) integrable function is determined only up to a set of measure zero, it can always be made Borelian by modification. The *weak* partial derivative with respect to x_j of f in $L_{\text{loc}}^1(D)$ exists iff there exists a function g in $L_{\text{loc}}^1(D)$ such that for all $\phi \in C_c^\infty(D)$ we have

$$\int_D g(x)\phi(x)\mathrm{d}x = -\int_D f(x)\frac{\partial \phi}{\partial x_j}(x)\mathrm{d}x. \tag{36}$$

This function g is then denoted by $\frac{\partial f}{\partial x_j}$; when $f \in C^1(D)$, it reduces to the strict partial derivative. The gradient operator in the weak sense is then defined as usual by

$$\nabla f = \left(\frac{\partial f}{\partial x_1}, \ldots, \frac{\partial f}{\partial x_d}\right).$$

Since $\frac{\partial f}{\partial x_j} \in L^1_{\mathrm{loc}}(D)$ when it exists, we can iterate the process to define partial derivatives of higher orders. In particular, the Laplacian Δf in the weak sense is a function $g \in L^1_{\mathrm{loc}}(D)$ such that for all $\phi \in C_c^\infty(D)$ we have

$$\int_D g(x)\phi(x)\mathrm{d}x = \int_D f(x)\Delta\phi(x)\mathrm{d}x. \tag{37}$$

When $f \in C^2(D)$, this reduces to the strict Laplacian defined in Section 1.4. Observe that by our definition, an equation like $\Delta f = g$ has a meaning only if both f and g are locally integrable. This will be sufficient for our purposes and we shall not treat the general case in which they are generalized functions.

If $\Delta f = 0$ in the weak sense, then Weyl's lemma (see Yosida (1980, Section II.7) asserts that there exists a harmonic function h such that $f = h$, m-almost everywhere. Thus, if f is also continuous then it is harmonic. We have omitted the words 'in D' several times in the last sentence. Indeed, from now on, D will be a fixed domain, arbitrary unless otherwise specified, and will be omitted from the notation. We begin with a classical result which shows that $-\frac{\Delta}{2}$ and $G(= G_D)$ acting on test functions are inverse to each other.

Proposition 2.8 *Let D be an arbitrary domain in $\mathbb{R}^d$, $d \geq 3$ or a Green-bounded domain in $\mathbb{R}^d$, $d = 1$ or 2. If $\phi \in C_c^\infty(D)$, then $G\phi \in C^\infty(D)$ and we have in D:*

$$\Delta(G\phi) = -2\phi; \tag{38}$$
$$G(\Delta\phi) = -2\phi. \tag{39}$$

Proof By (19), we have for $(x, y) \in D \times D$:

$$G(x, y) = g(x - y) + h(x, y),$$

where $h(\cdot, y)$ is harmonic in D for each $y \in D$. Now we extend the function ϕ from D to $\mathbb{R}^d$ by defining it to be zero in $\mathbb{R}^d \backslash D$. Then we have

$$\begin{aligned} G\phi(x) &= \int_{\mathbb{R}^d} g(x-y)\phi(y)\mathrm{d}y + \int_D h(x,y)\phi(y)\mathrm{d}y \\ &= \int_{\mathbb{R}^d} \phi(x-y)g(y)\mathrm{d}y + \int_D h(x,y)\phi(y)\mathrm{d}y. \end{aligned}$$

Since $g \in L^1_{\text{loc}}(\mathbb{R}^d)$, the first integral in the last member above belongs to C^∞. The second integral is harmonic in D, by direct differentiation. Since $\Delta_x\phi(x-y)$ has support in D, we may write

$$\begin{aligned} \Delta G\phi(x) &= \int_{\mathbb{R}^d} \Delta_x \phi(x-y)g(y)\mathrm{d}y \\ &= \left(\int_{B_r} + \int_{B_r^c}\right) \Delta_y \phi(x-y)g(y)\mathrm{d}y, \end{aligned} \tag{40}$$

where B_r is the ball $B(0,r)$, $r > 0$.

The above integral over B_r clearly converges to zero as $r \to 0$. We apply Green's second identity to the integral over B_r^c to obtain

$$\begin{aligned} &\int_{B_r^c} \phi(x-y)\Delta_y g(y)\mathrm{d}y + \int_{\partial B_r} \frac{\partial}{\partial n_y}\phi(x-y)g(y)\sigma(\mathrm{d}y) \\ &- \int_{\partial B_r} \phi(x-y)\frac{\partial g}{\partial n_y}(y)\sigma(\mathrm{d}y) \\ &= I_1(r) + I_2(r) + I_3(r), \end{aligned}$$

where $\frac{\partial}{\partial n_y}$ denotes the inner normal derivative at y. Since g is harmonic in $\mathbb{R}^d\backslash\{0\}$, $I_1(r) = 0$. Next we have for $|y| = r$:

$$g(y) = \begin{cases} C_d r^{2-d} & \text{if } d \geq 3, \\ \pi^{-1} \ln r^{-1} & \text{if } d = 2, \\ -r & \text{if } d = 1, \end{cases}$$

and $\sigma(\partial B_r)$ is given in Section 1.1. Hence we have

$$\overline{\lim_{r\to 0}}\, |I_2(r)| \leq \|\nabla\phi\|_\infty \overline{\lim_{r\to 0}}(\|1_{\partial B_r} g\|_\infty \sigma(\partial B_r)) = 0.$$

Finally, for $|y| = r$:

$$\begin{aligned} \frac{\partial g}{\partial n_y}(y) &= -\frac{\mathrm{d}}{\mathrm{d}r}\left(\frac{C_d}{r^{d-2}}\right) = \frac{(d-2)C_d}{r^{d-1}} = \frac{2}{\sigma(\partial B_r)} \quad \text{if } d \geq 3 \\ \frac{\partial g}{\partial n_y}(y) &= -\frac{\mathrm{d}}{\mathrm{d}r}\left(\frac{1}{\pi}\ln\frac{1}{r}\right) = \frac{1}{\pi r} = \frac{2}{\sigma(\partial B_r)} \quad \text{if } d = 2 \\ \frac{\partial g}{\partial n_y}(y) &= -\frac{\mathrm{d}}{\mathrm{d}r} r = 1 = \frac{2}{\sigma(\partial B_r)} \quad \text{if } d = 1. \end{aligned}$$

[Note that for $d = 1$, ∂B_r is a two-point set and its 0-dimensional measure is equal to 2 by an accepted convention]! Hence we have

$$\lim_{r\to 0} I_3(r) = \lim_{r\to 0} -\int_{\partial B_r} \phi(x-y)\frac{2\sigma(\mathrm{d}y)}{\sigma(\partial B_r)} = -2\phi(x),$$

by continuity. Collecting these results together, we obtain (38). Readers who are familiar with the use of Dirac's delta function will know that the preceding derivation amounts to a proof that $-\frac{1}{2}\Delta_x g(x-y)$ is the delta function so that heuristically

$$-\frac{\Delta}{2}G\phi(x) = -\frac{1}{2}\int \Delta_x g(x-y)\phi(y)\mathrm{d}y = \phi(x).$$

To prove (39), let D_n be regular bounded domains such that $D_n \uparrow\uparrow D$. We may suppose that the support of ϕ is in D_1. Applying (38) with D_n for D and $\Delta\phi$ for ϕ, we have

$$\Delta(G_{D_n}(\Delta\phi)) = -2\Delta\phi;$$

in other words, $G_{D_n}(\Delta\phi)+2\phi$ is harmonic in D_n. Since D_n is regular, $G_{D_n}(\Delta\phi) \in C_0(D_n)$ by Theorem 2.6 (iv). Hence $G_{D_n}(\Delta\phi)+2\phi$ has boundary value zero and is therefore identically zero in D_n by Corollary 1.11. For any positive Borel measurable function f with compact support in D_1, it is clear that

$$\begin{aligned} G_D f(x) &= E^x\left\{\int_0^{\tau_D} f(X_t)\mathrm{d}t\right\} = \lim_{n\to\infty} E^x\left\{\int_0^{\tau_{D_n}} f(X_t)\mathrm{d}t\right\} \\ &= \lim_{n\to\infty} G_{D_n} f(x), \end{aligned} \tag{41}$$

because $\tau_{D_n} \uparrow\uparrow \tau_D \leq \infty$. It follows that the same is true without the assumption of positivity, provided that $G_D f$ exists, which is the case for $f = \Delta\phi$ under the assumption on D. Thus we obtain in D

$$G_D(\Delta\phi) = \lim_n G_{D_n}(\Delta\phi) = -2\phi. \qquad \square$$

Although Proposition 2.8 is sufficient for our purposes later in the book, we shall extend equation (38) to the most general case. In fact, we shall prove that it is true whenever it 'makes sense'. This will justify the folklore that the Laplacian is the inverse of the Green operator. By contrast, we do not know what the most general form of (39) should be.

We need a lemma which introduces the notion of 'superharmonicity'.

Lemma 2.9 *Let D be an arbitrary domain in $\mathbb{R}^d$ $(d \geq 1)$ and G_D its Green operator. For each $\phi \in \mathcal{B}_+$ such that $G_D|\phi| \not\equiv \infty$, we have $G_D\phi \in L^1_{\mathrm{loc}}(D)$.*

Proof Let $x \in D$, $B = B(x,r) \subset\subset D$. Then

$$G_D\phi(x) = E^x\left\{\int_0^{\tau_B} \phi(X_t)\mathrm{d}t\right\} + E^x\left\{\int_{\tau_B}^{\tau_D} \phi(X_t)\mathrm{d}t\right\}.$$

Using the strong Markov property, followed by Proposition 1.12, as spelt out in the proof of Theorem 1.23, we can transform the second expectation above into

$$E^x\{G_D\phi(X(\tau_B))\} = \frac{1}{\sigma(S(x,r))}\int_{S(x,r)} G_D\phi(y)\sigma(\mathrm{d}y). \tag{42}$$

Since $\phi \geq 0$ it follows that $G_D\phi(x)$ is greater than or equal to the 'sphere-averaging' on the right-hand side of (42). A positive lower semi-continuous function having this property for each x in D and $B(x, r) \subset\subset D$ is said to be 'superharmonic in D'. Compare this property with the sphere-averaging property of a harmonic function in Theorem 1.9, (1.21). Note that the function is not assumed to be finite to begin with, in contrast to the h in Theorem 1.9 which is assumed to be finite and belong to $L^1_{\text{loc}}(D)$. Indeed, a positive superharmonic function in D has the fundamental property that if it is not identically equal to ∞ in D then it belongs to $L^1_{\text{loc}}(D)$. This is the assertion of the lemma. For a proof we refer readers to Chung (1982a, Section 4.5). □

Now we can state and prove the general form of Proposition 2.8.

Proposition 2.10 *Let D and G_D be as in Lemma 2.9. For any $\phi \in L^1_{\text{loc}}(D)$ such that $G_D|\phi| \not\equiv \infty$ in D, we have*

$$\Delta(G_D\phi) = -2\phi \tag{43}$$

in D in the weak sense.

Proof Let $\{D_n\}$ be bounded regular domains such that $D_n \uparrow\uparrow D$ (see Appendix to Chapter 1). Then $G_{D_n}\phi \equiv 0$ on D_n^c by regularity, and the Green function $G_{D_n}(\cdot,\cdot)$ exists by Theorem 2.5. Since $G_{D_n}|\phi| \leq G_D|\phi|$, it follows from Lemma 2.9 that for any $\psi \in C_c^\infty(D)$:

$$\begin{aligned}&\int_D \int_D G_{D_n}(x,y)|\phi(y)||\Delta\psi(x)|\mathrm{d}y\mathrm{d}x \\ &\qquad \leq \|\Delta\psi\|_\infty \int_K G_D|\phi|(x)\mathrm{d}x < \infty,\end{aligned} \tag{44}$$

where K is the compact support of ψ. Therefore, by Fubini's theorem and the symmetry of $G_{D_n}(x, y)$ (Theorem 2.6), we have

$$\begin{aligned}\int_D G_{D_n}\phi(x)\Delta\psi(x)\mathrm{d}x &= \int_D \left[\int_D G_{D_n}(x,y)\Delta\psi(x)\mathrm{d}x\right]\phi(y)\mathrm{d}y \\ &= \int_D [-2\psi(y)]\phi(y)\mathrm{d}y,\end{aligned}$$

where (39) is used with D_n for D and ψ for ϕ. When $n \to \infty$, $G_{D_n}\phi \to G_D\phi$ as in (41) and the first term above converges to $\int_D G_D\phi(x)\Delta\psi(x)\mathrm{d}x$ by dominated convergence, by virtue of (44). Thus we obtain

$$\int_D G_D\phi(x) \cdot \Delta\psi(x)\mathrm{d}x = \int_D -2\phi(x)\psi(x)\mathrm{d}x.$$

The fact that this equation holds for all test functions ψ implies the conclusion of our proposition. □

Our next object is the Sobolev space, which is well known in the theory of partial differential equations. For an arbitrary domain D in $\mathbb{R}^d$, $d \geq 1$, this space is defined as follows:

$$W^{1,2}(D) = \left\{ f \in L^2(D) : \frac{\partial f}{\partial x_j} \text{ exists and belongs to } L^2(D),\ 1 \leq j \leq d \right\}.$$

Here and in what follows, all partial derivatives are taken in the weak sense. In $W^{1,2}(D)$, we define a norm known as the *Dirichlet norm* as follows:

$$\|f\|_*^2 = \int_D f(x)^2 \mathrm{d}x + \int_D |\nabla f(x)|^2 \mathrm{d}x. \tag{45}$$

It is known that $W^{1,2}(D)$ is complete with respect to this norm; in other words, it is a Banach space, see e.g. Gilbarg and Trudinger (1977, Section 7.5). Clearly, $C_c^\infty(D) \subset W^{1,2}(D)$. The closure of $C_c^\infty(D)$ with respect to the Dirichlet norm is a subspace of $W^{1,2}(D)$ which we denote by $W_0^{1,2}(D)$. It can be proved that if $f \in W^{1,2}(D)$ and f has compact support in D, then $f \in W_0^{1,2}(D)$; see e.g. Gilbarg and Trudinger (1977, Section 7.5). In view of the importance of these spaces in analysis, we shall relate some salient developments in this book to them. In this section we treat the preliminaries.

Lemma 2.11 *Let D be a Green-bounded domain and G its Green operator. Then for $f \in L^1(D)$ we have*

$$\|Gf\|_1 \leq \|G1\|_\infty \|f\|_1,$$

while for $f \in L^2(D)$ we have

$$\|Gf\|_2 \leq \|G1\|_\infty \|f\|_2.$$

Proof Writing M for $\|G1\|_\infty$, we have by the symmetry of G, for $f \in L^1$:

$$\begin{aligned} \int G|f|(x)\mathrm{d}x &= \int\int G(x,y)|f|(y)\mathrm{d}y\mathrm{d}x \\ &= \int\left[\int G(y,x)\mathrm{d}x\right]|f|(y)\mathrm{d}y \leq M\|f\|_1 < \infty, \end{aligned}$$

where all integrals are over D. Hence we can apply Fubini's theorem to obtain the first inequality. If $f \in L^2$, we apply the Cauchy–Schwarz inequality first so that

$$\begin{aligned} \|Gf\|_2^2 &= \int\left[\int G(x,y)f(y)\mathrm{d}y\right]^2 \mathrm{d}x \\ &\leq \int\left(\int G(x,y)\mathrm{d}y\right)\left(\int G(x,y)f(y)^2\mathrm{d}y\right)\mathrm{d}x \\ &\leq M\int\int G(x,y)f(y)^2\mathrm{d}y\mathrm{d}x \leq M^2\int f(y)^2\mathrm{d}y = M^2\|f\|_2^2. \end{aligned}$$

□

Proposition 2.12 *Let D and G be as in Lemma 2.11. Then for any $\phi \in L^2(D)$:*

$$G\phi \in W_0^{1,2}(D). \tag{46}$$

Proof As before, D will be fixed in what follows and omitted from the notation when the context is clear. For any $\phi \in L^2$, we have $G\phi \in L^2$ by Lemma 2.11. We shall first prove (46) for any $\phi \in C_c^\infty$. Let D_n be C^2-domains with $D_n \uparrow\uparrow D$; see the Appendix to Chapter 1. Then, as in (41):

$$\forall x \in D: \ G_{D_n}\phi(x) \to G_D\phi(x)$$

and

$$G_{D_n}|\phi| \le G_D|\phi| \in L^2(D).$$

Hence by dominated convergence,

$$G_{D_n}\phi \to G_D\phi \quad \text{in } L^2(D). \tag{47}$$

Now by Widman's inequality (Widman (1967), Theorem 3), for $(x, y) \in \overline{D}_n \times \overline{D}_n$:

$$|\nabla G_{D_n}(x, y)| \le \frac{C}{|x-y|^{d-1}} \quad \text{if } d \ge 3 \text{ or } d = 1$$

and

$$|\nabla G_{D_n}(x, y)| \le \frac{C}{|x-y|} \ln \frac{A_n}{|x-y|} \quad \text{if } d = 2,$$

where A_n is twice the diameter of D_n. Since the right members of the above inequalities are uniformly integrable over D_n, we have for $x \in D_n$:

$$\frac{\partial}{\partial x_j} G_{D_n}\phi(x) = \int_{D_n} \frac{\partial}{\partial x_j} G_{D_n}(x, y)\phi(y)\mathrm{d}y, \quad 1 \le j \le d.$$

These functions are uniformly continuous in D_n, hence they have unique extensions to $\overline{D}_n$ which will be denoted by the same symbols; thus

$$\nabla G_{D_n}\phi \in C(\overline{D}_n).$$

Since D_n is regular, the function $G_{D_n}\phi$ as defined by (13) is zero on D_n^c. Using this we have $\nabla G_{D_n}\phi = 0$ in $(\overline{D}_n)^c$, although the gradient need not exist on ∂D_n, and even if it does, it may be discontinuous across ∂D_n. This warning remark is essential for the calculations below.

Let $m \le n$; then

$$\begin{aligned} &\int_D |\nabla G_{D_m}\phi - \nabla G_{D_n}\phi|^2 \mathrm{d}x \qquad (48)\\ &= \int_D |\nabla G_{D_m}\phi|^2 \mathrm{d}x - 2\int_D (\nabla G_{D_m}\phi) \cdot (\nabla G_{D_n}\phi)\mathrm{d}x + \int_D |\nabla G_{D_n}\phi|^2 \mathrm{d}x. \end{aligned}$$

Since $\nabla G_{D_m}\phi = 0$ in D_m^c, while ∂D_m has Lebesgue measure zero, the integrals on the right side of (48) may be taken over D_m, D_m and D_n respectively. It is sufficient to consider the second integral. By Green's second identity, that is equal to

$$-\int_{\partial D_m} G_{D_m}\phi(z)\frac{\partial}{\partial n_z}G_{D_n}\phi(z)\sigma(\mathrm{d}z) - \int_{D_m} G_{D_m}\phi(x)\Delta G_{D_n}\phi(x)\mathrm{d}x.$$

Note that the inner normal derivative exists on ∂D_m for $n = m$ as well as for $n > m$. The first integral above vanishes because $G_{D_m}\phi$ vanishes on ∂D_m. Hence the result is

$$0 - \int_{D_m} G_{D_m}\phi \cdot (-2\phi)\mathrm{d}x = 2\int_D \phi G_m\phi\mathrm{d}x.$$

Substituting into (48), we obtain

$$-2\int_D \phi G_{D_m}\phi\mathrm{d}x + 2\int_D \phi G_{D_n}\phi\mathrm{d}x = 2\int_D \phi(G_{D_m}\phi - G_{D_n}\phi)\mathrm{d}x.$$

Thus by the Cauchy–Schwarz inequality:

$$\|\nabla G_{D_m}\phi - \nabla G_{D_n}\phi\|_2^2 \le 2\|\phi\|_2\|G_{D_m}\phi - G_{D_n}\phi\|_2 \to 0$$

as $m, n \to \infty$ by (47). Hence $\{G_{D_n}\phi\}$ is a Cauchy sequence in $\|\cdot\|_*$ and so it converges to an element in $W_0^{1,2}(D)$, which must be $G_D\phi$. This proves (46) when $\phi \in C_c^\infty$.

Now for $\phi \in L^2(D)$, there exists a sequence $\{\phi_n\}$ in C_c^∞ such that $\|\phi_n - \phi\|_2 \to 0$. By Lemma 2.11 with $M = \|G1\|_\infty$, we have

$$\|G(\phi_m - \phi_n)\|_2 \le M\|\phi_m - \phi_n\|_2.$$

Hence $\{G\phi_n\}$ is a Cauchy sequence in L^2.

Next we have, as before, for any $\phi \in C_c^\infty$:

$$\int_D |\nabla G\phi|^2\mathrm{d}x = 2\int_D \phi G\phi\mathrm{d}x.$$

Thus,

$$\begin{aligned}\int_D |\nabla G(\phi_m - \phi_n)|^2\mathrm{d}x &= 2\int_D (\phi_m - \phi_n)(G\phi_m - G\phi_n)\mathrm{d}x \\ &\le 2\|\phi_m - \phi_n\|_2\|G(\phi_m - \phi_n)\|_2 \le 2M\|\phi_m - \phi_n\|_2^2.\end{aligned}$$

Hence $\{G\phi_n\}$ is a Cauchy sequence in $\|\cdot\|_*$. Therefore it converges to an element in $W_0^{1,2}$. But by Lemma 2.11,

$$\|G(\phi_n - \phi)\|_2 \le M\|\phi_n - \phi\|_2.$$

Hence $G\phi_n$ converges to $G\phi$ and therefore (46) is proved. □

We can now determine the infinitesimal generator A of the semigroup $\{P_t^D,\ t \geq 0\}$ in $L^2(D)$, for a Green-bounded domain D in $\mathbb{R}^d$, $d \geq 1$. Let us first give a brief review of the relevant theory, see e.g. Yosida (1980). For any strongly continuous semigroup $\{P_t,\ t \geq 0\}$ in a Banach space $\mathcal{S}$, the domain of its infinitesimal generator $\mathcal{D}(A)$ is the class of elements f in $\mathcal{S}$ such that

$$\lim_{t \downarrow 0} \frac{1}{t}(P_t f - f)$$

exists in the norm of $\mathcal{S}$, and Af is this limit element. Now suppose that the potential operator U for $\{P_t\}$ as defined in (12) is a bounded operator in $\mathcal{S}$, then we have

$$\mathcal{D}(A) = U[\mathcal{S}], \tag{49}$$

where the right member denotes the range of U. Moreover, the mapping U from $\mathcal{S}$ to $U[\mathcal{S}]$ is one-to-one and if $f = U\phi$, then

$$Af = -\phi. \tag{50}$$

Symbolically, this may be seen as follows:

$$\begin{aligned} A(U\phi) &= A\int_0^\infty P_t\phi \mathrm{d}t = \int_0^\infty \frac{\mathrm{d}}{\mathrm{d}t} P_t\phi \mathrm{d}t \\ &= \lim_{t\to\infty} P_t\phi - P_0\phi = -\phi; \end{aligned}$$

see e.g. Yosida (1980, page 241). Thus if $f \in \mathcal{D}(A)$, then

$$Af = -U^{-1}f, \tag{51}$$

where U^{-1} is the inverse of U.

We apply the general theory reviewed above to $\{P_t^D\}$, with $\mathcal{S} = L^2(D)$, and $U = G_D$ as given in (13). Since D is Green-bounded, G_D is a bounded operator in $L^2(D)$ by Lemma 2.11. Therefore we have as a special case of (49):

$$\mathcal{D}(A_2) = G_D[L^2(D)], \tag{52}$$

where A_2 denotes the infinitesimal generator of $\{P_t^D\}$ in $L^2(D)$.

Theorem 2.13 *$\mathcal{D}(A_2)$ is the class of f in $W_0^{1,2}(D)$ such that Δf exists in the weak sense and belongs to $L^2(D)$. If $f \in \mathcal{D}(A_2)$, then $A_2 f = \frac{\Delta}{2} f$.*

Proof Let us denote the class described in the theorem by $\mathcal{C}$. In view of (52), it is sufficient to identify $\mathcal{C}$ with $G_D[L^2(D)]$.

If $f \in G_D[L^2(D)]$, then $f = G_D\phi$ where $\phi \in L^2(D)$. Hence $f \in W_0^{1,2}(D)$ by Proposition 2.12. Since $L^2(D) \subset L^1_{\text{loc}}(D)$, we have by Proposition 2.10:

$$\Delta f = \Delta(G_D\phi) = -2\phi \in L^2(D).$$

Thus $f \in \mathcal{C}$.

Conversely, if $f \in \mathcal{C}$, then we set $\phi = -\frac{\Delta}{2} f \in L^2(D)$. Hence $G_D\phi \in W_0^{1,2}(D)$ by Proposition 2.12. By Proposition 2.10, we have

$$\Delta(G_D\phi) = -2\phi = \Delta f. \tag{53}$$

Both f and $G_D\phi$ belong to $W_0^{1,2}(D)$. A well-known property of the space $W_0^{1,2}(D)$ states that the only function h in the space satisfying $\Delta h = 0$ in the weak sense is $h \equiv 0$; see Gilbarg and Trudinger (1977, Corollary 8.2). Hence (53) implies that $f = G_D\phi \in G_D[L^2(D)]$.

We have therefore proved that $G_D[L^2(D)] = \mathcal{C}$, and that for each $f \in \mathcal{D}(A_2) = \mathcal{C}$:

$$f = G_D(-\frac{\Delta}{2} f).$$

Therefore as a special case of (50), we have

$$A_2 f = \frac{\Delta}{2} f. \qquad \square$$

We close this chapter by introducing an extension of (13) in Section 2.3. For an arbitrary domain D in $\mathbb{R}^d$ $(d \geq 1)$ and $\lambda \geq 0$, we define the operator G_D^λ by

$$G_D^\lambda f(x) = \int_0^\infty \mathrm{e}^{-\lambda t} P_t f(x) \mathrm{d}t = E^x \left\{ \int_0^{\tau_D} \mathrm{e}^{-\lambda t} f(X_t) \mathrm{d}t \right\}. \tag{54}$$

Thus G_D^0 is the previous G_D. We have, by an easy computation using polar coordinates, for $\lambda > 0$:

$$\begin{aligned} G_D^\lambda 1(x) &\leq \int_{\mathbb{R}^d} \int_0^\infty \frac{1}{(2\pi t)^{d/2}} \mathrm{e}^{-\lambda t - \frac{|x-y|^2}{2t}} \mathrm{d}t \mathrm{d}y \\ &= \sigma_{d-1}(1) \int_0^\infty \int_0^\infty \frac{1}{(2\pi t)^{d/2}} \mathrm{e}^{-\lambda t - \frac{r^2}{2t}} r^{d-1} \mathrm{d}r \mathrm{d}t < \infty. \end{aligned}$$

Hence we have as an analogue of (18):

$$\|G_D^\lambda 1\|_\infty < \infty. \tag{55}$$

This suggests that our previous results for a Green-bounded domain may be generalized to an arbitrary domain if we operate with G_D^λ for $\lambda > 0$. We shall deal with such a case in Section 3.4. Let us record the 'resolvent equation' (see Chung (1982a, page 83) as follows, for $\lambda > 0$, $\mu > 0$:

$$G_D^\mu - G_D^\lambda = (\lambda - \mu) G_D^\mu G_D^\lambda = (\lambda - \mu) G_D^\lambda G_D^\mu. \tag{56}$$

As a special case when $\mu \downarrow 0$, this implies

$$G_D^\lambda f = G_D^0 f - \lambda G_D^0 (G_D^\lambda f), \tag{57}$$

at least when all the three terms are well defined and finite. In fact, a little consideration shows that (57) holds with all terms finite if $G_D^0|f| < \infty$.

We can also define the extension of the Green function in (14) as follows:

$$G_D^\lambda(x,y) = \int_0^\infty e^{-\lambda t} p^D(t;x,y)\mathrm{d}t. \tag{58}$$

If $\lambda > 0$, then we have for all $(x,y) \in \mathbb{R}^d \times \mathbb{R}^d$, $d \geq 1$:

$$G_D^\lambda(x,y) \leq \int_0^\infty e^{-\lambda t} p(t;x,y)\mathrm{d}t < \infty,$$

as can be verified by calculus. Thus, even for $d = 1$ or 2, $G_D^\lambda(\cdot,\cdot)$ is a finite symmetric function. This shows the advantage of using G^λ rather than the previous $G = G^0$ in certain situations. An instance of this is described in Section 3.4. The following extension of Proposition 2.10 will be required.

Proposition 2.14 *Let D be an arbitrary domain in $\mathbb{R}^d$ $(d \geq 1)$ and G_D^λ as in (54) $(\lambda > 0)$. For any $\phi \in L^1_{\mathrm{loc}}(D)$ such that $G^\lambda|\phi| \not\equiv \infty$ in D, we have*

$$(\Delta - 2\lambda)G_D^\lambda \phi = -2\phi. \tag{59}$$

Proof We begin by proving (59) for a bounded D and $\phi \in C_c^\infty(D)$. Then

$$G_D^0|\phi| \leq (G_D^0 1)\|\phi\|_\infty < \infty$$

and by (55),

$$G_D^0(G_D^\lambda|\phi|) \leq (G_D^0 1)\|G_D^\lambda 1\|_\infty \|\phi\|_\infty < \infty.$$

Therefore (57) is valid when f there is replaced by ϕ. It then follows from Proposition 2.10 (for a bounded D) that

$$\Delta G_D^\lambda \phi = \Delta G_D^0(\phi - \lambda G_D^\lambda \phi) = -2(\phi - \lambda G_D^\lambda \phi), \tag{60}$$

which is (59) in this special case. For a general D we may now proceed as in the proof of Proposition 2.10. For any $\psi \in C_c^\infty(D)$, using (60) with ψ for ϕ, we obtain by Fubini's theorem and the symmetry of $G_{D_n}^\lambda(\cdot,\cdot)$:

$$\begin{aligned}\int_D G_{D_n}^\lambda \phi(x) \cdot \Delta\psi(x)\mathrm{d}x &= \int_D -2\psi(x)\phi(x)\mathrm{d}x + 2\lambda \int_D G_{D_n}^\lambda \psi(x) \cdot \phi(x)\mathrm{d}x \\ &= \int_D -2\phi(x)\psi(x)\mathrm{d}x + 2\lambda \int_D G_{D_n}^\lambda \phi(x) \cdot \psi(x)\mathrm{d}x.\end{aligned}$$

Here the last exchange of ϕ and ψ is again a consequence of the symmetry of $G_{D_n}^\lambda(\cdot,\cdot)$. Letting $n \to \infty$ we obtain the conclusion as before. □

Observe that we did not use the 'λ-Green function' $G_D^\lambda(\cdot,\cdot)$ for the general D in the above proof; thus, with hindsight, we could have absorbed Proposition 2.14 into Proposition 2.10.

For later use we add the following extension of (39) the proof of which is left to the reader.

If $\phi \in C_c^\infty(D)$, then $G_D^\lambda \phi \in C^\infty(D)$ and

$$G_D^\lambda(\Delta - 2\lambda)\phi = -2\phi. \tag{61}$$

Notes on Chapter 2

Since Brownian motion is a very special Markov process, its properties can usually be proved by a special argument or derived from a more general one. Lemma 2.1 is a typical example. It is true for any Feller process without continuity of paths or spatial homogeneity used in the proof given here; see Chung (1982a, page 73, Exercise 2). Thus, Theorem 2.2 is actually a particular case of a general theorem due to Chung (Chung 1986b) that a doubly Feller process killed off on an open set remains doubly Feller. This result solved an old problem discussed repeatedly in the earlier literature of Markov processes with no tangible conclusions; see e.g. Dynkin (1960). Moreover, the result is still true when a multiplicative functional satisfying certain conditions is attached to the process. This includes the case of the Feynman–Kac functional and so applies to the T_t in Theorem 3.17.

The construction of the transition density in Section 2.2 is due to Hunt (Hunt 1956). Once we have p^D, the natural way to *define* the Green function is by the formula (2.14), for all (x, y) in $\mathbb{R}^d \times \mathbb{R}^d$. Curiously enough, Hunt did not insist on this and chose to work in the classical framework. It is frustrating that despite the continuity of $p^D(t; \cdot, \cdot)$ no proof of the (extended) continuity of $G_D(\cdot, \cdot)$ for $\mathbb{R}^2$ is known. Indeed, a complete probabilistic treatment of the Green function for an arbitrary domain in $\mathbb{R}^2$ remains to be given (and would be a commendable thesis for a doctorate).

As alluded to in the text, we have abandoned the so-called Greenian domains in $\mathbb{R}^2$ for the more tractable Green-bounded domains. One reason for this is that we do not wish to muddy the water unnecessarily as far as our eventual applications are concerned. Thus our handling of Theorems 2.5 and 2.6 is direct and unencumbered by 'λ-potentials' (see (2.54)) used in other treatments such as Port and Stone (1978). The inequality (2.23), which will be required later, is apparently new.

A result like Proposition 2.10 deserves the most general formulation given here. This should be available in the literature of Schwartz distributions, but where can one find it?

Let us reiterate that the theory of Markov processes is much more than that of its transition semigroup. Therefore, questions regarding the infinitesimal generator and spectrum play a secondary role in this book. They are treated in Section 2.5 and again in Section 3.4 as a sop to readers reared in the tradition of older analysis ['isn't Wiener space also analysis?']. As Feller discovered in the late 1940s, the usual semigroup theory was of little use to his (Feller) processes. As will be seen from subsequent chapters, our principal results and proofs can be given in terms of probability without the intervention of any generator or eigenvalue. However, we have included a few topics from the other theories for the purposes of comparison and connection. For this purpose it becomes necessary to rely rather more heavily on the literature for certain prerequisites which cannot be treated fully within the confines of the present book. However, we prefer finger pointing to hand-waving wherever a reference is called for, and we do not cite unreadable books.

3. Schrödinger Operator

3.1 The Schrödinger Equation and Class J

Let D be a domain in $\mathbb{R}^d$ ($d \geq 1$). We consider the following equation:

$$\frac{\Delta}{2}u(x) + q(x)u(x) = 0, \quad x \in D, \tag{1}$$

where $\Delta = \sum_{i=1}^{d} \partial^2/\partial x_i^2$ is the Laplacian and q is a Borel measurable function on D. This equation is generally taken in the weak sense as discussed in Section 2.5. Thus (1) is satisfied when $u \in L^1_{\text{loc}}(D)$, $qu \in L^1_{\text{loc}}(D)$ and

$$\int_D u(x)\Delta\phi(x)\mathrm{d}x = -2\int_D q(x)u(x)\phi(x)\mathrm{d}x \tag{2}$$

for all $\phi \in C_c^{\infty}(D)$.

Equation (1) will be called Schrödinger's equation. It is one of the most important equations in mathematical physics, being a natural extension of Laplace's equation $\Delta u = 0$. The function q is referred to as a *'potential'* in physics, though this nomenclature is confusing in potential theory. We shall treat a more general class of functions q than those usually assumed in the theory of elliptic partial differential equations.

When $d = 1$, equation (1) is the reduced form of the Sturm–Liouville equation:

$$y''(t) + A(t)y'(t) + B(t)y(t) = 0, \quad t \in (a, b),$$

where $A, B \in \mathcal{B}_b((a, b))$. This may be given in the canonical form:

$$(ry')' + sy = 0, \quad r > 0, \ r, s \in \mathcal{B}((a, b)), \tag{3}$$

where $r(t) = \exp(\int_a^t A(u)\mathrm{d}u)$ and $s = rB$. If we set

$$x(t) = \int_a^t \frac{\mathrm{d}u}{r(u)}$$

and let $t(x)$ be the inverse of $x(t)$ and $u(x) = y[t(x)]$, then (3) becomes

$$\frac{\mathrm{d}^2 u}{\mathrm{d}x^2} + q(x)u(x) = 0, \text{ where } q(x) = (r^2 B)[t(x)].$$

Let q be a Borel measurable function from $\mathbb{R}^d$ to $\overline{\mathbb{R}^1}$. If q is only given in a domain D, we extend it to $\mathbb{R}^d$ by setting it to be zero in $\mathbb{R}^d \backslash D$.

We define the Kato class J as follows. Let g be as in (2.17), but without the constant factors:

$$g(u) = g(|u|) = \begin{cases} |u|^{2-d} & d \geq 3; \\ \ln \frac{1}{|u|} & d = 2; \\ |u| & d = 1. \end{cases} \tag{4}$$

Then $q \in J$ iff

$$\lim_{\alpha \downarrow 0} \left[\sup_{x \in \mathbb{R}^d} \int_{|y-x| \leq \alpha} |g(y-x)q(y)| \mathrm{d}y \right] = 0. \tag{5}$$

By calculus, (5) is satisfied if $q \equiv 1$, whence $L^\infty(\mathbb{R}^d) \subset J$. The product of a function in J by a function in $L^\infty(\mathbb{R}^d)$ is in J.

We say that $q \in J_{\text{loc}}$ iff for any bounded domain D, $1_D q \in J$. The following property of J is important.

Proposition 3.1 *If $q \in J$, then*

$$\sup_x \int_{|y-x| \leq 1} |q(y)| \mathrm{d}y < \infty. \tag{6}$$

Consequently, if $q \in J_{\text{loc}}$, then $q \in L^1_{\text{loc}}(\mathbb{R}^d)$.

Remark When $d = 1$ the fact that (6) implies (5) is trivial; thus, $q \in J$ iff (6) holds.

Proof Choose $0 < \alpha < 1$ such that

$$\sup_x \int_{|y-x| \leq \alpha} |g(y-x)q(y)| \mathrm{d}y \leq 1.$$

There exists an integer $N = N(d, \alpha)$ such that for any $x \in \mathbb{R}^d$, there exist points $x_1, \cdots, x_N$ in $\mathbb{R}^d$ such that

$$\overline{B}(x, 1) \subset \bigcup_{i=1}^N B(x_i, \alpha).$$

Hence for $d \geq 2$, we have for all $x \in \mathbb{R}^d$,

$$\int_{|y-x| \leq 1} |q(y)| \mathrm{d}y \leq \frac{1}{g(\alpha)} \sum_{i=1}^N \int_{|y-x_i| \leq \alpha} g(y-x_i)|q(y)| \mathrm{d}y \leq \frac{N}{g(\alpha)}.$$

For $d = 1$, we take $N = N(1, \frac{\alpha}{4})$, so that for any $x \in \mathbb{R}^1$, there exist $x_1, \cdots, x_N$ in $\mathbb{R}^1$ such that

$$\overline{B}(x,1) \subset \bigcup_{i=1}^{N} B(x_i, \frac{\alpha}{4}).$$

We let $z_i = x_i + \frac{\alpha}{2}$, $i = 1, \dots, N$. Then, for all $x \in \mathbb{R}^1$, we have

$$\begin{aligned}\int_{|y-x|\le 1} |q(y)|\mathrm{d}y &\le \sum_{i=1}^{N} \frac{4}{\alpha} \int_{|y-x_i|\le\frac{\alpha}{4}} |y-z_i||q(y)|\mathrm{d}y \\ &\le \sum_{i=1}^{N} \frac{4}{\alpha} \int_{|y-z_i|\le\alpha} |y-z_i||q(y)|\mathrm{d}y \le \frac{4N}{\alpha}. \qquad \square\end{aligned}$$

We now record a simple consequence of Proposition 3.1 for later use in Chapters 6 and 7.

Corollary to Proposition 3.1 *If $q \in J_{\mathrm{loc}}$, then*

$$\sup_{x\in\overline{D}} \int_D |g(y-x)q(y)|\mathrm{d}y < \infty$$

for any bounded Borel set D. Moreover the above supremum converges to zero as $m(D) \to 0$, provided that (the varying) D is restricted to be contained in a fixed ball B. Consequently if D is bounded, then the set of functions $\{g(x,\cdot)q(\cdot) : x \in D\}$ is uniformly integrable over D.

Proof Let A denote the diameter of D, so that $|y-x| \le A$ for $(x,y) \in \overline{D}\times\overline{D}$. Then for any $\alpha > 0$, on $\{(x,y) \in \overline{D}\times\overline{D} : |y-x| > \alpha\}$ we have, by the shape of $|u| \to |g(|u|)|$, (regardless of the value of d):

$$|g(y-x)| \le |g(\alpha)| \vee |g(A)|,$$

and consequently,

$$\int_D |g(y-x)q(y)|\mathrm{d}y \le \int_{|y-x|\le\alpha} |g(y-x)q(y)|\mathrm{d}y + (|g(\alpha)| \vee |g(A)|) \int_D |q(y)|\mathrm{d}y.$$

Since $q \in L^1(D)$ the first assertion of the corollary follows from (5). The second assertion follows from the fact that $q \in L^1(B)$. The third assertion follows from a characterization of uniform integrability. □

The following example shows the need for the restriction in the corollary.

Let $d = 3$. For each $n \ge 4$, there exist n^2 points $\{a_{n,k} : 1 \le k \le n^2\}$ on the sphere $\partial B(0,n)$ with $|a_{n,k} - a_{n,k'}| \ge \frac{\pi}{4}$ $(k \ne k')$. Let

$$q(y) = \sum_{n=4}^{\infty} \sum_{k=1}^{n^2} 1_{B(a_{n,k},r_n)}(y) \frac{1}{|y-a_{n,k}|},$$

where $r_n = \frac{1}{n(\ln n)^{1/2}}$. Then $q \in J$. For each $i \ge 4$, let

$$D_i = \bigcup_{n \geq i} \bigcup_{k=1}^{n^2} B(a_{n,k}, r_n).$$

Then $m(D_i) \to 0$ as $i \to \infty$. But

$$\sup_x \int_{D_i} \frac{|q(y)|}{|x-y|} \mathrm{d}y = +\infty.$$

The verification of the above assertions is left to the reader.

The next property of J plays an essential role in what follows.

Theorem 3.2 *Let D be a domain in $\mathbb{R}^d$, $d \geq 1$; and suppose that*

(i) *for $d = 1$, D is bounded and $q \in L^1(D)$;*

(ii) *for $d = 2$, D is Green-bounded and $q \in J \cap L^1(D)$;*

(iii) *for $d \geq 3$, $q \in J \cap L^1(D)$.*

Then $G_D q \in C_b(D)$. If, in addition, D is regular, then $G_D q \in C_0(D)$.

Proof We write G for G_D below.

(i) Let $D = (a, b)$; by the explicit formula for G given in (2.29), we have for $x \in \overline{D}$:

$$Gq(x) = \frac{2(x-a)}{b-a} \int_x^b (b-y)q(y)\mathrm{d}y + \frac{2(b-x)}{b-a} \int_a^x (y-a)q(y)\mathrm{d}y.$$

Hence $Gq \in C_b(\overline{D})$ and $Gq(x) = 0$ for $x = a$ or b; thus $Gq \in C_0(D)$.

(ii) and (iii) For each $n \geq 1$, let

$$G_n = G \wedge n.$$

There exists $\alpha_n > 0$ such that $G_n(x, y) = G(x, y)$ if $|y - x| \geq \alpha_n$, and $\alpha_n \to 0$ as $n \to \infty$. Hence we have, writing q_D for $1_D q$:

$$\begin{aligned}
\int_{|y-x|\geq\alpha_n} G(x,y)q_D(y)\mathrm{d}y &= \int_{|y-x|\geq\alpha_n} G_n(x,y)q_D(y)\mathrm{d}y \\
&= \int_{\mathbb{R}^d} G_n(x,y)q_D(y)\mathrm{d}y - \int_{|y-x|<\alpha_n} G_n(x,y)q_D(y)\mathrm{d}y.
\end{aligned}$$

Consequently,

$$\begin{aligned}
\int_{\mathbb{R}^d} G(x,y)q_D(y)\mathrm{d}y &= \int_{\mathbb{R}^d} G_n(x,y)q_D(y)\mathrm{d}y - \int_{|y-x|<\alpha_n} G_n(x,y)q_D(y)\mathrm{d}y \\
&\quad + \int_{|y-x|<\alpha_n} G(x,y)q_D(y)\mathrm{d}y.
\end{aligned}$$

Since $0 \leq G_n \leq G$, it follows, writing $G_n q_D$ for the first integral on the right-hand side, that

$$|Gq_D(x) - G_nq_D(x)| \leq 2 \int_{|y-x|<\alpha_n} G(x,y)|q_D(y)|\mathrm{d}y. \tag{7}$$

By Theorem 2.6, (recalling the new definition of g) we have for $d = 2$:

$$G(x,y) \leq \frac{1}{\pi} g^+(x,y) + C;$$

and for $d \geq 3$:

$$G(x,y) \leq Cg(x,y),$$

where C is a constant depending on D for $d = 2$, and on d only for $d \geq 3$. Hence in both cases the integral on the right-hand side of (7) is bounded by

$$C \int_{|y-x|<\alpha_n} g^+(x,y)|q(y)|\mathrm{d}y + C \int_{|y-x|<\alpha_n} |q_D(y)|\mathrm{d}y.$$

As $n \to \infty$, the first integral above converges to zero uniformly for $x \in \mathbb{R}^d$ because $q \in J$; the second does the same because $q_D \in L^1(\mathbb{R}^d)$. Hence by (7), G_nq_D converges to Gq_D uniformly in $\mathbb{R}^d$. But for each n, $x \to G_n(x,y)$ is bounded and continuous on D for each $y \in D$, by Theorem 2.6. Hence $G_nq_D \in C_b(D)$ by the dominated convergence theorem. Thus $Gq_D \in C_b(D)$.

When D is regular, $G_nq_D \in C_0(D)$ by the same theorem; therefore, $Gq_D \in C_0(D)$. □

Remark When D is a bounded domain and $q \in J_{\text{loc}}$, the conclusions of Theorem 3.2 hold in all dimensions, for in this case $q \in L^1(D)$ by Proposition 3.1.

The following example shows that even if q is bounded and integrable in an unbounded D, G_Dq may not be bounded.

Example 1 Let $d = 1$, $D = (0, \infty)$; $q(y) = 0$ if $0 < y \leq 1$ and $q(y) = y^{-2}$ if $1 < y < \infty$. Thus $q \in L^1(D) \cap L^\infty(D)$. We know that $G_D(x,y) = 2(x \wedge y)$ (see (2.30)). It is easy to show that $G_Dq(x) = 2x$ if $0 < x \leq 1$ and $G_Dq(x) = 2\ln x + 2$ if $x > 1$. Thus G_Dq is not bounded.

Let $d = 2$, $D = \{y \in \mathbb{R}^2 : |y| > 1\}$ and

$$q(y) = \frac{1}{|y|^2(1 + \ln|y|)^2}, \quad y \in D.$$

It is easy to verify that $G_D(x,y)$ has the same form as $G_{B(0,1)}(x,y)$ given in (2.28):

$$G_D(x,y) = \frac{1}{\pi} \ln \frac{|y||x - y^*|}{|x - y|}, \quad |x| > 1, \ |y| > 1,$$

where $y^* = |y|^{-2}y$; and $q \in L^\infty(D) \cap L^1(D) \subset J \cap L^1(D)$. Let $x_n \in D$ with $|x_n| \to \infty$, then by Fatou's lemma, we have

$$\begin{aligned}\underline{\lim}_n G_D q(x_n) &\geq \frac{1}{\pi}\int_{|y|>1} \underline{\lim}_n \left(\ln\frac{|y||x_n-y^*|}{|x_n-y|}\right)\cdot\frac{\mathrm{d}y}{|y|^2(1+\ln|y|)^2}\\ &= \frac{1}{\pi}\int_{|y|>1}\frac{\ln|y|}{|y|^2(1+\ln|y|)^2}\mathrm{d}y\\ &= 2\int_1^\infty \frac{\ln\, r}{r(1+\ln\, r)^2}\mathrm{d}r = 2\int_0^\infty \frac{u}{(1+u)^2}\mathrm{d}u=\infty.\end{aligned}$$

Thus $G_D q$ is not bounded in D.

We now give two typical examples of J.

Example 2 Suppose that for $d \geq 2$, $p > \frac{d}{2}$ and for $d = 1$, $p \geq 1$, we have

$$A = \sup_x \int_{|y-x|\leq 1} |q(y)|^p \mathrm{d}y < \infty,$$

then $q \in J$.

Let $0 < \alpha < 1$. If $d \geq 3$, then

$$\begin{aligned}\int_{|y-x|\leq\alpha}\frac{|q(y)|}{|y-x|^{d-2}}\mathrm{d}y &\leq \left(\int_{|y-x|\leq\alpha}|q(y)|^p\mathrm{d}y\right)^{1/p}\\ &\quad\times\left(\int_{|y-x|\leq\alpha}|y-x|^{(2-d)\frac{p}{p-1}}\mathrm{d}y\right)^{\frac{p-1}{p}}\\ &\leq A^{1/p}\left[\sigma_{d-1}(1)\int_0^\alpha r^{(2-d)\frac{p}{p-1}+d-1}\mathrm{d}r\right]^{\frac{p-1}{p}} \to 0,\end{aligned}$$

as $\alpha \downarrow 0$, since $(2-d)\frac{p}{p-1}+d-1 > -1$ when $p > \frac{d}{2}$.

If $d = 2$, then

$$\begin{aligned}&\int_{|y-x|\leq\alpha}\ln\frac{1}{|y-x|}|q(y)|\mathrm{d}y\\ &\leq \left(\int_{|y-x|\leq\alpha}|q(y)|^p\mathrm{d}y\right)^{1/p}\left(\int_{|y-x|\leq\alpha}\left(\ln\frac{1}{|y-x|}\right)^{\frac{p}{p-1}}\mathrm{d}y\right)^{\frac{p-1}{p}}\\ &\leq A^{1/p}\left[2\pi\int_0^\alpha\left(\ln\frac{1}{r}\right)^{\frac{p}{p-1}}r\mathrm{d}r\right]^{\frac{p-1}{p}} \to 0,\end{aligned}$$

as $\alpha \downarrow 0$, since $p > 1$.

If $d = 1$ and $p \geq 1$, then the condition that $A < \infty$ implies (6) which implies (5).

Example 3 *'Coulomb potential'* $d = 3$, $q(x) = \frac{1}{|x|} = g(x)$.

Let $\frac{3}{2} < p < 3$. Then for any $x \in \mathbb{R}^3$,

$$\int_{|y-x|\le 1} \frac{\mathrm{d}y}{|y|^p} \le \begin{cases} \frac{4\pi}{3}, & \text{if } |x| \ge 2; \\ \int_{|y|\le 3} \frac{\mathrm{d}y}{|y|^p} & \text{if } |x| < 2. \end{cases}$$

Hence $q \in J$ by Example 2. Note that $q \notin L^1(\mathbb{R}^3)$.

The following property of J is useful.

Lemma 3.3 *If $q \in J$, then for any $\alpha > 0$, $\beta > 0$ and real ν, we have*

$$\lim_{\lambda\to\infty} \left[\sup_x \lambda^\nu \int_{|y-x|>\alpha} \mathrm{e}^{-\lambda|y-x|^\beta} |q(y)|\mathrm{d}y \right] = 0. \tag{8}$$

Proof For $\lambda > 0$ and $x \in \mathbb{R}^d$, we have

$$\begin{aligned} &\lambda^\nu \int_{|y-x|>\lambda} \mathrm{e}^{-\lambda|y-x|^\beta} |q(y)|\mathrm{d}y \\ &\le \lambda^\nu \mathrm{e}^{-\frac{\lambda\alpha^\beta}{2}} \int_{\mathbb{R}^d} \mathrm{e}^{-\lambda|y|^\beta/2} |q(x+y)|\mathrm{d}y. \end{aligned} \tag{9}$$

It is easy to see that a ball of radius $n \ge 1$ in $\mathbb{R}^d$ can be covered by $A_d n^d$ balls of radius 1, where A_d is a constant depending only on d. Hence, it follows from Proposition 3.1 that for all n:

$$\sup_x \int_{|y|\le n} |q(x+y)|\mathrm{d}y \le A_d n^d \sup_x \int_{|y|\le 1} |q(x+y)|\mathrm{d}y \le A_d n^d M,$$

where M is the supremum in (6). Hence the integral on the right-hand side of (9) is trivially bounded in x and the lemma follows. □

Lemma 3.4 (i) $d = 1$. *For each $t > 0$, and all x and y, we have*

$$\int_0^t p(s; x, y)\mathrm{d}s \le \sqrt{t}.$$

(ii) $d = 2$. *There exists an absolute constant C such that for $|x - y| \le t \le \frac{1}{2}$, we have*

$$\int_0^t p(s; x, y)\mathrm{d}s \le Cg(x - y).$$

(iii) $d \ge 3$. *For each $t > 0$, and all x and y, we have*

$$\int_0^t p(s; x, y)\mathrm{d}s \le C_d g(x - y),$$

where C_d is the constant given in (2.17).

Proof (i) This follows from

$$p(s; x, y) \le \frac{1}{(2\pi s)^{1/2}}.$$

(ii) If we make the change of variables $t = |x-y|^2 u$ in (1.11), we obtain for all $d \geq 1$:

$$\int_0^t p(s;x,y)\mathrm{d}s = \frac{1}{|x-y|^{d-2}} \int_0^{\frac{t}{|x-y|^2}} p(u;0,a)\mathrm{d}u, \tag{10}$$

where a is any point in $\mathbb{R}^d$ with $|a| = 1$. Let

$$b_d = \int_0^1 p(u;0,a)\mathrm{d}u; \tag{11}$$

then $0 < b_d < \infty$. Now if $d = 2$ and $|x-y| \leq t \leq \frac{1}{2}$, then $|x-y|^2 \leq t$ and by (10) we have

$$\begin{aligned}\int_0^t p(s;x,y)\mathrm{d}s &\leq b_2 + \int_1^{\frac{t}{|x-y|^2}} \frac{1}{2\pi u}\mathrm{d}u \\ &= b_2 + \frac{1}{2\pi} \ln\left(\frac{t}{|x-y|^2}\right) \\ &\leq \left(\frac{b_2}{\ln 2} + \frac{1}{\pi}\right) \ln\left(\frac{1}{|x-y|}\right).\end{aligned}$$

This is the desired inequality.

(iii) This follows from (2.15). □

Lemma 3.5 *There is an absolute constant $C'_d > 0$ such that*

$$\int_0^t p(s;x,y)\mathrm{d}s \geq C'_d g(x-y), \tag{12}$$

provided

(i) $d = 1$, *and* $|x-y| \leq \sqrt{t} \leq 1$*; or*

(ii) $d = 2$, *and* $|x-y| \leq t \leq 1$*; or*

(iii) $d \geq 3$, *and* $|x-y| \leq \sqrt{t}$.

Proof (i) $d = 1$. For $s \geq \frac{t}{2}$, we have

$$\frac{|x-y|^2}{2s} \leq \frac{|x-y|^2}{t} \leq 1$$

by hypothesis; hence

$$\begin{aligned}\int_0^t p(s;x,y)\mathrm{d}s &\geq \frac{1}{\sqrt{2\pi}} \int_{t/2}^t \frac{\mathrm{e}^{-1}}{\sqrt{s}}\mathrm{d}s = \frac{\mathrm{e}^{-1}}{\sqrt{2\pi}}\left(1 - \frac{1}{\sqrt{2}}\right)\sqrt{t} \\ &\geq C'_1 g(x-y).\end{aligned}$$

(ii) $d = 2$. By (10) with $d = 2$, we have

$$\begin{aligned}\int_0^t p(s;x,y)\mathrm{d}s &\geq \frac{1}{2\pi}\mathrm{e}^{-1/2}\int_1^{\frac{t}{|x-y|^2}} \frac{1}{u}\mathrm{d}u = C_2'\ln\left(\frac{t}{|x-y|^2}\right) \\ &\geq C_2'\ln\left(\frac{1}{|x-y|}\right).\end{aligned}$$

In the first of the above inequalities we used the fact $t \geq |x-y|^2$, while in the second we used the fact that $t \geq |x-y|$.

(iii) $d \geq 3$. By (10) we have since $t \geq |x-y|^2$,

$$\int_0^t p(s;x,y)\mathrm{d}s \geq g(x-y)b_d,$$

where b_d is given in (11). Thus in this case we may take C_d' to be b_d. □

Theorem 3.6 *The class J is the class of $q \in \mathcal{B}$ satisfying the following condition:*

$$\limsup_{t\downarrow 0}\sup_x \int_0^t P_s|q|(x)\mathrm{d}s = 0. \tag{13}$$

Proof We have by Fubini's theorem:

$$\int_0^t P_s|q|(x)\mathrm{d}s = \int_{\mathbb{R}^d}\left[\int_0^t p(s;x,y)\mathrm{d}s\right]|q(y)|\mathrm{d}y.$$

Applying Lemma 3.5, we obtain

$$\int_0^t P_s|q|(x)\mathrm{d}s \geq \int_{|y-x|\leq\alpha} C_d' g(x-y)|q(y)|\mathrm{d}y$$

with $t \leq 1$, where $\alpha = \sqrt{t}$ if $d = 1$ or $d \geq 3$ and $\alpha = t$ if $d = 2$. Letting $t \downarrow 0$, whence $\alpha \downarrow 0$, in the above inequality, we see at once that the condition (13) implies the condition (5).

Conversely, suppose (5) is true. Let

$$\begin{aligned}I_1(\alpha;t,x) &= \int_{|y-x|\leq\alpha}\left[\int_0^t p(s;x,y)\mathrm{d}s\right]|q(y)|\mathrm{d}y, \\ I_2(\alpha;t,x) &= \int_{|y-x|>\alpha}\left[\int_0^t p(s;x,y)\mathrm{d}s\right]|q(y)|\mathrm{d}y,\end{aligned}$$

so that

$$\int_0^t P_s|q|(x)\mathrm{d}s = I_1(\alpha;t,x) + I_2(\alpha;t,x).$$

For $d = 1$, we have by Lemma 3.4(i), if $\alpha < 1$:

$$I_1(\alpha;t,x) \leq \sqrt{t}\int_{|y-x|\leq\alpha}|q(y)|\mathrm{d}y \leq \sqrt{t}\sup_x\int_{|y-x|\leq 1}|q(y)|\mathrm{d}y. \tag{14}$$

Assuming (5) is true, the above supremum is finite by (6).

For $d = 2$, we have by Lemma 3.4(ii), if $t \leq \frac{1}{2}$ and $\alpha \leq \frac{1}{2}$:

$$I_1(\alpha; t, x) \leq C \int_{|y-x|\leq\alpha} g(x-y)|q(y)|\mathrm{d}y.$$

For $d \geq 3$, the above inequality is also true by Lemma 3.4(iii) for $t > 0$ and any $\alpha > 0$, with C replaced by C_d. Therefore it follows from (14) for $d = 1$, and from the above inequality and (5) for $d \geq 2$, that for all $d \geq 1$ and any given $\varepsilon > 0$ there exists $\alpha = \alpha(\varepsilon)$ and $t_0 = t(\varepsilon)$ such that for all $0 < t < t_0$,

$$\sup_x I_1(\alpha(\varepsilon); t, x) \leq \varepsilon.$$

Now for this $\alpha(\varepsilon)$, we have

$$\begin{aligned} &\sup_x I_2(\alpha(\varepsilon); t, x) \\ &\leq \int_0^t \sup_x \int_{|y-x|>\alpha(\varepsilon)} (2s)^{-d/2} \mathrm{e}^{-(2s)^{-1}|x-y|^2} |q(y)|\mathrm{d}y\,\mathrm{d}s. \end{aligned}$$

By Lemma 3.3 with $\lambda = (2s)^{-1}$, this converges to zero as $t \downarrow 0$. Combining the above results we obtain (13), since ε is arbitrary. □

The advantage of using $p(t; x, y)$ instead of $g(x-y)$ should be crystal clear from the above discussion. Let us test it on the famous Coulomb potential of Example 3. In this case, by (2.15), since $C_3 = \frac{1}{2\pi}$, we have

$$q(x) = 2\pi \int_0^\infty p(t; x, 0)\mathrm{d}t,$$

and by the semigroup property, for all $x \in \mathbb{R}^3$:

$$\begin{aligned} P_s q(x) &= 2\pi \int_0^\infty p(s+t; x, 0)\mathrm{d}t = 2\pi \int_s^\infty p(t; x, 0)\mathrm{d}t \\ &\leq 2\pi \int_s^\infty \frac{\mathrm{d}t}{(2\pi t)^{3/2}} = \sqrt{\frac{2}{\pi}}\frac{1}{\sqrt{s}}. \end{aligned}$$

Thus we have

$$\sup_x \int_0^t P_s q(x)\mathrm{d}s \leq 2\sqrt{\frac{2t}{\pi}},$$

and (13) is true, whence $q \in J$. This should be compared with the previous proof!

We shall go one step further by using the Brownian motion $\{X_t, t \geq 0\}$ and recasting (13) in the 'process form' as follows:

$$\lim_{t\downarrow 0} \sup_x E^x \left\{ \int_0^t |q(X_s)|\mathrm{d}s \right\} = 0. \tag{15}$$

Although the new form might appear to be a merely capricious and unnecessary change of notation, this is by no means the case. The process is infinitely richer than its semigroup, as we shall demonstrate amply in the what follows. For the moment let us consider the possibility of substituting a random time for the t in (15): how would one do it in (13)? This leads naturally to the general orientation of the next section, but we must wait until Chapter 4 to appreciate the full significance.

3.2 Semigroup with Multiplicative Functional

We begin with a number of results which are not restricted to Brownian motion. Let $\mathcal{S}$ be a locally compact metric space and $X = \{X_t, t \geq 0\}$, a Markov process with state space $\mathcal{S}$. To be definite, we may suppose that X is a Hunt process (see Chung (1982a, Chapter 3)), although the initial results below are valid more generally. Previous notation such as $C_b(D)$ will be extended to $\mathcal{S}$ without comment. The standard definition of $C_0(S)$ is the class of continuous functions f on S such that $f(x)$ converges to zero as x leaves all compact subsets of S.

Let $q \in \mathcal{B}(\mathcal{S})$, and set

$$A(t) = \int_0^t q(X_s)\mathrm{d}s. \tag{16}$$

This is well defined if $q \geq 0$, and satisfies the additive property:

$$\forall s \geq 0,\ t \geq 0 :\ A(s+t) = A(s) + A(t) \circ \theta_s. \tag{17}$$

The following lemma, due to Khas'minskii (Khas'minskii 1959), plays an important role in what follows.

Lemma 3.7 *Let τ be an optional time of X satisfying the following condition:*

$$\forall t \geq 0 :\ \tau \leq t + \tau \circ \theta_t \text{ on } \{t < \tau\}. \tag{18}$$

Suppose that $q \geq 0$ and $E^x\{A(\tau)\} < \infty$ for all x. Then we have for each integer $n \geq 0$:

$$\sup_x E^x\{A(\tau)^n\} \leq n! \sup_x (E^x\{A(\tau)\})^n, \tag{19}$$

where the supremum is taken over $\mathcal{S}$. If

$$\sup_x E^x\{A(\tau)\} = \alpha < 1,$$

then

$$\sup_x E^x\{\mathrm{e}^{A(\tau)}\} \leq \frac{1}{1-\alpha}.$$

Remark Condition (18) is satisfied if τ is a constant or a hitting time (cf. (1.31)).

Proof Since $q \geq 0$, $A(\cdot)$ is positive increasing. Under our hypothesis we may suppose that $A(\tau) < \infty$ a.s. We have by calculus:

$$\frac{1}{n+1}A(\tau)^{n+1} = \int_0^\tau [A(\tau) - A(t)]^n \mathrm{d}A(t).$$

For $t < \tau$, we have by (18):

$$\begin{aligned} A(\tau) - A(t) &\leq A(t + \tau \circ \theta_t) - A(t) = \int_t^{t+\tau\circ\theta_t} q(X_s)\mathrm{d}s \\ &= \left[\int_0^\tau q(X_s)\mathrm{d}s\right] \circ \theta_t = A(\tau) \circ \theta_t. \end{aligned}$$

Hence by Fubini's theorem,

$$\begin{aligned} \frac{1}{n+1}E^x\{A(\tau)^{n+1}\} &\leq E^x\left\{\int_0^\tau [A(\tau)\circ\theta_t]^n \mathrm{d}A(t)\right\} \\ &= \int_0^\infty E^x\{t < \tau; [A(\tau)\circ\theta_t]^n q(X_t)\}\mathrm{d}t. \end{aligned}$$

This transformation is needed so that we may apply the Markov property. The last integral is then equal to

$$\begin{aligned} &\int_0^\infty E^x\{t < \tau;\ E^{X_t}[A(\tau)^n] q(X_t)\}\mathrm{d}t \\ &\leq \int_0^\infty E^x\{t < \tau; q(X_t)\}\mathrm{d}t \cdot \sup_x E^x\{A(\tau)^n\} \\ &= E^x\{A(\tau)\} \sup_x E^x\{A(\tau)^n\}. \end{aligned}$$

It follows that

$$\sup_x E^x\{A(\tau)^{n+1}\} \leq (n+1) \sup_x E^x\{A(\tau)\} \sup_x E^x\{A(\tau)^n\}.$$

Hence (19) is proved by induction on n.

The last assertion of the lemma is an immediate consequence of (19). □

Next we introduce the multiplicative functional:

$$e_q(t) = \exp\left[\int_0^t q(X_s)\mathrm{d}s\right] = \mathrm{e}^{A_q(t)}, \tag{20}$$

where $A_q(t)$ is the $A(t)$ in (16). When X is the Brownian motion this is known as the Feynman–Kac functional. It is well defined provided $A_q(t)$ is. This is the case under the condition (15), but then much more is true, as we shall now see.

Proposition 3.8 *Suppose (13) is true (where the supremum is now taken over all x in S). Then*

$$\lim_{t\downarrow 0} \sup_x E^x\{e_{|q|}(t)\} = 1 \tag{21}$$

and there exist positive constants C_0 and C_1 such that

$$\forall t > 0: \sup_x E^x\{e_{|q|}(t)\} \leq e^{C_0+C_1 t}. \tag{22}$$

Proof For sufficiently small t, we have $\sup_x E^x\{A_{|q|}(t)\} < 1$; hence by Lemma 3.7

$$1 \leq \sup_x E^x\{e_{|q|}(t)\} \leq \frac{1}{1-\sup_x E^x\{A_{|q|}(t)\}}.$$

Letting $t \downarrow 0$, we obtain (21) by (13).

Next, we have by (17) and the Markov property of X, for each x;

$$\begin{aligned} E^x\{e_{|q|}(s+t)\} &= E^x\{e_{|q|}(s)\cdot E^{X_s}[e_{|q|}(t)]\} \\ &\leq E^x\{e_{|q|}(s)\} \sup_x E^x\{e_{|q|}(t)\}. \end{aligned}$$

It follows that the function: $t \to \ln[\sup_x E^x\{e_{|q|}(t)\}]$ is subadditive. Since it is finite for sufficiently small t by (21), by elementary analysis there exist $C_0 \geq 0$, $C_1 \geq 0$ such that

$$\ln[\sup_x E^x\{e_{|q|}(t)\}] \leq C_0 + C_1 t.$$

Thus (22) is true. □

Corollary to Proposition 3.8 *We have for all* $t \geq 0$*:*

$$\sup_x E^x\{A_{|q|}(t)\} \leq C_0 + C_1 t. \tag{23}$$

Proof This follows from (22) by Jensen's inequality. □

Remark It follows from (15), (17), and the Markov property that the function $t \to E^x\{A_{|q|}(t)\}$ is finite and subadditive in $\mathbb{R}_+$; whence, there exist C_0 and C_1 for which (23) holds. But, curiously, this direct argument does not show that we may use the same constants C_0 and C_1 in (22).

It is trivial that (22) remains true if $|q|$ is replaced by q. To deal with (21), we note that by the Cauchy–Schwarz inequality:

$$1 = E^x\{e_{|q|/2}(t)e_{-|q|/2}(t)\}^2 \leq E^x\{e_{|q|}(t)\}E^x\{e_{-|q|}(t)\};$$

whence, it follows easily that

$$\lim_{t\downarrow 0}\sup_x E^x\{e_q(t)\} = \lim_{t\downarrow 0}\inf_x E^x\{e_q(t)\} = 1.$$

The next result will be needed shortly.

Proposition 3.9 *Under the same hypothesis as before, we have for every* $r \geq 1$*:*

$$\lim_{t\downarrow 0}\sup_x E^x\{|e_q(t)-1|^r\} = 0. \tag{24}$$

Proof We need the elementary inequality, valid for $z \in \mathbb{R}^1$, $r \geq 1$:

$$|e^z - 1|^r \leq (e^{|z|} - 1)^r \leq e^{r|z|} - 1.$$

Then, since $|A_q(t)| \leq A_{|q|}(t)$, we have

$$\begin{aligned} 0 &\leq E^x\{|e^{A_q(t)} - 1|^r\} \leq E^x\{e^{rA_{|q|}(t)} - 1\} \\ &= E^x\{e_{r|q|}(t)\} - 1. \end{aligned}$$

As $t \downarrow 0$, the last term above converges to zero uniformly in x by (21) with $|q|$ replaced by $r|q|$. □

From now on the condition (15) will be assumed in the general context of $\mathcal{S}$. Furthermore, we assume that the transition function P_t ($t > 0$) for the process $\{X_t\}$ has a density $p_t(x, y)$ with respect to a σ–finite measure m, which is symmetric and bounded. Thus, for $t > 0$, $x \in \mathcal{S}$, $B \in \mathcal{B}(\mathcal{S})$, we have:

$$P_t(x, B) = \int_B p_t(x, y) dm(y),$$

where

$$p_t(x, y) = p_t(y, x); \quad \sup_{x,y} p_t(x, y) < \infty. \tag{25}$$

Under the above assumptions, we define the operator T_t ($t \geq 0$) as follows:

$$T_t f(x) = E^x\{e_q(t) f(X_t)\}. \tag{26}$$

Clearly, T_t is linear and T_0 is the identity mapping. To see that $\{T_t,\ t \geq 0\}$ forms a semigroup, we have for $s \geq 0$, $t \geq 0$, and any admissible f:

$$\begin{aligned} T_{s+t} f(x) &= E^x\left\{e_q(s) \exp\left[\int_s^{s+t} q(X_u) du\right] f(X_{s+t})\right\} \\ &= E^x\{e_q(s)[e_q(t) f(X_t)] \circ \theta_s\} \\ &= E^x\{e_q(s) E^{X_s}[e_q(t) f(X_t)]\} \\ &= T_s(T_t f)(x). \end{aligned}$$

The semigroup $\{T_t : t \geq 0\}$ is called the Feynman–Kac semigroup.

We shall write L^p for $L^p(\mathcal{S}, m)$ in what follows. We recall the usual notation for norms of a mapping T from L^p into L^r:

$$\|T\|_{p,r} = \sup_{\|f\|_p \leq 1} \|Tf\|_r; \quad \|T\|_p = \|T\|_{p,p}.$$

Theorem 3.10 *$\{T_t,\ t \geq 0\}$ is a semigroup of linear positive operators in L^p, $1 \leq p \leq \infty$. For each such p, we have*

$$\|T_t\|_p \leq \|T_t\|_\infty \leq e^{C_0 + C_1 t} \tag{27}$$

where the constants C_0 and C_1 are as in (22). For each t, T_t is also a bounded operator from L^p ($1 \le p \le \infty$) into L^∞. There exists a symmetric and bounded Borel measurable density $u_t(\cdot,\cdot)$ such that for $1 \le p \le \infty$, and $f \in L^p$, we have

$$T_t f(x) = \int_S u_t(x,y) f(y) \mathrm{d}m(y). \tag{28}$$

In each L^p, $1 \le p \le \infty$, and in $C_0(S)$, $\{T_t\}$ is strongly continuous if $\{P_t\}$ is.

Proof Step 1: $L^\infty \to L^\infty$.
For all $f \in L^\infty$, $x \in S$,

$$|T_t f(x)| \le E^x[e_q(t)|f(X_t)|] \le \|f\|_\infty E^x[e_q(t)].$$

Hence by (22), T_t is a bounded operator in L^∞, and (27) is true for $p = \infty$.

Step 2: Symmetry of T_t.
We shall prove that for any two positive functions f_1 and f_2 on S, we have

$$\int_S f_1(x) T_t f_2(x) \mathrm{d}m(x) = \int_S f_2(x) T_t f_1(x) \mathrm{d}m(x). \tag{29}$$

By standard arguments, it is sufficient to verify this relation when $f_1 = 1_A$, $f_2 = 1_B$, where A and B are two bounded Borel sets. In fact, we can prove that for any $n \ge 0$:

$$\int_A E^x \left[\left(\int_0^t q(X_s)\mathrm{d}s \right)^n 1_B(X_t) \right] \mathrm{d}m(x)$$
$$= \int_B E^x \left[\left(\int_0^t q(X_s)\mathrm{d}s \right)^n 1_A(X_t) \right] \mathrm{d}m(x).$$

Note firstly that both members of the above are finite by the Corollary to Proposition 3.8, so that Fubini's theorem permits all rearrangements. Without real loss of generality, we may take $n = 3$ to lighten our visual load. The left member in the above is then equal to

$$3! \int_0^t \int_{s_1}^t \int_{s_2}^t \left\{ \int_S E^x[1_A(X_0) q(X_{s_1}) q(X_{s_2}) q(X_{s_3}) 1_B(X_t)] \mathrm{d}m(x) \right\} \mathrm{d}s_1 \mathrm{d}s_2 \mathrm{d}s_3.$$

The integrand under $\int_0^t \int_{s_1}^t \int_{s_2}^t$ is equal to

$$\begin{aligned} I_1 \equiv & \int_A \int_S \int_S \int_S \int_B p_{s_1}(x_0,x_1) q(x_1) p_{s_2-s_1}(x_1,x_2) q(x_2) p_{s_3-s_2}(x_2,x_3) \\ & q(x_3) p_{t-s_3}(x_3,x_4) \mathrm{d}m(x_0) \mathrm{d}m(x_1) \mathrm{d}m(x_2) \mathrm{d}m(x_3) \mathrm{d}m(x_4) \\ = & \int_B \int_S \int_S \int_S \int_A p_{t-s_3}(x_4,x_3) q(x_3) p_{s_3-s_2}(x_3,x_2) q(x_2) p_{s_2-s_1}(x_2,x_1) \\ & q(x_1) p_{s_1}(x_1,x_0) \mathrm{d}m(x_4) \mathrm{d}m(x_3) \mathrm{d}m(x_2) \mathrm{d}m(x_1) \mathrm{d}m(x_0) \\ \equiv & I_2, \end{aligned}$$

by the symmetry of $p_s(\cdot,\cdot)$ for each s. Now we change the variables $(x_4,\cdots,x_0)$ in I_2 into $(x_0,\cdots,x_4)$, and the indices $(t-s_3, s_3-s_2, s_2-s_1, s_1)$ into $(s_1'$, $s_2'-s_1', s_3'-s_2', t-s_3')$. Then we see by inspection that

$$\int_0^t\int_{s_1}^t\int_{s_2}^t I_1 \mathrm{d}s_1\mathrm{d}s_2\mathrm{d}s_3 = \int_0^t\int_{s_1'}^t\int_{s_2'}^t I_2 \mathrm{d}s_1'\mathrm{d}s_2'\mathrm{d}s_3'.$$

This establishes the desired equation for $n=3$. Dividing it by $n!$ and summing over all $n\geq 0$, we obtain

$$\int_A E^x[e_q(t)1_B(X_t)]\mathrm{d}m(x) = \int_B E^x[e_q(t)1_A(X_t)]\mathrm{d}m(x).$$

The summation is permitted by Proposition 3.8 and Fubini's theorem.

In what follows, we shall write $\int \phi \mathrm{d}m$ for $\int_S \phi(x)\mathrm{d}m(x)$.

Step 3: $L^1 \to L^1$.
For $f\in L^1$, we have by the symmetry relation (29),

$$\begin{aligned}\int |T_t f|\mathrm{d}m &\leq \int T_t|f|\mathrm{d}m = \int |f|T_t 1\mathrm{d}m\\ &\leq \|T_t\|_\infty \int |f|\mathrm{d}m = \|T_t\|_\infty\|f\|_1.\end{aligned}$$

Hence $T_t f\in L^1$ and $\|T_t\|_1\leq\|T_t\|_\infty$. Then by Step 1 and the Riesz convexity theorem (see Dunford and Schwartz (1958, VI 10.12)), T_t is a bounded operator in L^p $(1\leq p\leq\infty)$, and

$$\|T_t\|_p \leq \|T_t\|_\infty.$$

Hence by Step 1 and (22), (27) is true for $1\leq p\leq\infty$.

Step 4: $L^p\to L^\infty$, $1<p<\infty$.
Let $\frac{1}{p}+\frac{1}{p'}=1$. We have for $f\in L^p$:

$$\begin{aligned}|T_t f(x)| &\leq E^x[e_{p'q}(t)]^{1/p'}E^x[|f(X_t)|^p]^{1/p}\\ &\leq \sup_x E^x\{e_{p'q}(t)\}^{1/p'}\left[\sup_{x,y} p_t(x,y)\right]^{1/p}\|f\|_p.\end{aligned}$$

Hence by (22) and (25), T_t is a bounded operator from L^p to L^∞.

Step 5: $L^1\to L^2$.
By (29), for positive $f\in L^1$ and $\phi\in L^2$,

$$\begin{aligned}\int \phi T_t f\mathrm{d}m &= \int fT_t\phi\mathrm{d}m \leq \|T_t\|_{2,\infty}\|\phi\|_2\int f\mathrm{d}m\\ &= \|T_t\|_{2,\infty}\|f\|_1\|\phi\|_2.\end{aligned}$$

This shows that $T_t f\in L^2$ and

$$\|T_t\|_{1,2} \le \|T_t\|_{2,\infty}.$$

Step 6: $L^1 \to L^\infty$
Since $T_t = T_{t/2}T_{t/2}$, it follows from Steps 4 and 5 that T_t is a bounded operator from L^1 to L^∞.

The existence of a measurable and bounded density for T_t follows from Step 6 and a theorem due to Dunford and Pettis (see Simon (1982), Theorem A.1.1). Symmetry of the density $u_t(\cdot,\cdot)$ follows from the symmetry of T_t.

It remains to prove the assertion regarding the strong continuity of $\{T_t\}$.

Case L^1: Let $f_1 \in L^1$, $f_2 \in L^\infty$, then it follows from the symmetry of P_t and T_t that

$$\int f_2(T_t - P_t)f_1 \mathrm{d}m = \int f_1(T_t - P_t)f_2 \mathrm{d}m.$$

Setting $f_2 = \operatorname{sign}\{(T_t - P_t)f_1\}$ in the above, we obtain

$$\begin{aligned}\int |(T_t - P_t)f_1|\mathrm{d}m &\le \int |f_1||(T_t - P_t)f_2|\mathrm{d}m \\ &\le \sup_x E^x\{|e_q(t) - 1|\}\|f_1\|_1.\end{aligned}$$

By (24), this converges to zero as $t \downarrow 0$. Since

$$\|T_t f_1 - f_1\|_1 \le \int |(T_t - P_t)f_1|\mathrm{d}m + \int |P_t f_1 - f_1|\mathrm{d}m,$$

the strong continuity of $\{T_t\}$ in L^1 follows from that of $\{P_t\}$.

Case L^p, $1 < p < \infty$: Let $f \in L^p$ and $\frac{1}{p} + \frac{1}{p'} = 1$. Then

$$\begin{aligned}|T_t f(x) - P_t f(x)|^p &= |E^x\{(e_q(t) - 1)f(X_t)\}|^p \\ &\le E^x\{|e_q(t) - 1|^{p'}\}^{p/p'} E^x\{|f(X_t)|^p\}\end{aligned}$$

by Hölder's inequality. Hence we have

$$\int |T_t f - P_t f|^p \mathrm{d}m \le \sup_x E^x\{|e_q(t) - 1|^{p'}\}^{p/p'} \int p_t(x,y)|f(y)|^p \mathrm{d}m(y)\mathrm{d}(x).$$

By the symmetry of p_t, the second factor on the right-hand side of the above is simply $\int |f|^p \mathrm{d}m$, while the first factor converges to zero as $t \downarrow 0$ by Proposition 3.9. Thus $\{T_t\}$ is strongly continuous if $\{P_t\}$ is.

Case L^∞ *and* $C_0(S)$: Let $f \in L^\infty$ or $C_0(S)$, then we have

$$|T_t f(x) - P_t f(x)| \le E^x\{|e_q(t) - 1|\}\|f\|_\infty,$$

which converges to zero uniformly in all x as $t \downarrow 0$ by (24). Thus $\{T_t\}$ is strongly continuous if $\{P_t\}$ is. □

There is a basic connection between the two semigroups, $\{P_t\}$ and $\{T_t\}$, which transmits the properties of $\{P_t\}$ to $\{T_t\}$.

Proposition 3.11 *For each $t > 0$ and $f \in L^\infty(\mathcal{S})$, as $\delta \downarrow 0$, $P_\delta T_{t-\delta} f$ converges uniformly in $\mathcal{S}$ to $T_t f$.*

Proof For $0 < \delta < t$, we have by the Markov property:

$$\begin{aligned} P_\delta T_{t-\delta} f(x) &= E^x\{E^{X_\delta}[e_q(t-\delta)f(X_{t-\delta})]\} \\ &= E^x\left\{\exp\left[\int_\delta^t q(X_s)\mathrm{d}s\right] f(X_t)\right\} \end{aligned}$$

Hence by Cauchy–Schwarz inequality:

$$\begin{aligned} |T_t f(x) - P_\delta T_{t-\delta} f(x)| &\leq E^x\{e_q(t)[e_{|q|}(\delta) - 1]\}\|f\|_\infty \\ &\leq E^x\{e_{2q}(t)\}^{1/2} E^x\{e_{2|q|}(\delta) - 1\}^{1/2}\|f\|_\infty. \end{aligned}$$

This converges to zero uniformly in all x by (21). □

We recall the definition of the strong Feller property in Section 1.1, which may be generalized to the general case here.

Proposition 3.12 *If P_t maps $L^\infty(S)$ into $C_0(S)$ (or $C_b(S)$), then so does T_t.*

Proof Let $f \in L^\infty$; then if $0 < \delta < t$, $T_{t-\delta} f \in L^\infty$ by Theorem 3.10. Hence under the first hypothesis, $P_\delta T_{t-\delta} f \in C_0$. Therefore $T_t f \in C_0$ by Proposition 3.11. The result under the second hypothesis is proved in the same way. □

For certain purposes, it is useful to extend the above results to a product setting. Writing (x_1, x_2) for a generic point in the product space $\tilde{\mathcal{S}} = \mathcal{S} \times \mathcal{S}$, we define the product measure $\tilde{m} = m \times m$ and the product process $\tilde{X}_t = (X_t^{(1)}, X_t^{(2)})$, where $\{X_t^{(i)}\}$, $i = 1, 2$ are two (stochastically) independent copies of our previous $\{X_t\}$. Then the transition operator $\tilde{P}_t$ of $\tilde{X}_t$ has the product density function $\tilde{p}$ given by

$$\tilde{p}_t((x_1, x_2), (y_1, y_2)) \equiv p_t(x_1, y_1)p_t(x_2, y_2).$$

The classes of functions $C(\tilde{\mathcal{S}})$, $C_b(\tilde{\mathcal{S}})$ and $C_0(\tilde{\mathcal{S}})$ are defined using the product topology of $\tilde{\mathcal{S}} = \mathcal{S} \times \mathcal{S}$.

Lemma 3.13 *Suppose $m(\mathcal{S}) < \infty$. If P_t maps $L^\infty(\mathcal{S})$ into $C_0(\mathcal{S})$ (or $C_b(\mathcal{S})$), then $\tilde{P}_t$ maps $L^\infty(\tilde{\mathcal{S}})$ into $C_0(\tilde{\mathcal{S}})$ (or $C_b(\tilde{\mathcal{S}})$).*

Proof If $f(\cdot,\cdot) = f_1(\cdot)f_2(\cdot)$, where $f_i \in L^\infty(\mathcal{S})$, $i = 1, 2$, then

$$\tilde{P}_t f = (P_t f_1)(P_t f_2).$$

Hence the assertion is true. If $m(\mathcal{S}) < \infty$, then $\tilde{m}(\tilde{\mathcal{S}}) < \infty$ and $L^\infty(\tilde{\mathcal{S}}) \subset L^1(\tilde{\mathcal{S}})$. For any $f \in L^1(\tilde{\mathcal{S}})$, since m is σ-finite, by standard analysis, for any $\varepsilon > 0$ there exist $f_i^{(k)} \in L^\infty(\mathcal{S})$, $i = 1, 2$; $1 \leq k \leq n$ such that

$$\left\| f - \sum_{k=1}^{n} f_1^{(k)} f_2^{(k)} \right\|_1 \le \varepsilon;$$

consequently, by the boundedness condition in (25),

$$\left\| \tilde{P}_t f - \tilde{P}_t \left(\sum_{k=1}^{n} f_1^{(k)} f_2^{(k)} \right) \right\|_\infty \le \left(\sup_{x,y} p_t(x,y) \right)^2 \varepsilon.$$

The norms indicated above are those of $L^1(\tilde{\mathcal{S}}, \tilde{m})$ and $L^\infty(\tilde{\mathcal{S}})$, respectively. Since $\tilde{P}_t \left(\sum_{k=1}^{n} f_1^{(k)} f_2^{(k)} \right) = \sum_{k=1}^{n} \left(P_t f_1^{(k)} \right) \left(P_t f_2^{(k)} \right) \in C_0(\tilde{\mathcal{S}})$, the above approximation implies that $\tilde{P}_t f \in C_0(\tilde{\mathcal{S}})$. The proof for the case of C_b is similar. □

Next we set

$$\tilde{q}(x_1, x_2) = q(x_1) + q(x_2),$$

where q satisfies (15). Then, using self-explanatory notation, we have:

$$\tilde{e}_{\tilde{q}}(t) = \exp \left[\int_0^t \tilde{q}(\tilde{X}_s) \mathrm{d}s \right] = e_q^{(1)}(t) e_q^{(2)}(t),$$

and

$$\begin{aligned} \tilde{T} f &= E^\bullet \{ \tilde{e}_{\tilde{q}}(t) f(\tilde{X}_t) \} \\ &= E^\bullet \{ e_q^{(1)}(t) e_q^{(2)}(t) f(X_t^{(1)}, X_t^{(2)}) \}. \end{aligned}$$

If $f \in L^\infty(\tilde{\mathcal{S}})$ is of the product form $f_1 f_2$, where each $f_i \in L^\infty(\mathcal{S})$, $i = 1, 2$, we have by independence,

$$\tilde{T}_t(f_1 f_2)(x_1, x_2) = T_t^{(1)} f_1(x_1) T_t^{(2)} f_2(x_2).$$

By (28), the right-hand side of the above may be represented as

$$\int_{\mathcal{S}} u_t(x_1, y_1) f_1(y_1) \mathrm{d}m(y_1) \int_{\mathcal{S}} u_t(x_2, y_2) f_2(y_2) \mathrm{d}m(y_2).$$

Setting

$$\tilde{u}_t((x_1, x_2), (y_1, y_2)) = u_t(x_1, y_1) u_t(x_2, y_2),$$

we have

$$\tilde{T}_t f(x) = \int_{\tilde{\mathcal{S}}} \tilde{u}_t(x, y) f(y) \mathrm{d}\tilde{m}(y)$$

for the particular f. This representation then holds for any $f \in L^\infty(\tilde{\mathcal{S}})$ by a monotone class argument. □

Proposition 3.14 *Suppose $\tilde{P}_t$ maps $L^\infty(\tilde{\mathcal{S}})$ into $C_0(\tilde{\mathcal{S}})$ (or $C_b(\tilde{\mathcal{S}})$) for all $t > 0$. Then we have for each $t > 0$:*

$$u_t \in C_0(\mathcal{S} \times \mathcal{S}) \text{ (or } C_b(\mathcal{S} \times \mathcal{S})).$$

Proof We may apply Proposition 3.12 to $\tilde{P}_t$ and $\tilde{T}_t$ because the conditions on S and X are satisfied by $\tilde{S}$ and $\tilde{X}$. Under the present hypothesis, if $f \in L^\infty(\tilde{\mathcal{S}})$, then $\tilde{P}_t f \in C_0(\tilde{\mathcal{S}})$, hence also $\tilde{T}_t f \in C_0(\tilde{\mathcal{S}})$. Since $u_t(\cdot,\cdot) \in L^\infty(\tilde{\mathcal{S}})$ by Theorem 3.10, we have $\tilde{T}_t u_t \in C_0(\tilde{\mathcal{S}})$ for each $t > 0$. But u_t is a transition density hence

$$\begin{aligned} u_t(x,y) &= \int_{\mathcal{S}}\int_{\mathcal{S}} u_{t/3}(x,z)u_{t/3}(z,w)u_{t/3}(w,y)\mathrm{d}m(z)\mathrm{d}m(w) \\ &= \int_{\tilde{\mathcal{S}}} \tilde{u}_{t/3}((x,y),(z,w))u_{t/3}(z,w)\mathrm{d}\tilde{m}(z,w)) \\ &= (\tilde{T}_{t/3}u_{t/3})(x,y). \end{aligned}$$

Thus $u_t \in C_0(\tilde{\mathcal{S}})$. The proof for the case of C_b is similar. □

Proposition 3.15 *Suppose $m(\mathcal{S}) < \infty$. Then each T_t, $t > 0$, is a compact operator in $L^p(\mathcal{S})$, $1 \le p < \infty$.*

Proof By Theorem 3.10, the density $u_t(x,y)$ of T_t is bounded. The conclusion follows by the same argument as in Theorem 2.7 for P_t^D. □

For immediate applications in the next section, we need an important consequence of Proposition 3.8. Let us denote the class of functions $f \in \mathcal{B}(S)$ for which there exist constants $C_1 > 0$ and $C_2 > 0$ such that for all $x \in S$:

$$|f(x)| \le C_1 + C_2|q(x)| \tag{30}$$

by $\mathbb{F}(S,q)$.

Proposition 3.16 *Let $f \in \mathbb{F}(S,q)$. Then we have for $0 < t < \infty$:*

$$\int_0^t T_s|f|\mathrm{d}s \in L^\infty(S), \tag{31}$$

and

$$\lim_{t\downarrow 0} \| \int_0^t T_s|f|\mathrm{d}s\|_\infty = 0. \tag{32}$$

Furthermore, if $\{P_t\}$ has the strong Feller property, then we may replace $L^\infty(S)$ by $C_b(S)$ in (31).

Proof For (31) and (32), we need only consider the cases when f is a constant, and when f is q. In the first case, (31) and (32) follow from (22). In the second case, if $q \ge 0$, we have by pathwise integration and Fubini's theorem:

$$\begin{aligned} \int_0^t T_s q \mathrm{d}s &= E^\bullet\left\{\int_0^t \mathrm{e}^{\int_0^s q(X_r)\mathrm{d}r} q(X_s)\mathrm{d}s\right\} \\ &= E^\bullet\left\{\mathrm{e}^{\int_0^t q(X_r)\mathrm{d}r} - 1\right\} = E^\bullet\{e_q(t)\} - 1. \end{aligned}$$

The positivity of q is needed for Fubini's theorem. Note that for $q \in J$, if $\phi(t) = q(X_t)$ and $A(t) = \int_0^t \phi(r)\mathrm{d}r$, then almost surely $\phi \in L^1_{\mathrm{loc}}(\mathbb{R}^1)$ by virtue of (23). Hence $\mathrm{e}^{A(t)}$ is locally absolutely continuous in $\mathbb{R}^1$ and

$$\mathrm{e}^{A(t)} - 1 = \int_0^t \phi(s)\mathrm{e}^{A(s)}\mathrm{d}s$$

by the theory of Lebesgue integration. We have then for a general q:

$$\int_0^t T_s|q|\mathrm{d}s \le E^\bullet\{e_{|q|}(t)\} - 1.$$

Hence (31) follows from (22), and (32) from (21).

Under the assumption in the last sentence of the proposition, $\{T_t\}$ has the strong Feller property by Proposition 3.12; hence, for any $f \in \mathbb{F}(S,q)$, and $0 < \delta < t$, we have

$$\int_\delta^t T_s|f|\mathrm{d}s = T_\delta\left(\int_0^{t-\delta} T_s|f|\mathrm{d}s\right) \in C_b(S).$$

Therefore

$$\int_0^t T_s|f|\mathrm{d}s \in C_b(S)$$

by (32). □

Here, we note that $T_s|q| = E^\bullet\{e_q(s)|q|(X_s)\}$, not $E^\bullet\{e_{|q|}(s)|q|(X_s)\}$. This apparently minor difference will be a major one when the constant t in the above is replaced by an exit time in a later context. On the other hand, once (31) and (32) are proved, the results are of course true when $|q|$ is replaced by q.

When $f = q$ in (32), the latter represents an extension of the fundamental condition (13) from P_t to T_t. No wonder it is an important result.

3.3 Potential Operator and its Inverse

The general results in the preceding section will now be applied to the special case of Brownian motion. Let D be a domain in $\mathbb{R}^d$, $d \ge 1$, and let $q \in J$. Thus $S = D$, m is the Lebesgue measure, X is the killed Brownian motion denoted by X^D in Section 2.1, $\{P_t^D\}$ its transition semigroup and $p^D(t;\cdot,\cdot)$ its density introduced in Section 2.2. By Theorem 2.4, the last function is symmetric in $D \times D$, and it is trivial that for all $x \in D$, $y \in D$:

$$p^D(t;x,y) \le p(t;x,y) \le \frac{1}{(2\pi t)^{d/2}}, \tag{33}$$

where p is the density for the free Brownian motion. Thus our previous assumption (25) is satisfied. We continue to use the notation (20) for the Feynman–Kac multiplicative functional, and subject the function q to the condition (13) or (15). According to Lemma 3.5, this is equivalent to assuming $q \in J$. Both alternative characterizations of q are essential to the development here. The semigroup generated by X^D with the multiplicative functional $\{e_q(t)\}$ will be denoted as before by $\{T_t\}$ and referred to as the Feynman–Kac semigroup; thus for $t \geq 0$:

$$\begin{aligned} T_t f(x) &= E^x\left\{\exp\left[\int_0^t q(X_s^D)\mathrm{d}s\right] f(X_t^D)\right\} \\ &= E^x\{t < \tau_D; e_q(t) f(X_t)\}. \end{aligned} \tag{34}$$

Here we have adopted the convention of Section 2.1 that on $\{t \geq \tau_D\}$, $X_t^D = \partial$ together with the new convention that $f(\partial) = 0$ for any f. Thus the function f in (34) need only be defined in D, with $T_0 f = f$ in D. This is why we regard D, rather than $\mathbb{R}^d$, as the state space $\mathcal{S}$. However, we shall see that for certain questions the closure $\overline{D}$ of D must be considered. Note that if $z \in \partial D$, but z is not regular, then $T_0 f(z)$ is not necessarily equal to $f(z)$.

Let us recall the definition of an appropriate space for a domain D from Section 2.4. For $1 \leq p < \infty$, $L^p(D)$ is an appropriate space; if $m(D) < \infty$ and D is regular, $C_0(D)$ is also an appropriate space. By Theorem 2.7, $\{P_t^D, t > 0\}$ is a strongly continuous semigroup in each appropriate space for D. By Theorem 2.2, $\{P_t^D\}$ has the strong Feller property. Therefore, by the previous theorems, these properties can be transmitted to $\{T_t\}$. The following theorem constitutes a summary together with a reminder about eigenvalues as discussed in Theorem 2.7.

Theorem 3.17 *Let D be a domain in $\mathbb{R}^d$, and $\{T_t\}$ be defined as in (34). Then $\{T_t\}$ is a strongly continuous semigroup in each appropriate space for D. Each T_t is a bounded operator from $L^p(D)$, $1 \leq p \leq \infty$, to $L^\infty(D)$ and to itself, and*

$$\|T_t\|_p \leq \|T_t\|_\infty \leq \mathrm{e}^{C_0 + C_1 t}, \quad 0 \leq t < \infty, \tag{35}$$

where the constants C_0 and C_1 are the same as in (22). For each $t > 0$, T_t has the strong Feller property and possesses a symmetric density kernel $u_t \in C_b(D \times D)$ such that (28) holds. T_t maps $L^1(D)$ into $C_b(D)$.

Suppose in addition that $m(D) < \infty$, and D is regular. Then for each $t > 0$, $u_t \in C_0(D \times D)$; T_t is a bounded operator from $L^\infty(D)$ into $C_0(D)$. It is a compact operator in all the appropriate spaces, and has the same eigenvalues and eigenfunctions in all of them.

Proof These results follow from the general propositions Theorem 3.10, Propositions 3.11–3.15, and the properties of $\{P_t^D\}$ reviewed above, except those pertaining to a regular D. Note that in the general setting of Section 3.2, $\mathcal{S}$ is regarded as the whole space without boundary, while in this section we must consider ∂D, which is in general not empty and is vitally involved in the definition of $C_0(D)$.

Suppose that D is regular and $f \in L^1(D)$. We have

$$\begin{aligned} |P_t^D f(x)| &= |P_{t/2}^D(P_{t/2}^D f)(x)| = |E^x\{t/2 < \tau_D; (P_{t/2}^D f)(X_{t/2})\}| \\ &\leq P^x\{t/2 < \tau_D\}\|P_{t/2}^D f\|_\infty. \end{aligned}$$

As $x \to \partial D = (\partial D)_r$, this converges to zero by (1.35). Furthermore, we have

$$|P_t^D f(x)| \leq \int_D p(t; x, y)|f(y)|\mathrm{d}y.$$

Since for each $t > 0$ and $y \in \mathbb{R}^d$, $\lim_{|x|\to\infty} p(t; x, y) = 0$, it follows by dominated convergence and the fact that $f \in L^1(D)$ that $\lim_{|x|\to\infty} P_t^D f(x) = 0$. Combining this with the result when $x \to \partial D$, we have proved that $P_t^D f \in C_0(D)$. Hence $T_t f \in C_0(D)$ by Proposition 3.12.

If we further assume that $m(D) < \infty$, then since $L^\infty(D) \subset L^1(D)$, T_t maps $L^\infty(D)$ into $C_0(D)$.

To prove the assertion regarding u_t when D is regular, we apply Proposition 3.14 with $\mathcal{S} = D$, $\tilde{\mathcal{S}} = D \times D$, and $\tilde{P}_t = P_t^{D\times D}$. Let $\{X_t\}$ be the Brownian motion in $\mathbb{R}^{2d} = \mathbb{R}^d \times \mathbb{R}^d$, then $X_t = (X_t^{(1)}, X_t^{(2)})$, where $\{X_t^{(1)}\}$ and $\{X_t^{(2)}\}$ are two stochastically independent Brownian motion processes in $\mathbb{R}^d$. We have $\tau_{D\times D} = \tau_D^{(1)} \wedge \tau_D^{(2)}$, where $\tau_D^{(i)} = \inf\{t > 0 : X_t^{(i)} \notin D\}$, $i = 1, 2$. If $z = (z_1, z_2) \in \partial(D \times D) = (\partial D \times \overline{D}) \cup (\overline{D} \times \partial D)$, then either $z_1 \in \partial D$ or $z_2 \in \partial D$, hence we have

$$P^{(z_1,z_2)}(\tau_{D\times D} > 0) = P^{z_1}(\tau_D^{(1)} > 0)P^{z_2}(\tau_D^{(2)} > 0) = 0.$$

Thus $D \times D$ is regular in $\mathbb{R}^{2d}$. Since $m(D \times D) < \infty$, $u_t \in C_0(D \times D)$ by Proposition 3.14 □

The role of the additional assumption $m(D) < \infty$ in the second part of Theorem 3.17 deserves attention. Note also that $L^\infty(D)$ is not an appropriate space. Of course, when $m(D) < \infty$, it is trivial that T_t is a bounded operator from $L^\infty(D)$ to $L^p(D)$, $1 \leq p \leq \infty$; however, $\{T_t\}$ may not be strongly continuous in $L^\infty(D)$.

We now consider the following assumption:

$$\int_0^\infty \|T_t\|_\infty \mathrm{d}t < \infty. \tag{36}$$

By (35), this implies for all $p \in [1, \infty]$:

$$\int_0^\infty \|T_t\|_p \mathrm{d}t < \infty. \tag{37}$$

When $m(D) < \infty$, (36) and (37) are actually equivalent. For, suppose that (37) holds for a particular $p \in [1, \infty)$, then we have for $t \geq 0$:

$$\|T_{t+2}\|_\infty \leq \|T_1\|_{\infty,p}\|T_t\|_{p,p}\|T_1\|_{p,\infty};$$

whence, (37) implies that

$$\int_2^\infty \|T_t\|_\infty \mathrm{d}t < \infty. \tag{38}$$

Since $\int_0^2 \|T_t\|_\infty \mathrm{d}t < \infty$ by (35), (36) follows.

The significance of the condition (36) will be apparent later in Theorem 4.19, in conjunction with several equivalent conditions of fundamental importance in our study.

We introduce the potential operator for the semigroup $\{T_t\}$ as follows, for $f \in \mathcal{B}_+(\mathbb{R}^d)$:

$$Vf(x) = \int_0^\infty T_t f(x)\mathrm{d}t = E^x \left\{ \int_0^{\tau_D} e_q(t) f(X_t)\mathrm{d}t \right\}. \tag{39}$$

When $q \equiv 0$ and $T_t = P_t^D$, V reduces to the Green potential operator G_D given in (2.13). Thus V may be called the q-Green operator. The main purpose of this section is to show that V is related to the Schrödinger operator $\frac{\Delta}{2} + q$ as the classical Green operator G_D is to the (half-) Laplacian $\frac{\Delta}{2}$.

Under the assumption (36), V is a bounded operator in $L^p(D)$ for each $p \in [1, \infty]$; in particular, $V1 \in L^\infty(D)$.

Theorem 3.18 *Suppose that D is regular and $V1 \in L^\infty(D)$. Then for any $f \in \mathbb{F}(D, q)$, we have*

$$Vf \in C_0(D). \tag{40}$$

Proof We shall first prove that $V|f|$ is bounded. For each $x \in D$ and $t > 0$, we have

$$V|f|(x) = \int_0^t T_s|f|(x)\mathrm{d}s + V(T_t|f|)(x). \tag{41}$$

By (32), there exists $a > 0$ such that for each $0 < t \leq a$,

$$\| \int_0^t T_s|f|\mathrm{d}s\|_\infty \leq 1. \tag{42}$$

Integrating with respect to t from 0 to a on both sides of (41) and dividing by a, we have

$$\begin{aligned} V|f|(x) &\leq 1 + V\left(\frac{1}{a}\int_0^a T_t|f|\mathrm{d}t\right)(x) \\ &\leq 1 + a^{-1}\|V1\|_\infty, \end{aligned} \tag{43}$$

which is finite by assumption. Thus, $V|f|$ is bounded and so is Vf.

Now (41) also holds when $|f|$ is replaced by f. By Proposition 3.12, for each $t > 0$, $T_t(Vf) \in C_0(D)$. As $t \downarrow 0$, $\int_0^t T_s f\mathrm{d}s$ converges uniformly to zero by (32). Therefore $Vf \in C_0(D)$. □

Corollary to Theorem 3.18 *V has a symmetric density given by*

$$V(x, y) = \int_0^\infty u_t(x, y)\mathrm{d}t. \tag{44}$$

Proof Since $V1 \in L^\infty(D)$, this follows from (28), (39), and Fubini's theorem. □

The function $V(x, y)$ will be called the q-Green function for D.

The next two theorems establish the fundamental relationship between the two potentials V and G^λ (see (2.54)). Observe that if the function q implicit in the definition of V is the constant $-\lambda$, then V becomes G^λ. For the sake of clarity, the results will be presented first in the operator form and then in the density form.

Theorem 3.19 *Suppose that $q \in J$ and $V1 \in L^\infty(D)$. Then if $V[(|q|+\lambda)G^\lambda|f|] < \infty$ $(\lambda \geq 0)$, we have in D:*

$$Vf = G^\lambda f + V((q+\lambda)G^\lambda f). \tag{45}$$

If $G^\lambda[(|q|+\lambda)V|f|] < \infty$, then we have in D:

$$Vf = G^\lambda f + G^\lambda((q+\lambda)Vf). \tag{46}$$

Proof Both (45) and (46) will be proved by probabilistic calculations. Suppose that $V[(|q|+\lambda)G^\lambda|f|] < \infty$; then by the Markov property and Fubini's theorem, we have:

$$\begin{aligned}
&V((q+\lambda)G^\lambda f)(x)\\
&= E^x\left\{\int_0^{\tau_D} e_q(t)[q(X_t)+\lambda]E^{X_t}\left[\int_0^{\tau_D} \mathrm{e}^{-\lambda s} f(X_s)\mathrm{d}s\right]\right\}\mathrm{d}t\\
&= E^x\left\{\int_0^{\tau_D} e_q(t)[q(X_t)+\lambda]\mathrm{e}^{\lambda t}\left[\int_t^{\tau_D} \mathrm{e}^{-\lambda s} f(X_s)\mathrm{d}s\right]\right\}\mathrm{d}t\\
&= E^x\left\{\int_0^{\tau_D} \mathrm{e}^{-\lambda s} f(X_s)\left[\int_0^{s} e_{q+\lambda}(t)[q(X_t)+\lambda]\mathrm{d}t\right]\mathrm{d}s\right\}\\
&= E^x\left\{\int_0^{\tau_D} \mathrm{e}^{-\lambda s} f(X_s)[e_{q+\lambda}(s)-1]\mathrm{d}s\right\} = Vf(x) - G^\lambda f(x).
\end{aligned}$$

Next, suppose that $G^\lambda[(|q|+\lambda)V|f|] < \infty$, then we have

$$\begin{aligned}
&G^\lambda((q+\lambda)Vf)(x)\\
&= E^x\left\{\int_0^{\tau_D} \mathrm{e}^{-\lambda t}[q(X_t)+\lambda]E^{X_t}\left[\int_0^{\tau_D} e_q(s)f(X_s)\mathrm{d}s\right]\mathrm{d}t\right\}\\
&= E^x\left\{\int_0^{\tau_D} \mathrm{e}^{-\lambda t}[q(X_t)+\lambda]\int_t^{\tau_D} \exp\left[\int_t^{s} q(X_r)\mathrm{d}r\right] f(X_s)\mathrm{d}s\mathrm{d}t\right\}\\
&= E^x\left\{\int_0^{\tau_D} \mathrm{e}^{-\lambda s} f(X_s)\int_0^{s} [q(X_t)+\lambda]\exp\left[\int_t^{s} [q(X_r)+\lambda]\mathrm{d}r\right]\mathrm{d}t\mathrm{d}s\right\}\\
&= E^x\left\{\int_0^{\tau_D} \mathrm{e}^{-\lambda s} f(X_s)[e_{q+\lambda}(s)-1]\mathrm{d}s\right\} = Vf(x) - G^\lambda f(x).
\end{aligned}$$

□

Theorem 3.20 *Suppose that $q \in J$ and $V1 \in L^\infty(D)$. In addition, suppose that one of the following two conditions holds:*

(i) *$\lambda > 0$; D is any domain in $\mathbb{R}^d$ $(d \geq 1)$;*

(ii) *$\lambda = 0$; D is a Greenian domain in $\mathbb{R}^d$ $(d \geq 1)$.*

Then for each $x \in D$, we have

$$V(x,y) = G^\lambda(x,y) + \int_D V(x,z)[q(z)+\lambda]G^\lambda(z,y)\mathrm{d}z \tag{47}$$

for m-a.e. y in D. Moreover, we have for $(m \times m)$-a.e. (x,y) in $D \times D$:

$$V(x,y) = G^\lambda(x,y) + \int_D G^\lambda(x,z)[q(z)+\lambda]V(z,y)\mathrm{d}z. \tag{48}$$

Proof Under (i) or (ii), we have $G^\lambda(x,y) < \infty$ for $(x,y) \in D \times D$, $x \neq y$. On the other hand, for each $x \in D$, $V(x,y) < \infty$ for m-a.e. y in D (the set of y for which the statement holds depends on x). Now under (i), it is trivial that for any bounded Borel subset B of D, we have

$$G^\lambda 1_B \leq G^\lambda 1 \leq \int_0^\infty \mathrm{e}^{-\lambda t}\mathrm{d}t = \frac{1}{\lambda} < \infty.$$

Under (ii), if $d \geq 3$ (recall that any domain is then Greenian), we have

$$G1_B(x) \leq C_d \int_B \frac{\mathrm{d}y}{|x-y|^{d-2}}$$

where C_d is given in (2.17). It is well known that the right member in the above is bounded for all x (in $\mathbb{R}^d$). If $d = 2$, and D is Greenian, then $G1_B$ is bounded by a known result not proved here (see Port and Stone (1978, page 22)). If $d = 1$, and D is Greenian, then $D = (a,b)$, $D = (a,\infty)$ or $D = (-\infty,b)$ where $-\infty < a < b < \infty$. Using the explicit formula (2.29), (2.30) and a trivial analogue of (2.30), we can easily verify that $G1_B$ is bounded. Thus in all cases under (i) or (ii), we have

$$V[(|q|+\lambda)G^\lambda 1_B] \leq V(|q|+\lambda)\|G^\lambda 1_B\| < \infty$$

by Theorem 3.18. Therefore by Theorem 3.19, (45) holds when $f = 1_B$. This implies (47) since B is an arbitrary bounded Borel subset of D.

It is curious that we do not know how to verify that $G^\lambda[(|q|+\lambda)V1_B] < \infty$ under (i) or (ii), so as to deduce (48) in a way similar to the above. We must first invoke Fubini's theorem to infer that (47) holds for $(m \times m)$-a.e. (x,y) in $D \times D$, then interchange x and y in the equation, and use the symmetry of $G^\lambda(x,y)$ and $V(x,y)$ to obtain (48) under the same proviso. □

We note that the disparate analytic circumstances exposed in the above proof are inherent in the hidden incongruity of the hypothesis (36) and our assumption on D. For $d \geq 3$, $G_D 1$ need not even be finite, whereas $V1$ is bounded under

(36). More stringent hypotheses on D will be needed to make G^λ and V behave in a like manner.

From Propositions 2.8 and 2.10, we know that the operator $-\frac{\Delta}{2}$ acts as a left inverse of G_D. We shall generalize this relationship to the Schrödinger operator defined below:

$$S = -\left(\frac{\Delta}{2} + q\right)$$

and to the q-Green potential operator V. The operator S will be taken in the weak sense as explained in Section 3.1.

The first result below is a basic uniqueness theorem. For $q = 0$, this follows from the maximum (minimum) property for harmonic functions; see Corollary 1.11. The trivial example $\sin x$ in $(0, \pi)$ shows that no such principle holds in the Schrödinger case. Our proof depends on the use of Theorem 3.20.

Theorem 3.21 *Let D be a bounded regular domain in $\mathbb{R}^d$, $d \geq 1$, and $V1 \in L^\infty(D)$. If $\phi \in C(\overline{D})$ with $\phi = 0$ on ∂D, and $S\phi = 0$ in D, then $\phi = 0$ in D.*

Proof Consider

$$f = \phi - G(q\phi).$$

Since ϕ is bounded in D, $1_D q\phi \in J$. Hence by Theorem 3.2, $G(|q\phi|)$ is bounded in D, and so belongs to $L^1(D)$ because D is bounded. Therefore, by Proposition 2.10, we have in D:

$$\Delta f = \Delta\phi - \Delta G(q\phi) = -2q\phi + 2q\phi = 0.$$

Since $G(q\phi) \in C_0(D)$ by Theorem 3.2, so is f, and therefore f is harmonic by Weyl's lemma. Hence by the maximum principle we have $f = 0$ in D; in other words, $\phi = G(q\phi)$. Since D is bounded, $G|\phi| \leq \|\phi\|_\infty \|G1\|_\infty < \infty$. Therefore $V(|q|G|\phi|) < \infty$ by Theorem 3.18. Thus (45) with $\lambda = 0$ is true for $f = q\phi$:

$$V(q\phi) = G(q\phi) + V(qG(q\phi)) = \phi + V(q\phi).$$

Since $V(q\phi)$ is finite, $\phi = 0$. □

We note that in the above proof, the assumption that D is bounded may be generalized to the assumption that D is Green-bounded, provided we can prove that $G_D|q|$ is bounded. This is true by Theorem 4.3. Thus Theorem 3.21 may be extended to this case because Corollary 1.11 can be so extended (see the extension of Theorem 1.24 in Section 4.3). These extensions are significant but tend to belabour the exposition. It is for this reason that we confine ourselves here to the simplest case of a bounded domain. The next result shows the inverse roles of V and S acting on special classes of functions.

Theorem 3.22 *Under the assumptions of Theorem 3.21:*

(a) *if $f \in \mathbb{F}(D, q)$, then we have in D*

$$S(Vf) = f; \tag{49}$$

(b) *if* $f \in C_0(D)$ *and* $Sf \in \mathbb{F}(D, q)$*, then we have in* D

$$V(Sf) = f. \tag{50}$$

Proof (a) Both Vf and $V|f|$ are bounded in D by Theorem 3.18. Since D is bounded, $q \in L^1(D)$ by Proposition 3.1, hence $G(|q|V|f|) < \infty$ by Theorem 3.2, and (46) with $\lambda = 0$ holds by Theorem 3.20. Also $qVf \in L^1(D)$, and consequently by Proposition 2.10:

$$\Delta(Vf) = \Delta(Gf) + \Delta(G(qVf)) = -2f - 2qVf,$$

which is (49).

(b) Applying part (a) with f replaced by Sf, we have

$$S(V(Sf)) = Sf. \tag{51}$$

Since D is regular, $V(Sf) \in C_0(D)$ by Theorem 3.18. Hence by Theorem 3.21, (51) implies (50). □

For a given $f \in L^\infty(D)$, consider the inhomogeneous equation in D:

$$\left(\frac{\Delta}{2} + q\right)\phi = f.$$

It then follows from the preceding theorem that, under the stated assumptions on D and q, the unique (weak) solution with $\phi \in C_0(D)$ is given by $\phi = -Vf$. When $q \equiv 0$, V reduces to G_D and the equation is known as Poisson's equation.

3.4 Schrödinger Infinitesimal Generator

We shall now discuss the infinitesimal generators for the semigroup $\{T_t,\ t \geq 0\}$. We recall the definition of the appropriate spaces $L^p(D)$ for a bounded domain D $(1 \leq p < \infty)$, and $C_0(D)$ for a bounded regular domain D. The definition of the infinitesimal generator A in each case is given in Section 2.5. Its domain $\mathcal{D}(A)$ is the class of functions f in the appropriate space $\mathcal{S}$ for which there exists $g \in \mathcal{S}$ such that

$$\lim_{t \downarrow 0} \left\|\frac{1}{t}(T_t f - f) - g\right\| = 0;$$

and this g is denoted by Af. For any $\lambda > 0$, A is the infinitesimal generator of $\{T_t\}$ if and only if $A - \lambda I$ is the infinitesimal generator of the semigroup $\{T_t^\lambda\}$ where $T_t^\lambda = e^{-\lambda t}T_t$. If we take $\lambda > C_1$ where the constant C_1 is given in (27), then $\{T_t^\lambda\}$ satisfies the condition (36). Since the infinitesimal generator of $\{T_t^\lambda\}$ is just $A - \lambda I$ with the same definition domain as A, in questions relating to A and its spectrum, replacement of $\{T_t\}$ by $\{T_t^\lambda\}$ is immaterial. Hence we may as well assume that $\{T_t\}$ itself satisfies (36). Then V is a bounded operator in each appropriate space. The general result (2.49) becomes

$$\mathcal{D}(A) = V[\mathcal{S}]; \quad A = -V^{-1}. \tag{52}$$

When $\mathcal{S} = C_0(D)$, we have the following characterization.

Proposition 3.23 *Let D be bounded and regular, and $V1 \in L^\infty(D)$. The domain $\mathcal{D}(A_0)$ of the infinitesimal generator A_0 for $\{T_t\}$ acting on $C_0(D)$ is the class of $f \in C_0(D)$ such that*

$$Sf \in C_0(D).$$

If $f \in \mathcal{D}(A_0)$, then

$$A_0 f = -Sf. \tag{53}$$

Proof As a particular case of (52), we have

$$\mathcal{D}(A_0) = V[C_0(D)].$$

If $f \in C_0(D)$ and $Sf \in C_0(D)$, then $Sf \in \mathbb{F}(D, q)$. Hence by Theorem 3.22(b), $f = V(Sf) \in V[C_0(D)]$. Conversely, if $f = Vg$ where $g \in C_0(D)$, then $g \in \mathbb{F}(D, q)$, and so $f \in C_0(D)$ by Theorem 3.18. By Theorem 3.22(a), $Sf = S(Vg) = g \in C_0(D)$.

Since $f = V(Sf)$, (53) is a case of the general result (52). □

For the appropriate space $L^p(D)$, no such characterization is apparent, but the following partial result will be needed in Theorem 8.20. Let A_1 denote the infinitesimal generator of $\{T_t\}$ in $L^1(D)$.

Proposition 3.24 *Under the assumptions of Theorem 3.21, if $f \in C_0(D)$ and $Sf \in \mathbb{F}(D, q)$, then $f \in \mathcal{D}(A_1)$ and $A_1 f = -Sf$.*

Proof We have $\mathbb{F}(D, q) \subset L^1(D)$ by Proposition 3.1. Hence by applying (52) with $\mathcal{S} = L^1(D)$ we obtain $V(Sf) \in \mathcal{D}(A_1)$ and

$$A_1(V(Sf)) = -Sf.$$

But under our hypotheses we have $V(Sf) = f$ by Theorem 3.22(b). The conclusion follows. □

Before proving the next theorem we give a brief review of the required operator theory in $L^2(\mathbb{R}^d)$ $(d \geq 1)$ as a Hilbert space. We write

$$(f, g) = \int_{\mathbb{R}^d} f(x)g(x)\mathrm{d}x, \quad \|f\|_2^2 = (f, f),$$

as usual, but from here on we shall omit $\mathbb{R}^d$ from the notation. Let $\lambda > 0$, and G_λ be the G_D^λ defined in (2.54) with $D = \mathbb{R}^d$, which might also be denoted by U^λ as in (2.12). We now introduce a new operator, the 'square root' of G_λ, to be denoted by F_λ, as follows. For $\alpha > 0$ and $\lambda > 0$, we set

$$R_\lambda^\alpha(x) = \frac{1}{\Gamma(\frac{\alpha}{2})}\int_0^\infty \mathrm{e}^{-\lambda t} p(t;x,0) t^{\alpha/2}\frac{\mathrm{d}t}{t},$$

where p is as in (1.11). Observe that if we replace $p(t;x,0)$ by 1 in the right member of the above, the left member becomes $\lambda^{\alpha/2}$. This inspired the following celebrated equation due to Marcel Riesz:

$$R_\lambda^\alpha * R_\lambda^\beta = R_\lambda^{\alpha+\beta},$$

where $\beta > 0$ and $*$ denotes the usual convolution of functions. Furthermore, $R_\lambda^\alpha \in L^1$ by an easy computation. Now we define F_λ on L^2 by the explicit formula below, for m-almost all $x \in \mathbb{R}^d$:

$$(F_\lambda f)(x) = (R_\lambda^1 * f)(x) = \int_{\mathbb{R}^d} R_\lambda^1(x-y)f(y)\mathrm{d}y.$$

We then see that for $f \in L^2$:

$$G_\lambda f = R_\lambda^2 * f = (R_\lambda^1 * R_\lambda^1) * f = F_\lambda(F_\lambda f);$$

which shows that G_λ is the 'square' of F_λ. It is clear from the definition that the operator F_λ is bounded, linear, positive and symmetric, so that we have for all f and g in L^2:

$$(F_\lambda f, g) = (f, F_\lambda g), \quad (F_\lambda f, F_\lambda g) = (f, G_\lambda g)$$

and $|F_\lambda f| \le F_\lambda |f|$.

Next, the mapping G_λ from L^2 to L^2 is one-to-one (cf. (2.49) and (2.50)); hence so is F_λ because $F_\lambda F_\lambda = G_\lambda$. Therefore both the inverses G_λ^{-1} and F_λ^{-1} exist. For any $f \in C_c^\infty$, we have by (2.61):

$$f = G_\lambda g \text{ where } g = (\lambda I - \frac{\Delta}{2})f.$$

Thus $f = F_\lambda(F_\lambda g)$ and so

$$F_\lambda^{-1} f = F_\lambda g.$$

It follows that

$$\begin{aligned} \|F_\lambda^{-1} f\|_2^2 &= (F_\lambda g, F_\lambda g) = (g, G_\lambda g) \\ &= ((\lambda I - \frac{\Delta}{2})f, f) = \lambda(f,f) - \frac{1}{2}(\Delta f, f) \\ &= \lambda\|f\|_2^2 + \frac{1}{2}\||\nabla f|\|_2^2. \end{aligned} \tag{54}$$

Let us temporarily denote the last member of the above by $\|f\|_{\lambda *}^2$. For each $\lambda > 0$ this may be regarded as a norm equivalent to the Dirichlet norm defined in (2.45).

Recall the definition of $W^{1,2}(D)$ and $W_0^{1,2}(D)$ from Section 2.5. As before, when $D = \mathbb{R}^d$ it will be omitted from the notation. It is known that $W^{1,2} = W_0^{1,2}$, see Yosida (1980, page 58). Hence for each $f \in W^{1,2}$, there exists $f_n \in C_c^\infty$ such that $\|f_n - f\|_* \to 0$. It follows from (54) that as $n \to \infty$, $m \to \infty$:

$$\|F_\lambda^{-1} f_n - F_\lambda^{-1} f_m\|_2^2 = \|f_n - f_m\|_{\lambda *}^2 \to 0.$$

Since F_λ^{-1} is the inverse of a bounded linear operator, it is a closed operator. Hence we have $f \in \mathcal{D}(F_\lambda^{-1})$ and

$$F_\lambda^{-1} f = \lim_n F_\lambda^{-1} f_n.$$

Thus we have shown that

$$W^{1,2} \subseteq \mathcal{D}(F_\lambda^{-1}). \tag{55}$$

Furthermore it follows by applying (54) to f_n and passing to the limit that (54) holds also for any $f \in W^{1,2}$.

In fact, we have $W^{1,2} = \mathcal{D}(F_\lambda^{-1})$. This is best proved using Fourier transforms; since we do not need this result below, we will desist from this diversion and end the review here.

We are ready to prove the next theorem, due to Kato.

Theorem 3.25 *Let $q \in J$. For any number b in $(0, 1)$, there exists a number $a > 0$ such that for any $f \in W^{1,2}(\mathbb{R}^d)$:*

$$\int_{\mathbb{R}^d} |q(x)|(f(x))^2 \mathrm{d}x \le a \int_{\mathbb{R}^d} f(x)^2 \mathrm{d}x + b \int_{\mathbb{R}^d} |\nabla f(x)|^2 \mathrm{d}x. \tag{56}$$

Proof Set

$$C_\lambda^2 = \|G_\lambda |q|\|_\infty.$$

For $\phi \in L^2$, $\psi \in C_c^\infty$, we have $|q|^{1/2}\psi \in L^2$ by Proposition 3.1. Hence we have

$$\begin{aligned} (|q|^{1/2}|F_\lambda \phi|, \psi) &\le (F_\lambda|\phi|, |q|^{1/2}|\psi|) = (|\phi|, F_\lambda(|q|^{1/2}|\psi|)) \\ &\le \|\phi\|_2 \|F_\lambda(|q|^{1/2}|\psi|)\|_2 \end{aligned} \tag{57}$$

by the Cauchy–Schwarz inequality. Now by the review above, we have

$$\begin{aligned} \|F_\lambda(|q|^{1/2}|\psi|)\|_2^2 &= (|q|^{1/2}|\psi|, G_\lambda(|q|^{1/2}|\psi|)) \\ &\le (|q|^{1/2}|\psi|, G_\lambda(|q|)^{1/2} G_\lambda(|\psi|^2)^{1/2}) \\ &\le C_\lambda (|q|^{1/2}|\psi|, G_\lambda(|\psi|^2)^{1/2}) \\ &= C_\lambda (|\psi|, |q|^{1/2} G_\lambda(|\psi|^2)^{1/2}) \\ &\le C_\lambda \|\psi\|_2 \||q|^{1/2} G_\lambda(|\psi|^2)^{1/2}\|_2, \end{aligned}$$

where we have applied the Cauchy–Schwarz inequality to the integral operator G_λ in the first inequality above. Next we have

$$\begin{aligned}\||q|^{1/2}G_\lambda(|\psi|^2)^{1/2}\|_2^2 &= (|q|, G_\lambda(|\psi|^2)) = (G_\lambda|q|, |\psi|^2)\\ &\leq C_\lambda^2\|\psi\|_2^2;\end{aligned}$$

whence,

$$\|F_\lambda(|q|^{1/2}|\psi|)\|_2^2 \leq C_\lambda\|\psi\|_2 C_\lambda\|\psi\|_2 = C_\lambda^2\|\psi\|_2^2.$$

Substituting this into (57), we obtain

$$(|q|^{1/2}|F_\lambda\phi|, |\psi|) \leq C_\lambda\|\phi\|_2\|\psi\|_2.$$

This being true for all $\psi \in C_c^\infty$, it follows that $|q|^{1/2}|F_\lambda\phi| \in L^2$, and

$$\||q|^{1/2}|F_\lambda\phi|\|_2 \leq C_\lambda\|\phi\|_2.$$

In more explicit notation, we have proved that for any $\phi \in L^2$:

$$\int_{\mathbb{R}^d} |q(x)|(F_\lambda\phi(x))^2 dx \leq C_\lambda^2 \int_{\mathbb{R}^d} \phi(x)^2 dx.$$

If $f \in W^{1,2}$, then by (55) we have $\phi = F_\lambda^{-1}f \in L^2$ and $f = F_\lambda\phi$. Therefore by (54), the following is valid for f:

$$\begin{aligned}\int_{\mathbb{R}^d} |q(x)|(f(x))^2 dx &\leq C_\lambda^2 \int_{\mathbb{R}^d} (F_\lambda^{-1}f(x))^2 dx\\ &= C_\lambda^2\|f\|_{\lambda*}^2 = \lambda C_\lambda^2\|f\|^2 + \frac{1}{2}C_\lambda^2\||\nabla f|\|_2^2. \qquad (58)\end{aligned}$$

Comparing (58) with (56) we see that the theorem will be proved if we can prove that

$$\lim_{\lambda\to\infty} C_\lambda = 0. \qquad (59)$$

Now we have by (2.54) with $D = \mathbb{R}^d$, and Fubini's theorem:

$$G_\lambda|q|(x) = E^x\left\{\int_0^\infty e^{-\lambda t}|q(X_t)|dt\right\} = \int_0^\infty e^{-\lambda t}E^x\{|q(X_t)|\}dt. \qquad (60)$$

We recall from (23) that

$$\sup_x \int_0^t E^x\{|q(X_s)|\}ds \leq C_0 + C_1 t.$$

The crucial result (59) will now follow from the elementary lemma below, which is stated explicitly here for the sake of clarity.

Lemma *Suppose that for each x, the function $r(x;\cdot)$ is Borel measurable in $\mathbb{R}_+$ and that for all $t \geq 0$ we have:*

$$\begin{aligned} \sup_x \int_0^t r(x;s)\mathrm{d}s &\leq C_0 + C_1 t; \\ \limsup_{t\downarrow 0} \sup_x \int_0^t r(x;s)\mathrm{d}s &= 0. \end{aligned}$$

Then

$$\lim_{\lambda\to\infty} \sup_x \int_0^\infty \mathrm{e}^{-\lambda t} r(x;t)\mathrm{d}t = 0.$$

Applying the lemma to (60) with $r(x;t) = E^x\{|q(X_t)|\}$, we obtain (59). Let us not gloat over the second condition in the lemma which is precisely the defining condition (15) for J! □

Corollary to Theorem 3.25 *The assertion of Theorem 3.25 remains true when $\mathbb{R}^d$ is replaced by an arbitrary domain D in $\mathbb{R}^d$, provided that $f \in W_0^{1,2}(D)$.*

Proof This follows from the observation that $W_0^{1,2}(D) \subseteq W_0^{1,2}(\mathbb{R}^d)$. □

There is a complete characterization of the infinitesimal generator A_2 for $\{T_t\}$ acting in $L^2(D)$. The first step is an extension of Proposition 2.12. In order to handle an arbitrary D, we will use G_D^1 instead of G_D in what follows.

Proposition 3.26 *Let D be an arbitrary domain in $\mathbb{R}^d$. Then for any $f \in L^2(D)$, we have*

$$G_D^1 f \in W_0^{1,2}(D). \tag{61}$$

Proof For any domain D, we have $\|G_D^1\|_\infty < \infty$ by (2.55). Hence by the same argument as for Lemma 2.11, using $G_D^1(\cdot,\cdot)$ in (2.58), we see that G_D^1 is a bounded operator in $L^2(D)$ with $\|G_D^1\|_2 \leq \|G_D^1\|_\infty$. In other words, for any $f \in L^2(D)$, we have

$$G_D^1 f \in L^2(D) \text{ and } \|G_D^1 f\|_2 \leq \|G_D^1\|_\infty \|f\|_2. \tag{62}$$

Let $f \in L^2(D)$. We first assume that $f \geq 0$ and has a compact support. We take a sequence of bounded C^2 domains $D_n \uparrow\uparrow D$ with D_1 containing the support of f. By the monotone and dominated convergence theorems, we have as $n \to \infty$,

$$\|G_{D_n}^1 f - G_D^1 f\|_2 \to 0. \tag{63}$$

By (2.57) with $\lambda = 1$:

$$G_{D_n}^1 f = G_{D_n}(f - G_{D_n}^1 f).$$

Hence by Proposition 2.12 and (62) applied with D_n for D we have

$$G_{D_n}^1 f \in W_0^{1,2}(D_n) \subset W_0^{1,2}(D).$$

For $n \geq m \to \infty$, as in the proof of Proposition 2.12, we have by (63),

$$\begin{aligned}\int_D \nabla G^1_{D_n} f \cdot \nabla G^1_{D_m} f \mathrm{d}x &= \int_D \nabla G_{D_n}(f - G^1_{D_n} f) \cdot \nabla G_{D_m}(f - G^1_{D_m} f)\mathrm{d}x \\ &= -2\int_D (f - G^1_{D_n} f) G^1_{D_m} f \mathrm{d}x \\ &\to -2\int_D (f - G^1_D f) G^1_D f \mathrm{d}x.\end{aligned}$$

Thus $\{G^1_{D_n} f\}$ is a Cauchy sequence in $\{W_0^{1,2}(D), \|\cdot\|_*\}$, and so

$$G^1_D f \in W_0^{1,2}(D) \text{ and } \int_D |\nabla G^1_D f|^2 \mathrm{d}x = 2\int_D G^1_D f (G^1_D f - f)\mathrm{d}x. \tag{64}$$

Next for any $f \geq 0$ in $L^2(D)$, we can take a sequence $\{f_n\}$ of positive, Borel measurable functions with compact supports such that $f_n \uparrow f$ in D. Then as before (63) is true. Therefore (64) with $f_n - f_m$ for f implies that $\{G^1_D f_n\}$ is a Cauchy sequence in $\{W_0^{1,2}(D), \|\cdot\|_*\}$. It follows that (61) is true for any f in $L^2(D)$. □

The following theorem is an extension of Theorem 2.13 from the Laplacian to the Schrödinger case, and from a Green-bounded domain to an arbitrary one.

Theorem 3.27 *Let D be an arbitrary domain. Then we have*

$$\mathcal{D}(A_2) = \left\{ f \in W_0^{1,2}(D) : \left(\frac{\Delta}{2} + q\right) f \in L^2(D) \right\};$$

and if $f \in \mathcal{D}(A_2)$, then

$$A_2 f = \left(\frac{\Delta}{2} + q\right) f.$$

Proof D is fixed in what follows and will be omitted from the notation such as $L^2(D)$.

It follows from (27), with D for $\mathcal{S}$ and $|q|$ for q, that for all sufficiently large $\alpha > 0$, we have for $p \in [1, \infty]$:

$$\int_0^\infty \mathrm{e}^{-\alpha t} \|\tilde{T}_t\|_p \mathrm{d}t < \infty, \tag{65}$$

where $\tilde{T}_t$ is obtained from T_t by replacing q by $|q|$. From now on we consider only values of α for which (65) holds. The potential operator for the semigroup $\{\mathrm{e}^{-\alpha t}\tilde{T}_t, t \geq 0\}$ will be denoted by $\tilde{V}^\alpha$ and that for $\{\mathrm{e}^{-\alpha t} T_t,\ t \geq 0\}$ by V^α. Thus both V^α and $\tilde{V}^\alpha$ are bounded operators on L^p for $1 \leq p \leq \infty$, and $\mathcal{D}(A_p) = V^\alpha[L^p]$ for $1 \leq p < \infty$, by (2.49) and the remarks at the beginning of this section. We proceed to prove that $V^\alpha[L^2] \subseteq W_0^{1,2}$.

For $n \geq 1$, we set

$$q_n = \begin{cases} q - \alpha & \text{if } |q| \leq n; \\ n - \alpha & \text{if } q > n; \\ -n - \alpha & \text{if } q < -n. \end{cases}$$

When q is replaced by q_n, the resulting potential operator will be denoted by V^n. Since $q_n \leq |q| - \alpha$, we have by (65) for $f \in L^2$,

$$V^n|f| \leq \tilde{V}^\alpha|f| < \infty.$$

Since $q_n \to q - \alpha$ as $n \to \infty$, we have for all $x \in D$:

$$V^n f(x) = E^x \left\{ \int_0^{\tau_D} e_{q_n}(t) f(X_t) \mathrm{d}t \right\} \to V^\alpha f(x) \tag{66}$$

by dominated convergence. Since $\tilde{V}^\alpha$ is a bounded operator in L^2, $\tilde{V}^\alpha|f| \in L^2$, and for $x \in D$,

$$|V^n f(x) - V^\alpha f(x)|^2 \leq 4(\tilde{V}^\alpha|f|(x))^2.$$

Hence by dominated convergence, $V^n f \to V^\alpha f$ in L^2. If $f \in L^2$, then $f + (q_n + 1)V^n f \in L^2$. Since $V^n 1$ is bounded by (65) with $p = \infty$, Theorem 3.20 is applicable; hence it follows from (48) with q_n for q and $\lambda = 1$ that

$$V^n f = G_D^1[f + (q_n + 1)V^n f]$$

m-a.e.. Therefore $V^n f \in W_0^{1,2}$ by Proposition 3.26. Next we apply Proposition 2.14 to the preceding equation to infer that $\Delta(V^n f)$ exists, and

$$-\frac{\Delta}{2}(V^n f) = f + (q_n + 1)V^n f - V^n f = f + q_n V^n f.$$

For all $\phi \in C_c^\infty$, it follows from the above that

$$\begin{aligned} \frac{1}{2}\int_D \nabla(V^n f) \cdot \nabla\phi \mathrm{d}x &= -\frac{1}{2}\int_D \Delta(V^n f) \cdot \phi \mathrm{d}x \\ &= \int_D (f + q_n V^n f)\phi \mathrm{d}x. \end{aligned} \tag{67}$$

Since $V^n f \in W_0^{1,2}$, we have $|\nabla(V^n f)| \in L^2$, and $f + q_n V^n f \in L^2$.

Next, if $\phi \in W_0^{1,2}$, let $\phi_m \in C_c^\infty$ with $\|\phi_m - \phi\|_* \to 0$. It follows that as $m \to \infty$:

$$\frac{1}{2}\int_D (\nabla V^n f) \cdot \nabla\phi_m \mathrm{d}x \to \frac{1}{2}\int_D (\nabla V^n f) \cdot \nabla\phi \mathrm{d}x,$$

and

$$\int_D (f + q_n V^n f)\phi_m \mathrm{d}x \to \int_D (f + q_n V^n f)\phi \mathrm{d}x.$$

Thus (67) holds for $\phi \in W_0^{1,2}$. Setting $\phi = V^n f$ in (67), we obtain

$$\begin{aligned} \frac{1}{2}\||\nabla V^n f|\|_2^2 &= \frac{1}{2}\int_D |\nabla V^n f|^2 \mathrm{d}x = \int_D (f + q_n V^n f)V^n f \mathrm{d}x \\ &\leq \int_D f V^n f \mathrm{d}x + \int_D (|q| + \alpha)(V^n f)^2 \mathrm{d}x. \end{aligned}$$

By the Corollary to Theorem 3.25 applied to $V^n f$ for the f there, and with q replaced by $|q|+\alpha$, we have

$$\int_D (|q|+\alpha)(V^n f)^2 \mathrm{d}x \leq b\||\nabla V^n f|\|_2^2 + a\|V^n f\|_2^2$$

where $0 < b < \frac{1}{2}$ and $a > 0$. Hence, combining the two inequalities above, we have

$$\left(\frac{1}{2} - b\right) \||\nabla V^n f|\|_2^2 \leq \|f\|_2 \|V^n f\|_2 + a\|V^n f\|_2^2.$$

It follows that if $\{V^n f,\ n \geq 1\}$ is a bounded sequence in L^2, then it is also a bounded sequence in the Dirichlet norm $\|\ \|_*$. Since $V^n f \to V^\alpha f$ in L^2, this is the case.

We now proceed to prove that $\{V^n f,\ n \geq 1\}$ is a Cauchy sequence in $\|\ \|_*$, i.e.,

$$\|V^n f - V^m f\|_* \to 0 \text{ as } n, m \to \infty. \tag{68}$$

Setting $\phi = V^m f$ in (67), we have

$$\frac{1}{2}\int_D (\nabla V^n f) \cdot (\nabla V^m f) \mathrm{d}x = \int_D (f + q_n V^n f) V^m f \mathrm{d}x.$$

Hence we have by a simple computation:

$$\begin{aligned} a_{mn} &= \int_D |\nabla V^n f - \nabla V^m f|^2 \mathrm{d}x \\ &= \int_D [|\nabla V^n f|^2 - 2(\nabla V^n f)\cdot(\nabla V^m f) + |\nabla V^m f|^2] \mathrm{d}x \\ &= 2\int_D (q_n V^n f - q_m V^m f)(V^n f - V^m f) \mathrm{d}x. \end{aligned} \tag{69}$$

Now we have by the Cauchy–Schwarz inequality:

$$\begin{aligned} &\left(\int_D |q_n V^n f||V^n f - V^m f| \mathrm{d}x\right)^2 \\ &\leq \int_D (|q|+\alpha)(V^n f)^2 \mathrm{d}x \cdot \int_D (|q|+\alpha)(V^n f - V^m f)^2 \mathrm{d}x. \end{aligned}$$

It follows that

$$\begin{aligned} &\left(\int_D |q_n V^n f - q_m V^m f||V^n f - V^m f| \mathrm{d}x\right)^2 \\ &\leq 4\left(\sup_n \int_D (|q|+\alpha)|V^n f|^2 \mathrm{d}x\right) \int_D (|q|+\alpha)(V^n f - V^m f)^2 \mathrm{d}x. \end{aligned} \tag{70}$$

By the Corollary to Theorem 3.25, we have

$$\int_D (|q| + \alpha)|V^n f|^2 \mathrm{d}x \le b\||\nabla V^n f|\|_2^2 + a\|V^n f\|_2^2$$
$$\le (b \vee a)\|V^n f\|_*^2.$$

Since $\{V^n f\}$ is a bounded sequence in $\| \, \|_*$, the last term in the above is bounded in n, say by M. Then by (70), the left-hand side of (70) is bounded by

$$4M^2 \int_D (|q| + \alpha)(V^n f - V^m f)^2 \mathrm{d}x.$$

Using the Corollary to Theorem 3.25 again, this is less than

$$4M^2 \left(b \int_D |\nabla V^n f - \nabla V^m f|^2 \mathrm{d}x + a \int_D (V^n f - V^m f)^2 \mathrm{d}x \right).$$

Thus by (69),

$$a_{mn}^2 \le 16M^2 b a_{mn} + 16M^2 a \|V^n f - V^m f\|_2^2.$$

According to Theorem 3.25 and its corollary, we may choose b so that $16M^2 b < \frac{\varepsilon}{2}$, then for some $a(\varepsilon) > 0$, we have

$$a_{mn}^2 \le \frac{\varepsilon}{2} a_{mn} + 16M^2 a(\varepsilon)\|V^n f - V^m f\|_2^2.$$

Since $V^n f \to V^\alpha f$ in L^2, there exists an integer $n_0(\varepsilon)$ such that for $n > m > n_0(\varepsilon)$, the last term in the above $< \frac{\varepsilon^2}{2}$. Then we have

$$a_{mn}^2 \le \frac{\varepsilon}{2} a_{mn} + \frac{\varepsilon^2}{4}$$

which implies that

$$|a_{mn}| \le \varepsilon.$$

We have completed the proof of (68). Since $W_0^{1,2}$ is a Banach space with the norm $\| \, \|_*$, it follows that $V^n f$ converges in $\| \, \|_*$ to an element in $W_0^{1,2}$. The limit must be $V^\alpha f$ by (66), and so $V^\alpha f \in W_0^{1,2}$. This concludes the proof that $V^\alpha[L^2] \subseteq W_0^{1,2}$.

Letting $n \to \infty$ in (67), since $\nabla V^n f \to \nabla V^\alpha f$ in L^2, by what has just been proved, we obtain

$$\frac{1}{2} \int_D \nabla V^\alpha f \cdot \nabla \phi \mathrm{d}x = \int_D [f + (q - \alpha)V^\alpha f]\phi \mathrm{d}x.$$

By the very definition of $\Delta V^\alpha f$, this identifies it and yields the equation:

$$\left(\frac{\Delta}{2} + q - \alpha \right) V^\alpha f = -f,$$

or in previous notation:

$$(S+\alpha)V^\alpha f = f. \tag{71}$$

Thus we have proved that for any $f \in V^\alpha[L^2]$, $Sf \in L^2$ and $A_2 f = -Sf$. It follows from this and the fact that $V^\alpha[L^2] \subseteq W_0^{1,2}$ that

$$\mathcal{D}(A_2) \subseteq \{f \in W_0^{1,2} : Sf \in L^2\}.$$

To prove the opposite inclusion, let $f \in W_0^{1,2}$ and $Sf \in L^2$. We set

$$\phi = V^\alpha((S+\alpha)f); \quad \psi = f - \phi.$$

Then $\psi \in W_0^{1,2}$, $S\psi \in L^2$, and we have by (71):

$$(S+\alpha)\psi = 0.$$

Let $\psi_m \in C_c^\infty$ with $\|\psi_m - \psi\|_* \to 0$. Then we have

$$\int_D \nabla\psi \cdot \nabla\psi_m \mathrm{d}x = -\int_D (\Delta\psi)\psi_m \mathrm{d}x.$$

As $m \to \infty$, $\psi_m \to \psi$ and $\nabla\psi_m \to \nabla\psi$ both in L^2; hence we obtain

$$\int_D |\nabla\psi|^2 \mathrm{d}x = -\int_D (\Delta\psi)\psi \mathrm{d}x.$$

It follows that

$$\int_D \left[\frac{1}{2}|\nabla\psi|^2 - (q-\alpha)\psi^2\right] \mathrm{d}x = \int_D \psi(S+\alpha)\psi \mathrm{d}x = 0. \tag{72}$$

According to the Corollary to Theorem 3.25, there exists $a > 0$ such that

$$\int_D |q|\psi^2 \mathrm{d}x \le \frac{1}{2}\int [|\nabla\psi|^2 + a\psi^2]\mathrm{d}x.$$

Therefore

$$0 = \int_D \left[\frac{1}{2}|\nabla\psi|^2 - (q-\alpha)\psi^2\right] \mathrm{d}x \ge \int_D (\alpha - a)\psi^2 \mathrm{d}x,$$

which is impossible for sufficiently large α, unless $\int_D \psi^2 \mathrm{d}x = 0$, or $\psi = 0$ as an element of L^2. Thus $f = \phi \in V^\alpha[L^2] = \mathcal{D}(A_2)$ as was to be proved. □

The following supplement to Theorem 3.27 is essential.

Theorem 3.28 *The $\mathcal{D}(A_2)$ of Theorem 3.27 is dense in $W_0^{1,2}(D)$ with respect to the Dirichlet norm.*

Proof We begin, belatedly, to introduce the inner product induced by the Dirichlet norm defined in (2.45), as follows. For f and g in $W_0^{1,2}(D)$, we set

$$(f,g)_* = \int_D fg\mathrm{d}x + \int_D \nabla f \cdot \nabla g \mathrm{d}x; \tag{73}$$

thus $(f, f)_* = \|f\|_*^2$. Furthermore, for $\alpha > 0$, $q \in J$, f and g, as before, we set:

$$[f, g]_\alpha = \alpha \int_D fg\mathrm{d}x + \frac{1}{2}\int_D \nabla f \cdot \nabla g\mathrm{d}x - \int_D qfg\mathrm{d}x. \tag{74}$$

To see that the last integral in (74) is finite, we have by (56),

$$\left|\int_D qfg\mathrm{d}x\right|^2 \leq \int_D |q|f^2\mathrm{d}x \int_D |q|g^2\mathrm{d}x < \infty.$$

Next we have again by (56):

$$[f, f]_\alpha \geq (\alpha - a)\int_D f^2\mathrm{d}x + (\frac{1}{2} - b)\int_D |\nabla f|^2\mathrm{d}x. \tag{75}$$

Taking $b = \frac{1}{4}$ and denoting the corresponding a of Theorem 3.25 by a_1, we see from (75) that $[f, f]_\alpha \geq 0$ provided that $\alpha > a_1$. Then $[\cdot, \cdot]_\alpha$ is a positive-definite quadratic form in $W_0^{1,2}(D)$, and so $\|\cdot\|_\alpha = \sqrt{[\cdot,\cdot]_\alpha}$ is a norm on $W_0^{1,2}(D)$. We shall now show that this norm is equivalent to the Dirichlet norm. Using (56) again we have

$$[f, f]_\alpha \leq (\alpha + a_1)\int_D f^2\mathrm{d}x + \frac{3}{4}\int_D |\nabla f|^2\mathrm{d}x.$$

It follows from the above inequality and (75) with $a = a_1$ and $b = \frac{1}{4}$ that

$$\sqrt{(\alpha - a_1) \wedge \frac{1}{4}}\,\|f\|_* \leq \|f\|_\alpha \leq \sqrt{(\alpha + a_1) \vee \frac{3}{4}}\,\|f\|_*; \tag{76}$$

whence $\|\cdot\|_\alpha$ and $\|\cdot\|_*$ are equivalent. Therefore, $W_0^{1,2}(D)$ is also complete with $\|\cdot\|_\alpha$, and it is a Hilbert space with the inner product $[\cdot, \cdot]_\alpha$.

Now suppose that $g \in W_0^{1,2}(D)$ and

$$\forall \phi \in \mathcal{D}(A_2): \quad [\phi, g]_\alpha = 0. \tag{77}$$

Using the characterization of $\mathcal{D}(A_2)$, and proceeding as in the last part of the proof of Theorem 3.27 leading to (72) there, we see that the condition in (77) reduces to

$$\int_D (\alpha - \frac{\Delta}{2} - q)\phi \cdot g\mathrm{d}x = 0. \tag{78}$$

In particular, if $\phi = V^\alpha f$ where $f \in L^2(D)$ and α satisfies (65), so that (71) holds, then (78) reduces to

$$\int_D fg\mathrm{d}x = (f, g) = 0.$$

Since this is true for all $f \in L^2$, g must be the zero element in $L^2(D)$. We have therefore proved that the only element g in $W_0^{1,2}(D)$ that satisfies (77) is the zero element. Hence by a well-known property of Hilbert space (see e.g. Yosida (1980, page 82), Theorem 1), $\mathcal{D}(A_2)$ must be dense in $W_0^{1,2}(D)$ with respect to $\|\cdot\|_\alpha$, hence also with respect to $\|\cdot\|_*$. □

As an application of the preceding two theorems, we shall derive a formula for representing the 'tip' of the spectrum of the operator A_2:

$$\lambda_1 = \sup \mathrm{Spec}(A_2), \tag{79}$$

where $\mathrm{Spec}(A_2)$ is the spectrum set of A_2 (see Yosida (1980, page 209) for the definition of the spectrum set). According to general spectral theory, for any self-adjoint operator A in the Hilbert space H, we have

$$\lambda_1 = \sup\{(A\phi, \phi) : \ \phi \in \mathcal{D}(A), \ \|\phi\| = 1\}, \tag{80}$$

where $\|\cdot\|$ is the norm in H. In fact, the formula (80) is an easy consequence of the spectral resolution theorem for a self-adjoint operator (see Yosida (1980, page 313, Theorem 1)).

Proposition 3.29 *We have*

$$\lambda_1 = \sup\left\{\int_D \left[-\frac{|\nabla\phi|^2}{2} + q\phi^2\right] \mathrm{d}x : \ \phi \in C_c^\infty(D), \ \|\phi\|_2 = 1\right\}, \tag{81}$$

where $\|\phi\|_2^2 = \int_D \phi^2 \mathrm{d}x$.

Proof For the semigroup (T_t) defined in (26) and $\alpha > 0$ satisfying (65), the resolvent operator

$$U^\alpha = \int_0^\infty \mathrm{e}^{-\alpha t} T_t \mathrm{d}t = (\alpha I - A_2)^{-1}$$

being symmetric with domain $L^2(D)$, is self adjoint. Hence by a general result (see e.g. Yosida (1980, page 199, Theorem 1)), its inverse $\alpha I - A_2$, hence also A_2, is self-adjoint and (80) is true when $A = A_2$ with $\|\phi\| = \|\phi\|_2$. By Theorem 3.27, we have for $\phi \in \mathcal{D}(A_2)$:

$$(A_2\phi, \phi) = ((\frac{\Delta}{2} + q)\phi, \phi) = \int_D \left[-\frac{|\nabla\phi|^2}{2} + q\phi^2\right] \mathrm{d}x. \tag{82}$$

Thus (80) can be rewritten as

$$\lambda_1 = \sup\left\{\int_D \left[-\frac{|\nabla\phi|^2}{2} + q\phi^2\right] \mathrm{d}x : \ \phi \in \mathcal{D}(A_2), \ \|\phi\|_2 = 1\right\}. \tag{83}$$

The right member of (82) is just $-[\phi, \phi]_0$ in the notation of (74). The bilinear form $[f, g]_0$ is jointly continuous with respect to $\| \ \|_\alpha$ hence also with respect to $\| \ \|_*$ by (76). Hence by Theorem 3.28 we may replace $\mathcal{D}(A_2)$ in (83) first by $W_0^{1,2}(D)$, then by $C_c^\infty(D)$ because the latter is also dense in $W_0^{1,2}(D)$ with respect to $\| \ \|_*$ (readers will recall that this is how $W_0^{1,2}(D)$ was defined from $W^{1,2}(D)$). □

The formula (81) can be found in the literature (see Courant and Hilbert (1953, VI(43), page 446)), but only for a more special class of q. If $q \in L^2_{\text{loc}}(D)$, in particular if q is bounded, then for all $\phi \in C_c^\infty(D)$ we have $|\int_D q^2\phi^2 \mathrm{d}x| < \infty$ and so $(\frac{\Delta}{2} + q)\phi \in L^2(D)$. Therefore by Theorem 3.27, $\mathcal{D}(A_2) \supset C_c^\infty(D)$ and Proposition 3.29 becomes 'trivial'. The finesse required in the case $q \in J$ is the gist of Theorem 3.28.

Notes on Chapter 3

For an account (in fact more than one) of the historical background of the true Schrödinger equation and Feynman's path integrals in quantum mechanics, see Kac (1959). A point of nomenclature is in order: a 'path integral' is not an ordinary integral evaluated along a (hidden generic) path such as $\int_0^t q(X_s)\mathrm{d}s$, which is the staple of this book, but is Feynman's way of expressing an integral over the probability space, namely an expectation E, written by him as $\int \cdots \mathrm{d}(\text{path})$. Actually his integral is complex-valued and nobody but nobody has made rigorous probabilistic sense of it yet.

Kac (1951) took our q to be negative and evaluated (in one dimension) the expectation $E^0\{e_q(t);\ a < X(t) < b\} = T_t 1_{a,b}(0)$ (see (3.26)) as $\int_a^b \phi(t,x)\mathrm{d}x$ where ϕ satisfies the 'time-dependent' equation:

$$\frac{\partial \phi}{\partial t} = \frac{1}{2}\frac{\partial^2 \phi}{\partial x^2} + q(x)\phi.$$

Thus $\phi(t,x)$ is the $u_t(0,x)$ in (3.28). The method he used is the forerunner of the calculations in our Lemma 3.7 (what else can one do with an exponential other than use its power series?). Since Kac only considered $\mathbb{R}^d$, there was no need to introduce an optional (stopping) time which he 'hated' (private communication). But even for a general domain D, he might conceivably have achieved his aim by using $p^D(t;x,y)$ (see Section 2.2) instead of $p(t;x,y)$, which latter he always wrote out in its explicit form as physicists love to do.

The class of functions denoted here by J, rather than the clumsy K_d (in $\mathbb{R}^d$!), is associated with the names Stummel, Kato, and Fujiwara. The example given after the Corollary to Proposition 3.1 is new and due to Zhao. Theorem 3.2 is also new at least in its explicit form; it will be improved considerably in Theorem 4.3. Example 1 is due to V. Papanicolaou. The inclusion of the unbounded Coulomb potential (Example 3) in the class of admissible q is apparently of prime concern to mathematical physicists.

The remarkable Theorem 3.6 is due to Aizenman and Simon (1982). The proof has been reorganized and elaborated to bring out the equivalence of the two conditions (5) and (13) more visibly. Readers will observe how much easier it is to deduce (5) from (13) than the other way around. This makes one wonder how far a probabilist who had chanced upon the condition (13) might have gone before needing some part of (5) such as Proposition 3.1? The equivalence given in Theorem 3.6 is extended to a more general class of Markov processes in Zhao (1991) and is applied to the perturbation problems for the pseudo-differential operator $|-\Delta|^{d/2}$ and the relativistic Hamiltonian operator $\sqrt{-\Delta+m^2}-m$.

The deliberate exposition in Section 3.2 of the additive and multiplicative functionals A_q and e_q is intended to show the advantage of viewing certain mathematical objects from a properly generalized standpoint. Nothing is gained and much can be lost by a hasty treatment of such basic structures. For a further extension of multiplicative functionals attached to doubly Feller processes, see Chung (1986b). Another extension is discussed in Chung and Rao (1988).

The functional analysis of the semigroup $\{T_t\}$ in Theorem 3.10 is largely culled from Aizenman and Simon (1982). We note that the symmetry of T_t is far from obvious; indeed even that of P_t^D in Theorem 2.4 is a lucky event. Hunt who first proved the latter was aware of this and used a trick to explain it. When D is the whole space $\mathbb{R}^d$, the invariance relation $mP = m$ lends itself to an intuitive argument suggested by the time-reversibility of the Brownian motion process; see Chung (1982a). For the symmetry of Green's function, see Chung (2001).

The fundamental relationship between the Green kernel G and the Schrödinger kernel V was described in Chung (1986a) for $\lambda = 0$. Formally, this generalizes the resolvent equation (2.56).

Theorem 3.25 is due to Kato (1966), but the proof of the key relation (59) is new. Moreover, he treated only the case $d \geq 3$; our unified treatment for $d \geq 1$ may well be new. This is an instance where the condition (13) proves superior to (5). The explicit analytic definition of the operator F_λ, rather that its adoption from general operator theory, was recommended by Falkner. Theorem 3.27 is given in Simon (1979) for $D = \mathbb{R}^d$. The methods used here are more direct and elementary, minimizing the reliance on operator theory. Of course, in these matters what is simple to one might well be complex to another. We have taken pains to spell out certain details which are often left to the reader. In the same spirit, we have given full coverage to Theorem 3.28 and Proposition 3.29, for both of which we are indebted to Falkner for generous assistance.

4. Stopped Feynman–Kac Functional

4.1 Harnack Inequality and Gauge Theorem

Let D be a domain in $\mathbb{R}^d$, and for a function $q \in \mathcal{B}^d$, let

$$e_q(\tau_D) = \exp\left(\int_0^{\tau_D} q(X_t)\mathrm{d}t\right), \tag{1}$$

where $\{X_t\}$ is the Brownian motion in $\mathbb{R}^d$, and τ_D is the exit time from D defined in Section 1.5. The random variable in (1) is well defined if and only if $\int_0^{\tau_D} q(X_t)\mathrm{d}t$ is well defined, almost surely. This is trivially the case if $q \geq 0$, or if q is bounded and $\tau_D < \infty$. To see that this is also the case when $q \in J$ and $\tau_D < \infty$, we need the Corollary to Proposition 3.8 which implies that for each $t > 0$,

$$\int_0^t |q(X_s)|\mathrm{d}s < \infty$$

a.s.; thus, the same is true when t is replaced by τ_D provided the latter is finite a.s. As before, 'a.s.' will be omitted in what follows when the context is obvious. Under these circumstances we have $0 < e_q(\tau_D) < \infty$; and in fact for each $x \in \mathbb{R}^d$, if $P^x\{\tau_D < \infty\} > 0$, then

$$0 < E^x\{\tau_D < \infty; e_q(\tau_D)\} \leq \infty. \tag{2}$$

Let $f \in \mathcal{B}_+(\partial D)$, and set

$$u(D, q, f; x) = E^x\{\tau_D < \infty; e_q(\tau_D) f(X(\tau_D))\}. \tag{3}$$

This function is well defined, positive and Borel measurable (see a similar discussion in Section 2.3), but may take the value ∞. If $m(D) < \infty$, we may omit '$\tau_D < \infty$' in (3) by Theorem 1.17. Note that when D is bounded and $q \equiv 0$, the function $u(D, 0, f; \cdot)$ reduces to $H_D f$ in the notation of Theorem 1.23. Thus we are dealing with the extension of a harmonic function to a more general setting.

We shall say that the *Harnack inequality* holds for (D, q) iff for each compact subset K of D, there exists a constant $c > 0$ which depends only on D, K, and q such that for any $x \in K$ and $x' \in K$, we have

$$u(D, q, f; x) \leq c\, u(D, q, f; x'). \tag{4}$$

It follows from this that if the function $u(D, q, f; \cdot)$ is not identically ∞ in D, then it is finite everywhere in D. Also, if it is not identically 0 in D, then it is strictly positive in D. The Harnack inequality is the analogue of a result in classical potential theory given in Theorem 1.14, from which the name is derived. For our purposes, however, there is a more important result which strengthens it when f is bounded rather than positive, to the effect that, if $u(D, q, f; \cdot)$ is not identically ∞ in D, then it is bounded in $\overline{D}$. Such a result will be called a *gauge theorem*, for reasons which will become apparent in its applications. It turns out that whereas the Harnack inequality holds for an arbitrary domain in $\mathbb{R}^d$ $(d \geq 1)$, when $q \in J$, the gauge theorem is true under somewhat stronger hypotheses to be specified below.

In this section, we treat the Harnack inequality and the gauge theorem for a domain D with $m(D) < \infty$, and a bounded Borel measurable q. Historically, this was the first case considered; it presents the probabilistic method at its simplest. Moreover, part of the essential argument in this case can be carried over without change under more general hypotheses; see Chapter 5.

Theorem 4.1 *Let D be a domain in $\mathbb{R}^d$. If $q \in L^\infty(\mathbb{R}^d)$, then the Harnack inequality holds for (D, q).*

Proof We write $u(x)$ for $u(D, q, f; x)$, and suppose that there exists $x_0 \in D$ such that $u(x_0) < \infty$. Let K be a compact subset of D; by enlarging K, if necessary, we may suppose that $x_0 \in K$. For any $r > 0$ we define

$$T(r) = \inf(t > 0 : \ |X_t - X_0| \geq r).$$

It follows from Proposition 1.18 that there exists $\delta > 0$, which depends only on $Q \equiv \sup_x |q(x)|$ and d, such that for all $x \in \mathbb{R}^d$:

$$\frac{1}{2} \leq E^x[\exp(-QT(2\delta))]; \quad E^x[\exp(QT(2\delta))] \leq 2. \tag{5}$$

In fact, the two expectations in (5) do not depend on x by the spatial homogeneity of Brownian motion. Now let

$$2r = \rho(K, \partial D) \wedge 2\delta; \tag{6}$$

then for any $s < 2r$, since $T(s) < \tau_D$ under P^{x_0}, we have by the strong Markov property:

$$\begin{aligned} \infty > u(x_0) &= E^{x_0}\left[e_q(T(s))u(X(T(s)))\right] \\ &\geq E^{x_0}\left[\exp(-QT(s))u(X(T(s)))\right]. \end{aligned} \tag{7}$$

The isotropic property of the Brownian motion implies that the random variables $T(s)$ and $X(T(s))$ are stochastically independent for each s under P^{x_0} (see Chung (1982a, page 149)). Hence, we obtain from (5) and (7) that

$$u(x_0) \geq \frac{1}{2} E^{x_0}\left[u(X(T(s)))\right]. \tag{8}$$

The expectation on the right-hand side above is the area average of the values of u on the boundary of $B(x_0, s)$ by Proposition 1.21. Hence if we integrate with respect to the radius, we obtain

$$\int_0^{2r} E^{x_0}\left[u(X(T(s)))\right]\sigma(s)ds = \int_{B(x_0,2r)} u(y)dy, \tag{9}$$

where $\sigma(s) = \sigma(\partial B(x_0, s))$. It follows from (8) and (9) that

$$u(x_0) \geq \frac{1}{2v_d(2r)} \int_{B(x_0,2r)} u(y)dy, \tag{10}$$

where $v_d(r) = m(B(x_0, r)) = \int_0^r \sigma(s)ds$. cf. Section 1.1 for notation.

Next, let $x \in B(x_0, r)$ so that $\rho(x, \partial D) \geq r$ by (6). Then, for $0 < s < r$, we have:

$$\begin{aligned} u(x) &= E^x[e_q(T(s))u(X(T(s)))] \leq E^x[\exp(QT(s))u(X(T(s)))] \\ &= E^x[\exp(QT(s))]E^x[u(X(T(s)))] \leq 2E^x[u(X(T(s)))] \end{aligned}$$

by independence and (5). Integrating as before, we obtain

$$u(x) \leq \frac{2}{v_d(r)} \int_{B(x,r)} u(y)dy. \tag{11}$$

Since $B(x, r) \subset B(x_0, 2r)$ and $u \geq 0$, (10) and (11) together yield

$$u(x) \leq 2^{d+2}u(x_0). \tag{12}$$

To proceed we need a geometrical lemma known as the 'chain argument,' a precise statement of which is given below.

Lemma *Let D_0 be a bounded domain in $\mathbb{R}^d$ $(d \geq 1)$ and let*

$$\overline{D}_0 \subset \bigcup_{i=1}^{N} B\left(x_i, \frac{r}{2}\right). \tag{13}$$

Then for any two distinct points x and x' in $\overline{D}_0$, there exists a subset $\{x_{i_j}, 1 \leq j \leq l\}$, $1 \leq l \leq N$, of the set of centers of the balls satisfying the following conditions:

$$|x - x_{i_1}| < \frac{r}{2}; \; |x_{i_j} - x_{i_{j+1}}| < r, \; 1 \leq j < l; \; |x_{i_l} - x'| < \frac{r}{2}.$$

For a proof of this result, see Chung (1982a, page 205, Exercise 3).

To apply the Lemma, let D_0 be a domain such that $K \subset D_0 \subset\subset D$. It is sufficient to prove (4) when K is replaced by $\overline{D}_0$; indeed we may as well suppose that $K = \overline{D}_0$. Then the covering (13) exists by the Heine–Borel theorem. If $u(x) \equiv +\infty$ for all $x \in K$, then (4) is trivial with $c = 1$. Hence we may suppose that there exists x' in K for which $u(x') < \infty$. Now apply the Lemma to an arbitrary x in K, $x \neq x'$. It then follows by successive use of (12) with x_0 replaced by $x_{i_1}, \cdots, x_{i_l}$ and the given x' that

$$u(x) \leq 2^{(d+2)(l+1)} u(x').$$

Hence firstly $u(x) < \infty$ for all $x \in K$, and secondly

$$u(x) \leq 2^{(d+2)(N+1)} u(x')$$

where now x' as well as x is arbitrary in K. We have thus proved (4) with $c = 2^{(d+2)(N+1)}$. □

The next result is a *gauge theorem*.

Theorem 4.2 *Let D be a domain with $m(D) < \infty$, let $q \in L^\infty(\mathbb{R}^d)$, and let $f \in L_+^\infty(\partial D)$. If $u(D, q, f; \cdot) \not\equiv \infty$ on D, then it is bounded on $\overline{D}$.*

Proof Since q is bounded, it follows from Proposition 1.18 that for any $\varepsilon > 0$, there exists $\delta > 0$ such that for any open subset E of D with $m(E) < \delta$, we have

$$\sup_{x \in \overline{E}} E^x[e_{|q|}(\tau_E)] \leq 1 + \varepsilon. \tag{14}$$

Let K be a compact subset of D such that $m(E) < \delta$ where $E = D \backslash K$. Note that E need not be a connected set; in fact, it cannot be connected when $d = 1$.

For $x \in \overline{E}$, we set

$$u_1(x) = E^x\ [\tau_E < \tau_D;\ e_q(\tau_D) f(X(\tau_D))],$$

and

$$u_2(x) = E^x\ [\tau_E = \tau_D;\ e_q(\tau_D) f(X(\tau_D))]. \tag{15}$$

We have by the strong Markov property:

$$\begin{aligned} u_1(x) &= E^x \left\{ \tau_E < \tau_D;\ e_q(\tau_E) E^{X(\tau_E)} \left[e_q(\tau_D) f(X(\tau_D)) \right] \right\} \\ &= E^x \left\{ \tau_E < \tau_D;\ e_q(\tau_E) u(X(\tau_E)) \right\}. \end{aligned} \tag{16}$$

On the set $(\tau_E < \tau_D)$, we have $X(\tau_E) \in K$. Since $u \not\equiv \infty$ in D, by Theorem 4.1 u is bounded on K; let $\|u\|_K \equiv \sup_{x \in K} |u(x)| < \infty$. Then we have by (14) and (16):

$$u_1(x) \leq E^x \left[e_{|q|}(\tau_E) \right] \|u\|_K \leq (1 + \varepsilon) \|u\|_K. \tag{17}$$

On the other hand, we have by (14),

$$u_2(x) \leq E^x \left[e_{|q|}(\tau_E) f(X(\tau_E)) \right] \leq (1 + \varepsilon) \|f\|, \tag{18}$$

where $\|f\|$ denotes the sup-norm of f.

Combining the last two inequalities, we have

$$u(x) \leq (1 + \varepsilon)(\|u\|_K + \|f\|). \tag{19}$$

Since $\overline{D} \backslash \overline{E} \subset K$, (19) holds trivially for $x \in \overline{D} \backslash \overline{E}$; therefore it holds for all $x \in \overline{D}$. □

Fig. 4.1.

4.2 Properties of $G_D q$

We begin with an essential improvement on an earlier result, Theorem 3.2. This will now be proved by a totally different method based on the probabilistic characterization of J given in Theorem 3.6.

Theorem 4.3 *Let D be a Green-bounded domain in $\mathbb{R}^d$ $(d \geq 1)$, and $q \in J$. Let G_D denote the Green operator defined in (2.13).*

(i) *For any number $b > 0$, there exists a number $a > 0$ depending only on q and b such that*

$$G_D|q| \leq aG_D 1 + b. \tag{20}$$

Consequently, for a fixed $q \in J$, we have $\|G_D q\|_\infty \to 0$ as $\|G_D 1\|_\infty \to 0$.

(ii) *$G_D q \in L^\infty(\mathbb{R}^d) \cap C(D)$, and for any $z \in (\partial D)_r$ we have*

$$\lim_{x \to z} G_D q(x) = 0. \tag{21}$$

(iii) *Under the additional hypothesis:* (a) *$q \in L^1(D)$, or* (b) *$m(D) < \infty$, we have*

$$\lim_{|x| \to \infty} G_D q(x) = 0. \tag{22}$$

Hence $G_D q \in C_0(D)$ provided D is regular.

Proof In the proof of (i) we may suppose that $q \geq 0$. For any $s > 0$, let Λ_n denote the indicator of the set $\{\tau_D > ns\}$, $n \geq 0$. The following inequality holds:

$$\int_0^{\tau_D} q(X_t)dt \leq \sum_{n=0}^{\infty} \Lambda_n \int_{ns}^{(n+1)s} q(X_t)dt. \tag{23}$$

This can be derived directly, but for those readers who care a little for history we note that it is a case of Abel's lemma known in old mathematics! It follows from (23) by taking expectations, using Fubini's theorem and the Markov property that

$$G_D q(x) \leq \sum_{n=0}^{\infty} E^x \left\{ \tau_D > ns;\ E^{X(ns)} \left[\int_0^s q(X_t)dt \right] \right\}.$$

Now for any given $b > 0$, by (3.15) we can choose a number $s > 0$ such that

$$\sup_x E^x \left\{ \int_0^s q(X_t)dt \right\} \leq b.$$

We then have

$$G_D|q|(x) \leq \sum_{n=0}^{\infty} P^x\{\tau_D > ns\} b \leq \left[1 + E^x \left(\frac{\tau_D}{s} \right) \right] b,$$

which is (20) with $a = b/s$.

From here on we no longer suppose that $q \geq 0$. It follows at once from (20) that $G_D q$ is bounded in $\mathbb{R}^d$. Since P_t^D $(t > 0)$ has the strong Feller property by Theorem 2.2, $P_t^D(G_D q)$ is continuous in D. By (2.13),

$$G_D q - P_t^D G_D q = \int_0^t P_s^D q ds. \tag{24}$$

Since $|P_s^D q| \leq P_s|q|$, as $t \downarrow 0$, the right member above converges to zero uniformly in $\mathbb{R}^d$ by the equivalent definition of J given in (3.13). Therefore $P_t^D G_D q$ converges to $G_D q$ uniformly in $\mathbb{R}^d$ and so $G_D q$ is continuous in D. Next, if $z \in (\partial D)_r$, we have

$$\lim_{x \to z} |P_t^D G_D q(x)| \leq \lim_{x \to z} (P_t^D 1(x)) \|G_D q\|_\infty = 0$$

by (1.35). Hence (21) is true also by the uniform convergence.

Under the hypothesis (a) in (iii), we use Lemma 2.11 to obtain $G_D q \in L^1(D)$. It then follows as in the proof of Theorem 3.17 that

$$\lim_{|x| \to \infty} P_t^D G_D q(x) = 0,$$

and so (22) follows by uniform convergence as before. Under the hypothesis (b) in (iii), $G_D q \in L^\infty(D) \subset L^1(D)$, and so the same proof is valid. □

The proof of Theorem 4.3 is an illuminating case of the power of probabilistic methods. We recall Theorem 3.2 which is the special case of Theorem 4.3 when D is bounded, and which is proved by simple analysis. It is curious that the extension of Theorem 3.2 to the case of an unbounded D, albeit of finite measure, is so much harder; in fact, no 'purely analytic' proof has yet been found. One difficulty is that $q \in J$ does not imply $q \in L^1(D)$ for such a D; another difficulty is that in $\mathbb{R}^2$, the Green function $G_D(x,y)$ is not very amenable to analysis. Hence it is not easy to see what to do with the analytic formula $\int_D G_D(x,y)q(y)\mathrm{d}y$. The first proof of the boundedness of $G_D q$ was given in Chung (1986a); the proof given above is a slicker version.

We now cite the following examples, due to V.Papanicolaou.

Example 1 $D = \{(x,y) \in \mathbb{R}^2 : \ x > 1, 0 < y < \frac{1}{x^2}\}$, $q((x,y)) = x$.
It can be shown (although the proof is rather tedious) that $1_D q \in J$; clearly $m(D) < \infty$, but $q \notin L^1(D)$.

Example 2 $d \geq 3$, $\beta = \frac{1}{d-2}$.

$$D = \{(x_1, \cdots, x_d) \in \mathbb{R}^d : \ 1 < x_1, \ 0 < x_j < \frac{1}{x_1^\beta}, \ 2 \leq j \leq d\}$$

and

$$q((x_1, \cdots, x_d)) = |x_1|^\beta.$$

Then $m(D) < \infty$, $q \notin L^1(D)$. It can be shown that $1_D q \in J$.

We now also include an example in which D is Green-bounded, $q \in J$, but (21) is false.

Example 3 Let $D = \{(x,y) \in \mathbb{R}^2 : \ |x| < 1\}$ and $q \equiv 1$. Let $x_k = (0,k)$, $k \geq 1$; then $x_k \in D$ and $\lim_k |x_k| = \infty$, but

$$G_D 1(x_k) \equiv E^0[\tau_{(-1,1)}] > 0$$

by the independence of the two components of the Brownian motion in $\mathbb{R}^2$.

Remark It is a general result that if D is unbounded, $m(D) < \infty$, $q \geq 0$, and $q \notin L^1(D)$, then there exist bounded subdomains D_n such that $m(D_n) \to 0$ but $\int_{D_n} q(x)\mathrm{d}x \to \infty$. This is worth mentioning in case an analytic proof of (i) and (ii) above can be found.

Since $1 \in J$, one immediate consequence of Theorem 4.3 is the following corollary, which is important enough to be stated explicitly as a supplement to Theorem 1.17. Elementary as the result seems, no simpler proof is known when D is unbounded.

Corollary to Theorem 4.3 *If $m(D) < \infty$, then*

$$\lim_{|x| \to \infty} E^x\{\tau_D\} = 0,$$

and consequently

$$\forall t > 0 : \ \lim_{|x| \to \infty} P^x\{\tau_D > t\} = 0. \tag{25}$$

Next we need some supplements to previous results concerning harmonic functions. We recall that for $D \subset \mathbb{R}^d$, $\overline{D}$ and ∂D are subsets of $\mathbb{R}^d$ without the point '∞'; whereas $x \to \infty$ simply means $|x| \to \infty$.

Supplement to Theorem 1.23 *Assume that (25) is true and that*

$$\lim_{\substack{x \in \partial D \\ x \to \infty}} f(x) = f(\infty) \tag{26}$$

exists, then

$$\lim_{\substack{x \in \overline{D} \\ x \to \infty}} H_D f(x) = f(\infty). \tag{27}$$

Proof Since by hypothesis f is bounded on ∂D, $f(\infty)$ is finite. For any $\varepsilon > 0$, there exists $r_1 > 0$ such that if $x \in \partial D$ and $|x| > r_1$, then

$$|f(x) - f(\infty)| < \frac{\varepsilon}{2}.$$

Choose $s > 0$ so that

$$P^0\{\tau_{B(0,r_1)} \leq s\} < \frac{\varepsilon}{8M},$$

where $M = \|f\|$, and furthermore by (25) there exists $r_2 > 0$ such that

$$\sup_{|x| \geq r_2} P^x\{\tau_D > s\} < \frac{\varepsilon}{8M}.$$

It now follows as in the proof of Theorem 1.23 that for $x \in \overline{D}$, $x > \max(2r_1, r_2)$ we have

$$P^x\{\tau_{B(x,r)} < \tau_D\} \leq P^x\{\tau_{B(x,r)} \leq s\} + P^x\{\tau_D > s\} \leq \frac{\varepsilon}{4M};$$

and

$$\begin{aligned} &E^x\{\tau_D \leq \infty;\ |f(X(\tau_D)) - f(\infty)|\} \\ &\quad \leq P^x\{\tau_D \leq \tau_{B(x,r)}\}\frac{\varepsilon}{2} + 2MP^x\{\tau_{B(x,r)} < \tau_D\} \leq \varepsilon. \end{aligned}$$

This implies (27). □

The notation $C(A)$ for $A \subset \mathbb{R}^d$ in Section 1.1 may be extended to $A \subset \overline{\mathbb{R}^d}$. Thus for any unbounded D, $C(\overline{D} \cup \{\infty\})$ is the class of finite continuous (hence bounded!) functions on the closed subset $\overline{D} \cup \{\infty\}$ of $\overline{\mathbb{R}^d}$. For bounded D, of course, this should be replaced by $C(\overline{D})$.

Extension of Theorem 1.24 *Let D be any domain, $h \in C(\overline{D} \cup \{\infty\})$, and h be harmonic in D. Then for $x \in D$:*

$$h(x) = E^x\{h(X(\tau_D))\}, \tag{28}$$

provided that on $\{\tau_D = \infty\}$ we put $h(X(\tau_D)) = h(\infty)$.

Remark The interpretation given in the proviso above is purely a symbolic convenience. In $\mathbb{R}^d$, $d = 1$ or 2, $X(t)$ does not have any limit as $t \to \infty$ because X is a recurrent process. In $\mathbb{R}^d$, $d \geq 3$, it is known that $\lim_{t\to\infty} |X(t)| = \infty$ so that on $\{\tau_D = \infty\}$ it is logical to set $X(\tau_D) = \infty$, and $h(X(\tau_D)) = h(\infty)$ by the hypothesis on h.

Proof We must first improve the second assertion in Theorem 1.24 by dropping the assumption that D is regular. To be precise, suppose that D is a bounded domain, $h \in C(\overline{D})$ and h is harmonic in D; we shall prove that (28) is true. Let D_n be regular bounded domains such that $D_n \uparrow\uparrow D$ (see Appendix to Chapter 1). Then by Theorem 1.24, we have for $x \in D_n$:

$$h(x) = E^x\{h(X(\tau_{D_n}))\}.$$

For a fixed x in D, this representation then holds for all sufficiently large n. Letting $n \to \infty$, we have $X(\tau_{D_n}) \to X(\tau_D)$ and $h(X(\tau_{D_n})) \to h(X(\tau_D))$, by the continuity of X and that of h in $\overline{D}$. Hence (28) follows by bounded convergence.

Now let $x_0 \in D$ and $B_n = B(x_0, n)$ $(n \geq 1)$. Let $D_n = D \cap B_n$. Then D_n is open but not necessarily connected. For each $x \in D$, let C_n be the component of D_n which contains x. Then C_n is a bounded domain and $h \in C(\overline{C_n})$. We have just proved that for this x and all n:

$$h(x) = E^x\{h(X(\tau_{C_n}))\}. \tag{29}$$

In what follows, x is fixed. On $\{\tau_D = \infty\}$, we have for all n, $\tau_{C_n} < \tau_D$, hence $X(\tau_{C_n}) \in \partial B_n$ because $(\partial C_n)\backslash(\partial D) \subset \partial B_n$, and consequently, $\lim_{n\to\infty} |X(\tau_{C_n})| = \infty$. Therefore we have

$$\begin{aligned} \lim_{n\to\infty} E^x\{\tau_D = \infty;\ h(X(\tau_{C_n}))\} &= E^x\{\tau_D = \infty;\ \lim_{n\to\infty} h(X(\tau_{C_n}))\} \\ &= E^x\{\tau_D = \infty;\ h(\infty)\} \end{aligned}$$

by the hypothesis on h. On $\{\tau_D < \infty\}$, there exists $N = N(w)$ such that $\tau_{C_n} = \tau_D$ for all $n \geq N$, hence we see that

$$\lim_{n\to\infty} E^x\{\tau_D < \infty;\ h(X(\tau_{C_n}))\} = E^x\{\tau_D < \infty;\ h(X(\tau_D))\}.$$

The conclusion (28) follows from (29) by adding these two relations. □

It is clear from the proof that when $\tau_D < \infty$ a.s., then (28) is true when $h \in C_b(\overline{D})$, without any condition on $h(x)$ as $x \to \infty$.

4.3 Fundamental Properties of the Gauge

For a given D and q, let us denote the function in (3) by u_f; when $f \equiv 1$, u_1 will be abbreviated to u:

$$u(x) = E^x\{\tau_D < \infty;\ e_q(\tau_D)\}.$$

This is well defined if $q \in J$ as affirmed at the beginning of Section 4.1, and it may also be well defined in other cases. It is defined and ≤ 1 for any negative (≤ 0) $q \in \mathcal{B}^d$, which is therefore a 'trivial' case. We shall call it the *gauge* (function) for (D, q), and when it is bounded in D, we shall say that (D, q) is *gaugeable.* Thus, by Theorem 4.2, (D, q) is gaugeable when $m(D) < \infty$, $q \in L^\infty(\mathbb{R}^d)$, and $u(x) < \infty$ for some $x \in D$. We shall prove later that the same is true more generally if $q \in J$. Furthermore, it is also true, e.g. when D is an arbitrary domain in $\mathbb{R}^d$, $d \geq 3$, and $q \in J \cap L^1(D)$. These extensions are postponed to Chapter 5 for technical reasons. In this section we shall first derive some general properties of the gauge.

It is important to observe that on $\{\tau_D = \infty\}$, $\int_0^{\tau_D} q(X_t)dt$ is not even defined for a general $q \in J$; therefore the condition $\{\tau_D < \infty\}$ in the definition of u is indispensable.

For a fixed $q \in J$ but a variable domain D, we shall use the alternative notation u_D for the gauge for (D, q). This function is defined in $\mathbb{R}^d$. For $x \in (\overline{D})^c \cup (\partial D)_r$, it is trivial that $u_D(x) = 1$. Let x be an irregular boundary point, so that $P^x\{\tau_D > 0\} = 1$ by Proposition 1.3. Hence we have by the Markov property and (3.21):

$$\begin{aligned} u_D(x) &= \lim_{t \downarrow 0} E^x\{t < \tau_D < \infty;\ e_q(\tau_D)\} \\ &= \lim_{t \downarrow 0} E^x\{t < \tau_D;\ e_q(t) u_D(X_t)\} \\ &\leq \sup_{x \in D} u_D(x) \sup_{x \in D} \lim_{t \downarrow 0} E^x\{e_{|q|}(t)\} \\ &= \sup_{x \in D} u_D(x). \end{aligned} \tag{30}$$

Exactly the same argument shows that

$$u_D(x) \geq \inf_{x \in D} u_D(x).$$

It follows from the above that the infimum and supremum of u_D over $\overline{D}$ are equal to the infimum and supremum of u_D over D, respectively.

Proposition 4.4 *For any domain $E \subset D$, we have*

$$[\inf_{x \in D} u_D(x) \wedge 1] u_E \leq u_D \leq [\sup_{x \in D} u_D(x) \vee 1] u_E. \tag{31}$$

Proof We have $\tau_E \leq \tau_D$ a.s. Hence by the strong Markov property:

$$\begin{aligned} u_D(x) &= E^x\{\tau_E < \tau_D;\ e_q(\tau_E)u_D(X(\tau_E))\} + E^x\{\tau_E = \tau_D < \infty;\ e_q(\tau_E)\} \\ &\leq [\sup_{x\in D} u_D(x)]E^x\{\tau_E < \tau_D;\ e_q(\tau_E)\} + E^x\{\tau_E = \tau_D < \infty;\ e_q(\tau_E)\}. \end{aligned}$$

The second inequality in (31) follows from this. The first inequality follows by a similar argument. □

Proposition 4.5 *If $G_D(q)$ is bounded below, then*

$$\inf_{x\in\mathbb{R}^d} u_D(x) > 0. \tag{32}$$

If (32) is true and (D,q) is gaugeable, then (E,q) is gaugeable for any domain $E \subset D$.

Proof By Jensen's inequality,

$$u_D(x) \geq \exp\{G_D(q)(x)\},$$

and consequently

$$\inf_{x\in\mathbb{R}^d} u_D(x) \geq \exp(\inf_{x\in\mathbb{R}^d} G_D(q)) > 0.$$

The second assertion then follows from the first inequality in (31). □

Corollary to Proposition 4.5 *If D is Green-bounded and $q \in J$, then the gauge is bounded away from 0, i.e. (32) holds.*

Proof The statement follows from Theorem 4.3(i) and Proposition 4.5. □

Next, we prove that gaugeability implies the condition (3.36) of Section 3.3, namely (33) below. This will be needed in the following sections for the uniqueness assertion in Theorem 4.7(iv) and the equivalence assertion in Theorem 4.19. However, an alternate treatment without using it will be given later in Section 4.5.

Theorem 4.6 *Let $m(D) < \infty$ and $q \in J$. If (D,q) is gaugeable, then*

$$\int_0^\infty \|T_t\| dt < \infty, \tag{33}$$

where the operator norm is taken in $L^\infty(D)$.

Proof By the Corollary to Proposition 4.5, there exists a constant $c > 0$ such that

$$E^x\{e_q(\tau_D)\} \geq c \text{ for all } x \in D.$$

Hence for each $x \in D$ and $t > 0$:

$$\begin{aligned} T_t 1(x) &= E^x[t < \tau_D;\ e_q(t)] \\ &\leq c^{-1}E^x[t < \tau_D;\ e_q(t)E^{X(t)}[e_q(\tau_D)]] \\ &= c^{-1}E^x[t < \tau_D;\ e_q(\tau_D)], \end{aligned} \tag{34}$$

where the last equation follows by the Markov property. The assumption of gaugeability implies that the right member of (34) converges to zero as $t \to \infty$. Hence for each x,

$$\lim_{t\to\infty} T_t 1(x) = 0.$$

But (34) also implies that

$$T_t 1(x) \leq c^{-1} u_D(x).$$

The gaugeability means that the function $u_D(x)$ is bounded, whence integrable over D because $m(D) < \infty$. Therefore we have by bounded convergence, as $t \to \infty$:

$$\|T_t 1\|_1 = \int_D T_t 1(x)\mathrm{d}x \to 0. \tag{35}$$

By Theorem 3.17, T_t is a bounded operator in $L^1(D)$, and also a bounded operator from $L^1(D)$ to $L^\infty(D)$. Hence we have for $t > 1$,

$$\begin{aligned}\|T_t\|_\infty &= \|T_t 1\|_\infty = \|T_1(T_{t-1}1)\|_\infty \\ &\leq \|T_1\|_{1,\infty}\, \|T_{t-1}1\|_1. \end{aligned} \tag{36}$$

It follows from (35) and (36) that $\|T_t\|_\infty \to 0$ as $t \to \infty$; hence there exists $t_0 > 0$ such that $\|T_{t_0}\|_\infty < 1$. Dropping the subscript on the norm, we have

$$\begin{aligned}\int_0^\infty \|T_t\|\mathrm{d}t &\leq \sum_{n=0}^\infty \int_0^{t_0} \|T_{nt_0}\|\; \|T_t\|\mathrm{d}t \\ &\leq \sum_{n=0}^\infty \|T_{t_0}\|^n \int_0^{t_0} \|T_t\|\mathrm{d}t < \infty,\end{aligned}$$

where we have used (3.35) in the last integral above. □

4.4 Dirichlet Boundary Value Problem for the Schrödinger Equation

We now come to one of the major results of this chapter, namely, the solution of the Dirichlet boundary value problem for the Schrödinger equation. We first treat the case of a bounded domain.

Theorem 4.7 *Suppose that D is bounded, $q \in J_{\text{loc}}$ and (D, q) is gaugeable. For any $f \in \mathcal{B}_b(\partial D)$ and $x \in D$ we set:*

$$\begin{aligned}u(x) = u_f(x) &= E^x\{e_q(\tau_D)f(X(\tau_D))\}, \\ h(x) = h_f(x) &= E^x\{f(X(\tau_D))\}.\end{aligned} \tag{37}$$

Then we have

(i) *u satisfies the following integral equation:*

$$u = h + G(qu), \tag{38}$$

where G is the Green operator for D.

(ii) *$u \in C_b(D)$ and u is a weak solution (see (3.2)) of the Schrödinger equation in D:*

$$(\frac{\Delta}{2} + q)u = 0. \tag{39}$$

(iii) *If f is continuous at $z \in (\partial D)_r$, then*

$$\lim_{\overline{D}\ni x \to z} u(x) = f(z). \tag{40}$$

In particular, if D is regular, then $u \in C(\overline{D})$ with $u = f$ on ∂D.

(iv) *If D is regular, and $f \in C(\partial D)$, then u is the unique solution of* (39) *such that $u \in C(\overline{D})$ and $u = f$ on ∂D.*

(v) *If D is regular, $u \in C(\overline{D})$ and (39) holds in D, then $u = u_u$ in $\overline{D}$ where u_u is the u_f in (37) with f replaced by u (restricted to ∂D).*

Proof (i) By hypothesis, $u_{|f|}$ is bounded; hence by Theorem 4.3(ii),

$$G(|q|u_{|f|}) < \infty. \tag{41}$$

This allows us to apply Fubini's theorem in the following calculation. Consider

$$\begin{aligned}\phi(t,\omega) &= 1_{\{t<\tau_D\}} q(X_t) \exp\left[\int_t^{\tau_D} q(X_s)\mathrm{d}s\right] f(X(\tau_D)),\\ \psi(t,\omega) &= 1_{\{t<\tau_D\}} |q|(X_t) \exp\left[\int_t^{\tau_D} q(X_s)\mathrm{d}s\right] |f|(X(\tau_D)).\end{aligned}$$

We have by the Markov property and the Fubini–Tonelli theorem

$$\begin{aligned}\int_0^\infty E^x\{\psi(t,\omega)\}\mathrm{d}t &= E^x\left\{\int_0^{\tau_D} |q|(X_t)E^{X_t}[e_q(\tau_D)|f|(X(\tau_D))]\right\}\\ &= E^x\left\{\int_0^{\tau_D} |q|(X_t)u_{|f|}(X_t)\mathrm{d}t\right\}\\ &= G(|q|u_{|f|})(x).\end{aligned}$$

Therefore under the condition (41), Fubini's theorem allows us to obtain a similar equality

$$\int_0^\infty E^x\{\phi(t,\omega)\}\mathrm{d}t = G(qu_f)(x),$$

and to evaluate the double integral below in either order:

$$\begin{aligned}\int_0^\infty E^x\{\phi(t,\omega)\}dt &= E^x\{\int_0^\infty \phi(t,\omega)dt\} \\ &= E^x\{[e_q(\tau_D)-1]f(X(\tau_D))\};\end{aligned}$$

in other words, $G(qu_f) = u_f - h_f$. For the pathwise integration see a similar discussion in the proof of Proposition 3.16.

(ii) By Theorem 1.23, h is bounded and harmonic in D. By gaugeability, u is bounded in D. Since $q \in J_{\text{loc}}$, $1_D qu \in J$; hence by Theorem 4.3(ii), $G(qu) \in C_b(D)$. Therefore $u \in C_b(D)$ by (38). Since D is bounded, $q \in L^1(D)$ by Proposition 3.1, hence $qu \in L^1(D)$; while $G(|qu|) < \infty$. Hence Proposition 2.10 is applicable, and we obtain

$$\Delta u = \Delta h + \Delta G(qu) = -2qu.$$

(iii) By Theorem 4.3(ii), applied with $1_D qu$ for the q there, we have $\lim_{x\to z} G(qu) = 0$. Since $\lim_{x\to z} h(x) = f(z)$ by Theorem 1.23, (40) follows from (38).

(iv) Under the hypotheses, if we define u on ∂D to be f, then $u \in C(\overline{D})$, by (iii). In particular, when $f \equiv 0$ on ∂D, then $u \in C_0(D)$. By virtue of Theorem 4.6, gaugeability implies the condition (33). Consequently, Theorem 3.21 is applicable and yields the conclusion $u \equiv 0$ in $\overline{D}$. This is equivalent to the asserted uniqueness.

(v) This follows from (iv). □

The following result is a special case of Theorem 4.7 in which properties of strict differentiability are handled. A function ϕ is said to be (locally) Hölder continuous in a domain D, denoted by $\phi \in H(D)$, iff for any ball $B \subset\subset D$, there exist $c > 0$ and $\alpha \in (0, 1]$ such that

$$|\phi(x) - \phi(y)| \le c|x-y|^\alpha,\ x \in B,\ y \in B.$$

The next proposition is well known in analysis.

Proposition 4.8 *Let D be a bounded domain with Green operator G_D and D' any subdomain of D.*

(a) *If ϕ is locally bounded in D', then $G_D\phi \in C^1(D')$.*

(b) *If $\phi \in H(D')$, then $G_D\phi \in C^2(D')$, and $\Delta G_D\phi = -2\phi$ in D'.*

For a proof of Proposition 4.8, see e.g. Port and Stone (1978, page 115).

Theorem 4.9 *Under the assumptions of Theorem 4.7, if $q \in H(D')$, where D' is a subdomain of D, then the function u in (37) belongs to $C^2(D')$, and the equation (39) holds in the strict sense in D'.*

Proof By Theorem 4.7(ii), $qu \in C(D')$; hence, by Proposition 4.8(a), $G_D(qu) \in C^1(D')$. Since $h \in C^\infty(D)$ by Theorem 1.9, it follows from (38) that $u \in C^1(D')$. This implies by simple analysis that $qu \in H(D')$. Hence, by Proposition 4.8(b),

$G_D(qu) \in C^2(D')$. Using this in (38) again, we obtain $u \in C^2(D')$, and so (39) is true pointwise in D'. □

We now extend Theorem 4.7 to a more general case. It is instructive to scrutinize the arguments to see where improvements are needed.

Theorem 4.10 *Suppose that D is Green-bounded, $q \in J$ and (D, q) is gaugeable. Let f, u, and h be as in (37). Then the assertions* (i), (ii), *and* (iii) *of Theorem 4.7 are true. Moreover, if $m(D) < \infty$ or $q \in L^1(D)$ and (26) is assumed, then*

$$\lim_{\overline{D} \ni x \to \infty} u(x) = f(\infty). \tag{42}$$

Proof Firstly, under the present hypothesis, (41) is true by Theorem 4.3(ii), which yields the assertion (i) as before.

Next, it is no longer true that $q \in L^1(D)$ in general, but fortunately $qu \in L^1_{\text{loc}}(D)$ because u is bounded by hypothesis and $q \in L^1_{\text{loc}}(D)$. Since $G(qu)$ is bounded, it certainly belongs to $L^1_{\text{loc}}(D)$. Thus, by Proposition 2.10, the assertion (ii) follows. Note that the harmonicity of h in the present case is covered by Theorem 1.23.

The assertion (iii) is proved now by Theorem 4.3(ii). The new assertion (42) is proved by the Supplement to Theorem 1.23, and (22) in Theorem 4.3. □

What happens to the assertions (iv) and (v) in Theorem 4.7? They will be reformulated and established by a new method involving several new results of independent interest. So far as our previous method is concerned, it is possible to extend Theorem 3.21 to the case where D is regular but unbounded with $m(D) < \infty$, and $\phi \in C_0(D)$. Instead of Theorem 3.2, we must now use Theorem 4.3; Theorem 3.18 is still applicable by virtue of Theorem 4.6. The details are left to the reader.

4.5 Representation Theorem

The definition below is meaningful for a general $q \in \mathcal{B}$, although we shall soon need $q \in J$.

Definition *A function ϕ is q-harmonic in D iff $\phi \in C(D)$ and satisfies the Schrödinger equation:*

$$\left(\frac{\Delta}{2} + q\right)\phi = 0 \tag{43}$$

in the weak sense in D.

In Section 5.5 we will prove that the hypothesis of continuity of ϕ imposed above may be omitted, provided an equivalent version of ϕ is used. This fine point is not needed in this section. The definition is an extension of harmonicity defined in Sections 1.4 and 2.5. Observe that when $q \equiv 0$, the above definition of a 0-harmonic function renders it harmonic in the classic sense, owing to Weyl's lemma. Standard analysis shows that q-harmonicity is a local property; in other words, ϕ is q-harmonic in D if and only if it is q-harmonic in a neighborhood of each point in D.

In Section 4.4 we demonstrated the existence of q-harmonic functions in gaugeable domains. In particular, when $q \leq 0$, any domain is gaugeable and so the results there are all true. Indeed, in this case we may consider a general Borel measurable q; however, we shall not concern ourselves with this facile generalization. The next result, which may be referred to as 'local gaugeability', is more interesting. We note that, by definition, $q \in J$ is equivalent to $|q| \in J$; by contrast, the gaugeability of $(D, |q|)$ implies that of (D, q) but is not implied by the latter. This makes a serious difference in what follows.

Proposition 4.11 *For any $q \in J$, there exists $\delta > 0$ such that if $m(D) < \delta$, then $(D, |q|)$ is gaugeable.*

Proof By Theorem 4.3, there exists $\delta > 0$ such that

$$\sup_x G_D|q|(x) < \varepsilon < 1.$$

Hence, by Lemma 3.7 with $\tau = \tau_D$, we have

$$\sup_x E^x\{e_{|q|}(\tau_D)\} \leq \frac{1}{1-\varepsilon} < \infty.$$

□

By virtue of this proposition, all the results of Theorem 4.7 are true for a sufficiently small ball. In particular, there is a unique local solution of (39), or in our new terminology, a unique q-harmonic function with prescribed boundary values. This result of local solvability, at least for a more special q, is known in the theory of partial differential equations but is proved here by probability methods; furthermore, the solution is given by the explicit formula (37). No analogue of this exists in classical analysis.

We now proceed to derive the representation of q-harmonic function in a general gaugeable domain. We assume first that $(D, |q|)$ is gaugeable, in which case the representation can be constructed step by step by means of a Taylor series. The following lemma is the key.

Lemma 4.12 *Let $m(D) < \infty$, $q \in J$, $f \in L^\infty(\partial D)$. For integers $n \geq 0$ and $x \in D$ let:*

$$f_n(x) = \frac{1}{n!}E^x\left\{\left(\int_0^{\tau_D} q(X_t)dt\right)^n f(X(\tau_D))\right\}. \tag{44}$$

Then f_n is bounded in $\mathbb{R}^d$ for each n, and

$$f_{n+1} = G(qf_n), \tag{45}$$

where $G = G_D$ is the Green operator.

Proof By (3.19) and Theorem 4.3:

$$\sup_x |f_n(x)| \leq \left(\sup_x E^x \left\{ \int_0^{\tau_D} |q(X_t)| \mathrm{d}t \right\} \right)^n \|f\| = \|G|q|\|^n \|f\| < \infty.$$

This bound is needed to justify the calculations below. Using the notation of Lemma 3.7, we set $\tau = \tau_D$ and

$$A(t) = \int_0^t q(X_s)\mathrm{d}s = \int_0^t q^+(X_s)\mathrm{d}s - \int_0^t q^-(X_s)\mathrm{d}s,$$

where $q = q^+ - q^-$ is the usual decomposition. Thus $t \to A(t)$ is a process of locally bounded variation, and integration with respect to it is defined path by path. On the set $\{t < \tau\}$, we have $\tau = t + \tau \circ \theta_t$, $X(\tau) = X(\tau) \circ \theta_t$, and

$$A(\tau)^n \circ \theta_t = (A(\tau) - A(t))^n.$$

Hence by the Markov property we have

$$\begin{aligned} G(qf_n)(x) &= \frac{1}{n!} E^x \left\{ \int_0^\tau f_n(X_t) \mathrm{d}A(t) \right\} \\ &= \frac{1}{n!} E^x \left\{ \int_0^\tau E^{X_t}\{A(\tau)^n f(X(\tau))\} \mathrm{d}A(t) \right\} \\ &= \frac{1}{n!} E^x \left\{ \int_0^\tau (A(\tau) - A(t))^n \mathrm{d}A(t) f(X(\tau)) \right\} \\ &= \frac{1}{(n+1)!} E^x \{A(\tau)^{n+1} f(X(\tau))\} = f_{n+1}(x). \end{aligned}$$

□

Proposition 4.13 *Let D be a regular Green-bounded domain, $q \in J$, and $(D, |q|)$ be gaugeable. Let $u \in C_b(\overline{D})$ and u be q-harmonic in D. Then we have the representation:*

$$u(x) = E^x\{e_q(\tau_D)u(X(\tau_D))\}, \ x \in \overline{D}. \tag{46}$$

Proof Set in $\overline{D}$:

$$h = u - G(qu). \tag{47}$$

Since $qu \in L^1_{\mathrm{loc}}(D)$ and $G(|qu|) < \infty$ by Theorem 4.3, Proposition 2.10 is applicable and we have in D:

$$\Delta h = \Delta u - \Delta G(qu) = -2qu + 2qu = 0.$$

Although u is only given in $\overline{D}$, we may define it to be zero on $(\overline{D})^c$; then since u is bounded, $1_D qu \in J$, and consequently by Theorem 4.3, $G(qu) = G_D(1_D qu) \in C_b(\mathbb{R}^d)$. Thus $h \in C_b(\overline{D})$, and so h is harmonic in D by Weyl's lemma. By the Extension of Theorem 1.24 given in Section 4.2, (28) holds. Note that since D is Green-bounded, we have $\tau_D < \infty$ a.s., and so the continuity of h at ∞ is not needed here. Since D is regular, $G(qu) = 0$ on ∂D, and so $h = u$ on ∂D. We have therefore the representation:

$$h(x) = E^x\{u(X(\tau_D))\}, \quad x \in \overline{D}. \tag{48}$$

Setting $h_0 = h$, $u_0 = u$, we define inductively for $n \geq 0$:

$$h_{n+1} = G(qh_n), \quad u_{n+1} = G(qu_n).$$

Then by (47),

$$u_0 = h_0 + G(qu_0) = h_0 + u_1.$$

A simple induction shows that

$$u_n = h_n + u_{n+1},$$

and consequently for all $n \geq 0$,

$$u_0 = \sum_{j=0}^{n} h_j + u_{n+1}. \tag{49}$$

Applying Lemma 4.12 with $f \equiv u$ on ∂D, so that $f_0 = h_0$, we obtain, using the notation of Lemma 4.12:

$$h_n(x) = \frac{1}{n!} E^x\{A(\tau)^n u(x(\tau))\}. \tag{50}$$

Next we apply Lemma 4.12 again with q replaced by $|q|$, and $f \equiv 1$ on ∂D. We set $B(t) = \int_0^t |q(X_s)| \mathrm{d}s$, and

$$e_n(x) = \frac{1}{n!} E^x\{B(\tau)^n\}. \tag{51}$$

Then $e_1 = G(|q|)$, and for $n \geq 1$:

$$e_{n+1} = G(|q| e_n). \tag{52}$$

Let $\|u_0\| = M < \infty$; then $|u_1| \leq e_1 M$, and a simple induction shows that

$$|u_n| \leq e_n M. \tag{53}$$

The condition of gaugeability of $(D, |q|)$, together with Fubini's theorem for a series of positive terms, yields that

$$\sum_{n=0}^{\infty} e_n(x) = E^x\{\sum_{n=0}^{\infty} \frac{1}{n!} B(\tau)^n\} = E^x\{\mathrm{e}^{B(\tau)}\}$$

is bounded in x. Actually, it is sufficient here that the sum be finite for all $x \in D$. Then of course $\lim_{n\to\infty} e_n(x) = 0$, and consequently by (53), $\lim_{n\to\infty} u_n(x) = 0$. Therefore we obtain by (49):

$$\begin{aligned} u_0(x) &= \sum_{j=0}^{\infty} h_j(x) = \sum_{j=0}^{\infty} \frac{1}{j!} E^x\{A(\tau)^j u(X(\tau))\} \\ &= E^x\{\mathrm{e}^{A(\tau)} u(X(\tau))\}. \end{aligned}$$

This is the desired representation. □

It is curious to note that the above proof requires the gaugeability of $(D, |q|)$, although the result is true under the weaker assumption of the gaugeability of (D, q). This improvement will be achieved by a detour via Proposition 4.11, using a method of successive balayage to extend the domain. This approach is reminiscent of similar procedures of analytic continuation in complex analysis.

Proposition 4.14 *Suppose that for all $x \in D$:*

$$P^x\{\tau_D < \infty\} = 1, \tag{54}$$

(D, q) is gaugeable and (32) holds. Let $D = \cup_{i=1}^{k} D_i$, where each D_i is a regular domain contained in D. Suppose that $u \in C(\overline{D})$, and that for each i $(1 \le i \le k)$ we have

$$u(x) = E^x\{e_q(\tau_{D_i}) u(X(\tau_{D_i}))\}, \quad x \in D_i. \tag{55}$$

Then (55) is true when D_i is replaced by D.

Proof By Proposition 4.5, $\left(\cup_{i=1}^{j} D_i, q\right)$ is gaugeable for $1 \le j \le k$. Hence, by induction on j, it is sufficient to prove the proposition when $k = 2$. Thus $D = D_1 \cup D_2$. We may suppose that $x \in D_1$. Set $T_0 = 0$, and for $n \ge 1$;

$$\begin{aligned} T_{2n-1} &= T_{2n-2} + \tau_{D_1} \circ \theta_{T_{2n-2}} \\ T_{2n} &= T_{2n-1} + \tau_{D_2} \circ \theta_{T_{2n-1}} \end{aligned}$$

We shall first prove from the assumption (55) above, without using gaugeability, that

$$u(x) = E^x[e_q(T_m) u(X(T_m))], \quad m \ge 0. \tag{56}$$

This is true for $m = 0$. Suppose that

$$u(x) = E^x\left[e_q(T_{2n}) u(X(T_{2n}))\right].$$

Then on $\{T_{2n} < \tau_D\}$ we have $X(T_{2n}) \in D_1$; hence by (55):

$$\begin{aligned} u(x) = &\; E^x\{T_{2n} = \tau_D,\ e_q(T_{2n}) u(X(T_{2n}))\} \\ &+ E^x\left\{T_{2n} < \tau_D,\ e_q(T_{2n}) E^{X(T_{2n})}\left[e_q(\tau_{D_1}) u(X(\tau_{D_1}))\right]\right\}. \end{aligned} \tag{57}$$

On $\{T_{2n} = \tau_D\}$, we have $T_{2n} = T_{2n+1}$. Hence the first term on the right-hand side of (57) is equal to

$$E^x[T_{2n} = \tau_D,\ e_q(T_{2n+1}) u(X(T_{2n+1}))].$$

By the definition of T_{2n+1} and the strong Markov property, the second term on the right-hand side of (57) is equal to

$$E^x\{T_{2n} < \tau_D,\ e_q(T_{2n+1})u(X(T_{2n+1}))\}.$$

Adding these two terms in (57), we obtain (56) with $m = 2n + 1$. Similarly we can prove that if (56) holds for $m = 2n + 1$, then it holds for $m = 2n + 2$. Hence it holds for all $m \geq 0$ by induction.

Since $T_m \uparrow$ and $T_m \leq \tau_D$, we have $\lim_{m\to\infty} T_m = S \leq \tau_D$. If $S < \tau_D$, then since $X(T_{2n-1}) \in \partial D_1$ and $X(T_{2n}) \in \partial D_2$ for all n, we have by continuity

$$X(S) \in (\partial D_1) \cap (\partial D_2) \subset (D_1 \cup D_2)^c;$$

thus $S \geq \tau_D$, which is a contradiction. Therefore almost surely

$$\lim_{m\to\infty} T_m = \tau_D. \tag{58}$$

Now by assumption the gauge for (D, q) is finite, which means that $e_q(\tau_D)$ is integrable with respect to P^x. Hence we have

$$\begin{aligned} E^x\{e_q(\tau_D)|\mathcal{F}_{T_m}\} &= e_q(T_m)E^{X(T_m)}[e_q(\tau_D)] \\ &\geq e_q(T_m)c, \end{aligned} \tag{59}$$

where $c > 0$ is the infimum in (32). Consequently, $\{e_q(T_m),\ m \geq 1\}$ is uniformly integrable with respect to P^x (see Chung, (1974, page 328, Exercise 2)), hence, so is the sequence $\{e_q(T_m)u(X(T_m)),\ m \geq 1\}$ because u is bounded. We may therefore let $m \to \infty$ in (56), and use (58) and the continuity of u in $\overline{D}$ to obtain

$$u(x) = E^x[e_q(\tau_D)u(X(\tau_D))]. \qquad \square$$

We are ready to state the representation theorem for q-harmonic functions.

Theorem 4.15 *Let D be Green-bounded, $q \in J$, and (D, q) be gaugeable. Let u be q-harmonic in D and $u \in C_b(\overline{D})$. Then we have the representation (46).*

Proof Since D is Green-bounded, (54) is true; by Theorem 4.3(i) and Proposition 4.5, (32) is also true. Assume first that D is bounded and regular. Choose δ as in Proposition 4.11. It is possible to 'decompose' D into a finite number of regular subdomains D_i, not necessarily disjoint, such that $D = \cup_{i=1}^k D_i$, and $m(D_i) < \delta$, $1 \leq i \leq k$. Then $(D_i, |q|)$ is gaugeable by Proposition 4.11 so that (55) holds true by Proposition 4.13 for each i. Therefore (46) is proved by Proposition 4.14.

In the general case, there exist regular bounded domains $D_n \uparrow\uparrow D$. We have just proved that for $x \in D_n$:

$$u(x) = E^x\{e_q(\tau_{D_n})u(X(\tau_{D_n}))\}. \tag{60}$$

Fix x and let $n \to \infty$. Then $\tau_{D_n} \uparrow \tau_D < \infty$, hence $e_q(\tau_{D_n}) \to e_q(\tau_D)$ and $u(X(\tau_{D_n})) \to u(X(\tau_D))$. Now since the gauge for (D, q) is finite, we have for each x:

$$E^x\{e_q(\tau_D)|\mathcal{F}_{\tau_{D_n}}\} = e_q(\tau_{D_n})E^{X(\tau_{D_n})}\{e_q(\tau_D)\}.$$

As before in (59), this implies that the sequence $\{e_q(\tau_{D_n}),\ n \geq 1\}$ is uniformly integrable, and therefore we can let $n \to \infty$ in (60) to obtain (46). □

One consequence is the following important uniqueness theorem, which was proved for a bounded D in Theorem 4.7(iv) by quite different methods, and was omitted in Theorem 4.10.

Corollary to Theorem 4.15 *Under the conditions of Theorem 4.15, a q-harmonic function which is bounded and continuous in $\overline{D}$ and vanishes on ∂D is identically zero in $\overline{D}$.*

Thus, the assertion in part (iv) of Theorem 4.7 remains true under the more general assumptions of Theorem 4.10. This extension from a bounded domain to an unbounded one, even under the specified restriction, is far from trivial. Note that by our convention made at the outset, the closure of an unbounded set does not include the point at infinity, and so the above result does not require any knowledge of the behavior of the function at infinity, apart from its boundedness. Probabilistically, this is understandable because when the exit time is finite the path does not go to infinity.

Theorems 4.10 and 4.15 extend previous results from a bounded D to a Green-bounded D. In a similar vein we state the following result which may be regarded as an extension of Theorems 3.21 and 3.22.

Extension of Theorems 3.21 and 3.22(a) *Let D be Green-bounded and regular, $q \in J$, and $V1 \in L^\infty(D)$. For any $f \in \mathbb{F}(D, q)$, the unique (weak) solution ϕ in $C_b(D)$ of the equation*

$$\left(\frac{\Delta}{2} + q\right)\phi = f$$

with the boundary condition

$$\phi|_{\partial D} = 0$$

is given by $\phi = -Vf$ in D.

The proof is similar to that of Theorem 3.22, but we must use Theorem 4.3 instead of Theorem 3.2, and observe that Proposition 2.10 is applicable because $qVf \in L^1_{\text{loc}}(D)$. The uniqueness follows from the Corollary to Theorem 4.15. Note that no assertion is made about the behavior of $\phi(x)$ as $|x| \to \infty$.

We observe that the representation result has been proved for a bounded and regular domain in Theorem 4.7(v), where it is deduced from the uniqueness result in (iv). The latter depends on Theorem 3.21. Here we established the more general representation result Theorem 4.15 by a different method, and deduce the corresponding uniqueness result from it. The new method is constructive and reveals the basic structure of the exponential in the gauge.

4.6 Equivalence Theorem for Gaugeability

We begin with a lemma based on an argument in the proof of Proposition 4.14.

Lemma 4.16 *Let D_1 and D_2 be two bounded regular domains with $D_1 \cap D_2 \neq \emptyset$; $q \in J_{\text{loc}}$, and (D_i, q) is gaugeable for $i = 1, 2$. Let $D = D_1 \cup D_2$. Suppose there exists a q-harmonic function u in D which is strictly positive and continuous in $\overline{D}$, then (D, q) is gaugeable.*

Proof By Theorem 4.7(v) or Theorem 4.15, for such a function u and each $i = 1, 2$, the equation (55) holds. Hence (56) follows as shown there. The hypotheses on u then imply that for each $m \geq 1$ and $x \in D$,

$$[\inf_{x\in\overline{D}} u(x)] E^x\{e_q(T_m)\} \leq \sup_{x\in\overline{D}} u(x).$$

It follows from (58), Fatou's lemma and the Corollary to Proposition 4.5 that

$$E^x[e_q(\tau_D)] \leq \frac{\sup_{x\in\overline{D}} u(x)}{\inf_{x\in\overline{D}} u(x)} < \infty. \qquad \square$$

We proceed to develop a number of conditions which are equivalent to the boundedness of gauge in a domain. The first is a very general sufficient condition.

Theorem 4.17 *Let D be an arbitrary domain in $\mathbb{R}^d$, $d \geq 1$; and $q \in J_{\text{loc}}$. Suppose that there exists a q-harmonic function u in D satisfying*

$$0 < \inf_{x\in D} u(x) \leq \sup_{x\in D} u(x) < \infty. \tag{61}$$

Then (D, q) is gaugeable.

Proof There exists a sequence of bounded domains $\{D_n\}$ such that $\overline{D}_n \subset D$, $D_n \uparrow D$; and for each n, D_n is the finite union of balls B so small that each (B, q) is gaugeable by Proposition 4.11. Then by repeated application of Lemma 4.16, (D_n, q) is gaugeable for each n, and consequently by Theorem 4.7(v) or Theorem 4.15:

$$u(x) = E^x[e_q(\tau_{D_n}) u(X(\tau_{D_n}))], \ x \in D_n.$$

It follows as before that

$$E^x[e_q(\tau_{D_n})] \leq \frac{\sup_{x\in D} u(x)}{\inf_{x\in D} u(x)}. \tag{62}$$

Since $\tau_{D_n} \uparrow \tau_D \leq \infty$ a.s., we have by Fatou's lemma

$$E^x\{\tau_D < \infty;\ e_q(\tau_D)\} \leq \underline{\lim}_{n\to\infty} E^x[e_q(\tau_{D_n})].$$

This is bounded by (61) and (62). □

Theorem 4.18 *Let D be a Green-bounded domain in $\mathbb{R}^d$ $(d \geq 1)$ and $q \in J$. Then the existence of a q-harmonic function u satisfying (61) is not only sufficient, but also necessary for the gaugeability of (D, q).*

Proof Suppose $u(x) = E^x[e_q(\tau_D)]$ is bounded in D. For any bounded domain $E \subset\subset D$, we have by the strong Markov property:

$$u(x) = E^x[e_q(\tau_E)u(X(\tau_E))].$$

By Theorem 4.7(ii) u is q-harmonic in E. This being true for each E, u is q-harmonic in D because q-harmonicity is a local property. It is bounded away from 0 by the Corollary to Proposition 4.5. Thus the gauge is a q-harmonic function whose existence is claimed. □

Theoretically, the necessary and sufficient condition for gaugeability given in Theorem 4.18 is the most significant one. We shall now give a number of other conditions which may be more operational.

Let us first introduce the formal definition of the so-called Dirichlet eigenvalues of the Schrödinger operator in a domain D with $m(D) < \infty$. A real number λ is said to be an eigenvalue of the Schrödinger operator $\frac{\Delta}{2} + q$ in a domain D iff there exists a nonzero function $\phi \in W_0^{1,2}(D)$ such that

$$\left(\frac{\Delta}{2} + q\right)\phi = \lambda\phi.$$

By Theorem 3.27, this definition is equivalent to that of an eigenvalue of the generator A_2 of the Feynman–Kac semigroup $\{T_t\}$ in $L^2(D)$. Therefore, according to Theorem 3.17, the Dirichlet eigenvalues of $\frac{\Delta}{2} + q$ in D are identical to the eigenvalues of the generator A_p of $\{T_t\}$ in $L^p(D)$ $(1 \leq p < \infty)$, and to those of A_0 in $C_0(D)$ if $m(D) < \infty$ and D is regular.

By Theorem 3.10 and Proposition 3.15, for each $t > 0$, T_t is a symmetric and compact operator in $L^2(D)$. Therefore, a classical result (see e.g. Dunford and Schwartz (1958, Volume 1, page 579)) states that the spectrum Spec(T_1) is a countable set contained in $[0, \|T_1\|]$. According to the spectral resolution theorem for self-adjoint operators (see Yosida (1980, page 313)), we have

$$e^{[\mathrm{Spec}(A)]} = \mathrm{Spec}(T_1)\backslash\{0\}. \tag{63}$$

Hence Spec(A) consists of the eigenvalues

$$\infty > \lambda_1 > \lambda_2 > \cdots > \lambda_n > \cdots > -\infty.$$

Let $\lambda(D, q)$ denote λ_1, the first eigenvalue. Then we have as in (3.79),

$$\lambda(D,q) = \sup[\mathrm{Spec}(A)],$$

where A is the generator of $\{T_t\}$ in one of the appropriate spaces for D.

We recall the representation of $\lambda(D,q)$ given in (3.81).

Theorem 4.19 *Let D be a domain in $\mathbb{R}^d$ $(d \geq 1)$ with $m(D) < \infty$ and $q \in J$. The following conditions are equivalent to one another.*

(i) *(D,q) is gaugeable; in other words, the gauge $u(\cdot) = E^{\cdot}\{e_q(\tau_D)\}$ is bounded in $\mathbb{R}^d$.*

(ii) *The semigroup $\{T_t,\ t \geq 0\}$ satisfies the condition (33).*

(iii) *$V1$ is bounded in $\mathbb{R}^d$.*

(iv) *$V|q|$ is bounded in $\mathbb{R}^d$.*

(v) *$\lambda(D,q) < 0$.*

Proof We proved that (i) implies (ii) in Theorem 4.6. The fact that (ii) implies (iii) is trivial because

$$V1 = \int_0^\infty T_t 1 \mathrm{d}t.$$

(iii) implies (iv) by (3.43) with $f = q$.

If (iv) is true, then the following calculation is valid:

$$\begin{aligned} Vq(x) &= \int_0^\infty T_t q(x)\mathrm{d}t = \int_0^\infty E^x\{t < \tau_D; e_q(t)q(X_t)\}\mathrm{d}t \\ &= E^x\left\{\int_0^{\tau_D} e_q(t)q(X_t)\mathrm{d}t\right\} = E^x\{e_q(\tau_D) - 1\}. \end{aligned} \tag{64}$$

This follows from Fubini's theorem and requires $V|q| < \infty$ rather than $Vq < \infty$! Thus (iv) implies (i), and we have completed the circle from (i) to (iv).

By the spectral radius theorem (see Yosida (1980, VIII.2)) and (3.79), we have

$$\lim_{t\to\infty} \frac{\ln \|T_t\|}{t} = \lambda(D,q). \tag{65}$$

If $\lambda(D,q) < 0$, then there exist $t_0 > 0$ and $\beta > 0$ such that for all $t \geq t_0$,

$$\|T_t\| \leq \mathrm{e}^{-\beta t}. \tag{66}$$

By Proposition 3.8

$$\|T_t\| = \|T_t 1\| = \sup_x E^x\{e_q(t)\} \leq \mathrm{e}^{C_0 + C_1 t}$$

with $C_1 > 0$, and consequently

$$\sup_{0\leq t\leq t_0} \|T_t\| \leq \mathrm{e}^{C_0 + C_1 t_0} < \infty. \tag{67}$$

Therefore (33) is true by (66) and (67). Conversely, if (33) is true, then there exist $t_0 > 0$ and $\beta > 0$ such that $\|T_{t_0}\| \le e^{-\beta}$, and so $\|T_{nt_0}\| \le e^{-n\beta}$ by the semigroup property. Hence the limit in (65) must be strictly less than zero. We have proved the equivalence of (ii) and (v). □

We have in fact proved that each point of the theorem is equivalent to each of the following further points.

(vi) *There exists* $t_0 > 0$ *such that* $\|T_{t_0}\| < 1$.

(vii) *There exist* $t_0 > 0$ *and* $\beta > 0$ *such that for all* $t \ge t_0$,

$$\|T_t\| \le e^{-\beta t}.$$

The last property is known as 'exponential decay'.

We note that the above proof actually shows that for each given x in D, $V1(x) < \infty$ if and only if $u(x) < \infty$. Therefore the gauge theorem may be stated in terms of $V1$; in other words:

$$V1 \not\equiv \infty \text{ in } D \text{ implies } V1 \text{ is bounded in } \overline{D}.$$

In this form the result becomes a major extension and strengthening of the result in classical potential theory which asserts that if a positive superharmonic function in a domain is finite at any point, then it is locally integrable (see Chung (1982a)).

The condition (v) above in terms of the spectrum of the operator looks elegant and is beloved by the traditional analysts, *qua* vested interest. It is clear that this condition plays only a peripheral role here. However, more vital involvements with eigenvalues will be found in Chapter 8.

We note that some of the implications in Theorem 4.19 hold under more general assumptions. If D is Green-bounded, then (iii) implies (iv) by the same proof; conversely (iv) implies (iii) by Theorem 3.20, if we integrate (3.47) with $\lambda = 0$ with respect to y over D. Finally if D satisfies the assumption (54), then (iv) implies (i) by (64).

We close this chapter with a numerical example of gaugeability.

Example Let $D = B(0, r)$ be a ball in $\mathbb{R}^d$ $(d \ge 1)$, and let $q \in L^\infty(D)$ with $|q| \le Q$. From martingale theory, we know the following exact formula:

$$G_D 1(x) = E^x\{\tau_D\} = \frac{r^2 - |x|^2}{d}. \tag{68}$$

This can also be verified by the results in Chapters 1 and 2, because $G_D 1$ is the unique solution $\phi \in C_0(D)$ of the equation $\frac{\Delta}{2}\phi = -1$; see Chung (1982a, pages 195–196). Hence we have

$$\|G_D 1\|_\infty = \sup_x E^x\{\tau_D\} = \frac{r^2}{d};$$

and consequently

$$\|G_D|q|\|_\infty \leq \frac{Qr^2}{d}.$$

It follows from the proof of Proposition 4.11 that $(D, |q|)$ is gaugeable provided

$$\frac{Qr^2}{d} < 1 \text{ or } r < \sqrt{\frac{d}{Q}}.$$

This estimate can be extended to any domain D using the remark following Theorem 1.17. Given the number $m(D)$, the radius r of the ball with volume equal to $m(D)$ is determined by

$$\frac{\pi^{d/2}}{\Gamma(\frac{d}{2}+1)} r^d = m(D)$$

(see Section 1.1). A little computation then shows that $(D, |q|)$ is gaugeable provided

$$m(D) < \frac{(\pi d)^{d/2}}{\Gamma(\frac{d}{2}+1)} \frac{1}{Q^{d/2}}.$$

For $d = 1$, this improves on an ancient result of de la Vallee-Poussin, to the effect that any nonzero solution of the differential equation

$$\phi'' + 2q\phi = 0$$

in $\mathbb{R}^1$ cannot have two consecutive zeros in an interval of length less then $\frac{2}{\sqrt{Q}}$. His estimate gives $\frac{1}{\sqrt{2Q}}$.

We recall that gaugeability implies that the Dirichlet boundary value problem for the Schrödinger equation has a unique solution. The classical analytic approach to this kind of problem depends on the 'eigen connection', i.e. condition (v) of Theorem 4.19. An estimate for $\lambda_1(D, q)$ can be derived from the Sobolev and Poincaré inequalities (which look rather complicated to us); see Chung, Li and Williams (1986).

Notes on Chapter 4

The organization of the two main chapters, Chapters 4 and 5, is motivated by historical as well as didactic considerations, Theorems 4.1, 4.2, 4.7 and 4.18, and parts of Theorem 4.19 were first published in Chung and Rao (1981), after a preliminary announcement in 1980. The function q (commonly called the 'potential term' in old texts) was assumed to be Borel measurable and bounded, in other words, in $\mathcal{B}_b^d$. While this class of functions is contained in J it is much broader than the usual classes of q treated in works on partial differential equations. From our standpoint the case $-q \in \mathcal{B}_+^d$ is trivial; the case $q \in \mathcal{B}_+^d$ is of some interest and was previously studied in Khas'minskii (1959). For further historical comments see the introduction to Chung and Rao (1981) and Chung (1985b).

The Gauge Theorem, so named despite other competing usage of the term 'gauge', is a new kind of result. Its intuitive content is yet another manifestation of the incredibly weird behavior of the Brownian paths, when one pauses to reflect on the arbitrariness of the distribution of values of q, as well as that of the shape of D. This aesthetic appeal is complemented by its immediate utility, because the boundedness of the gauge is absolutely essential for the rest of the analytic manipulations.

Theorem 4.3 contains Theorem 3.2 which was retained for the sake of comparison. The major departure in the method of proof, making use of time rather than space, is a noteworthy example of the probabilistic approach. For a discursive account containing an earlier version of the result, see Chung (1988): another version was given in Chung's Pisa lectures (Spring 1989) and recorded in Mancino (1989). The latest shortcut to (20) was found by Zhao during the final revision of the present chapter.

Theorem 4.6 stemmed from Chung (1983a), which treats the Dirichlet problem for an integrable but not bounded boundary function. This topic deserves further study but is omitted here.

The extension of Theorem 4.17(ii) and (iii) from $q \in \mathcal{B}_b$ to $q \in J$, under the condition (v) of Theorem 4.19, was given in Aizenman and Simon (1982). Actually, we now know that condition (v) is equivalent to condition (i) in the same theorem, i.e. the gaugeabilty of (D, q). Then u_f is bounded and the original proof of Theorem 4.7 given in Chung and Rao (1988) can be carried over to q in J, without any change.

The uniqueness part (iv) in Theorem 4.7 was proved in Chung and Rao (1981), using Ito's stochastic calculus for a Hölder-continuous q. This is a point of contact with the popular martingaling methods in similar problems such as the Neumann boundary value problem. A challenging question is whether this kind of technique can be extended to handle nonsmooth functions.

It is important to separate the representation theorem from the boundary value problem. Historically, George Green set out around 1828 to represent a harmonic function by its boundary values, and did so by solving a particular boundary value problem which was physically evident (for a grounded conductor). Theorem 4.15 was first given in Chung's Pisa lectures mentioned above, see Mancino (1989). An earlier hint of this approach was given in Chung (1982a, page 206), Exercise 7.

It must be emphasized here that in our framework of the Brownian motion process (alias Wiener space) we have not only solved the Dirichlet boundary value problem for Schrödinger's equation but also produced a simple, elegant representation of the solution by means of the u_f in (37). There is no counterpart to this in the non-probabilistic theory, and this is a point often overlooked by non-probabilists.

The part of Theorem 4.19 stating that (i) and (iii) are equivalent was given in Chung and Rao (1981), while the equivalence of (iii) and (v) is just elementary semigroup theory. On the other hand, it takes longer to introduce the entity $\lambda(D, q)$ in an appropriate manner. This was probably why we omitted the eigen connection in the earlier paper, causing anguish to some readers who looked in vain for their 'ground state'. An instructive case of this kind occurred in Aizenman and Simon (1982), p. 212, where they lamented the absence of a 'key condition' in the prior work by Chung and Rao (1981), without realizing that their pet key is nothing but the condition (v) in Theorem 4.19 here, which is superseded by our gaugeability. Mathematicians, like politicians, can be quite parochial.

5. Conditional Brownian Motion and Conditional Gauge

5.1 Conditional Brownian Motion

In this section, we develop the notion of conditional Brownian motion introduced by Doob in 1957 (see Doob (1984)) for a general boundary theory. Here it will be used as a tool to sharpen our previous results to include the exit place as well as the exit time. Our treatment is essentially self-contained.

Let D be a domain in $\mathbb{R}^d$ ($d \geq 1$) and $h > 0$ a harmonic function in D.

We define

$$p_h^D(t; x, y) = h(x)^{-1} p^D(t; x, y) h(y), \qquad t > 0,\ x, y \in D,$$

where p^D is the transition density of the killed Brownian motion (see Section 2.2). We show that p_h^D satisfies the conditions for a transition probability density as p^D. The Borel measurability and semigroup property are trivial. The condition

$$\int_D p_h^D(t; x, y) dy \leq 1,$$

is equivalent to (1) below.

Lemma 5.1 *For any $t > 0$, $x \in D$, we have*

$$E^x[t < \tau_D;\ h(X_t)] \leq h(x). \tag{1}$$

Proof Take a sequence of bounded and regular domains $\{D_n\}$ such that $D_n \uparrow\uparrow D$.

By Theorem 1.24, for each $n \geq 1$, $y \in D_n$, we have

$$h(y) = E^y[h(X(\tau_{D_n}))].$$

Thus, if $x \in D_n$:

$$\begin{aligned} E^x[t < \tau_{D_n};\ h(X_t)] &= E^x\{t < \tau_{D_n};\ E^{X_t}[h(X(\tau_{D_n}))]\} \\ &= E^x[t < \tau_{D_n};\ h(X(\tau_{D_n}))] \\ &\leq E^x[h(X(\tau_{D_n}))] = h(x). \end{aligned}$$

Letting $n \to \infty$ in the above, we obtain inequality (1) by monotone convergence. □

According to Theorem 1.1 and the discussion preceding it, p_h^D determines a Markov process on the state space $D_\partial = D \cup \{\partial\}$, where ∂ is the extra point needed in the definition of the transition probabilities. This process is called the 'h-conditioned Brownian motion' or 'h-Brownian motion'. Its *lifetime* is defined to be $T_{\{\partial\}} = \tau_D$. The process remains at ∂ in $[T_{\{\partial\}}, \infty)$ on $\{T_{\{\partial\}} < \infty\}$. For simplicity of notation, we continue to use X_t to denote a generic random variable of the conditional process, but use P_h^x and E_h^x to indicate the associated quantities. Thus, we have for any $B \in \mathcal{B}(D)$:

$$\begin{aligned} P_h^x(X_t \in B) &= \frac{1}{h(x)} \int_B p^D(t; x, y) h(y) \mathrm{d}y \\ &= \frac{1}{h(x)} E^x\{t < \tau_D,\ X_t \in B; h(X_t)\}. \qquad (2) \\ P_h^x(X_t = \partial) &= 1 - P_h^x(X_t \in D) = 1 - \frac{1}{h(x)} E^x\{t < \tau_D;\ h(X_t)\}. \end{aligned}$$

We begin with some elementary properties of h-Brownian motion.

Proposition 5.2 *For $t > 0$, if $\Phi \geq 0$ is an $\mathcal{F}_t$-measurable function, then*

$$E_h^x[t < \tau_D;\ \Phi] = h(x)^{-1} E^x[t < \tau_D;\ \Phi \cdot h(X_t)],\ x \in D. \qquad (3)$$

Proof By a routine argument based on the monotone class theorem, it suffices to prove (3) for the special case when Φ is the indicator of the set $(X_{t_1} \in B_1, \cdots, X_{t_n} \in B_n)$, where $0 < t_1 < t_2 < \cdots < t_n \leq t$, $B_i \in \mathcal{B}(D)$, $i = 1, \cdots, n$.

For $n = 1$, (3) reduces to (2). Suppose that (3) is true for $n - 1$, $n \geq 2$. By the Markov property of h-Brownian motion followed by that of Brownian motion, we have

$$\begin{aligned} &P_h^x[t < \tau_D,\ X_{t_1} \in B_1, \cdots, X_{t_n} \in B_n] \\ &= E_h^x\{t_1 < \tau_D,\ X_{t_1} \in B_1;\ P_h^{X_{t_1}}[t - t_1 < \tau_D, \\ &\quad X_{t_2 - t_1} \in B_2, \cdots, X_{t_n - t_1} \in B_n]\} \\ &= h(x)^{-1} E^x\{t_1 < \tau_D,\ X_{t_1} \in B_1;\ E^{X_{t_1}}[t - t_1 < \tau_D, \\ &\quad X_{t_2 - t_1} \in B_2, \cdots, X_{t_n - t_1} \in B_n;\ h(X_{t - t_1})]\} \\ &= h(x)^{-1} E^x\{t < \tau_D,\ X_{t_1} \in B_1, \cdots, X_{t_n} \in B_n;\ h(X_t)\}. \end{aligned}$$

Thus (3) is true for n and the induction is complete. □

For any optional time τ, the tribe $\mathcal{F}_{\tau+}$ was defined in Section 1.2. We now define $\mathcal{F}_{\tau-}$ to be the tribe generated by $\mathcal{F}_{0+}$ and the class of sets:

$$(A_t \cap (t < \tau) :\ t \geq 0,\ A_t \in \mathcal{F}_t).$$

It is easy to see that $\mathcal{F}_{\tau-} \subset \mathcal{F}_{\tau+}$. The left tribe $\mathcal{F}_{\tau-}$ plays a fundamental role in the general theory of processes; see Chung (1982a, Section 1.3). Its emergence here is a cause for reflection by the attentive reader.

Proposition 5.3 *For any optional time T and any $\mathcal{F}_{T+}$-measurable function $\Phi \geq 0$, we have*

$$E_h^x[T < \tau_D; \Phi] = h(x)^{-1} E^x[T < \tau_D; \Phi \cdot h(X_T)]. \tag{4}$$

Proof If T is countably valued and Φ is $\mathcal{F}_T$-measurable, then (4) follows easily from (3). Next, let $T_n \downarrow\downarrow T$, where T_n is countably valued. Then

$$E_h^x[T_n < \tau_D; \Phi] = h(x)^{-1} E^x[T_n < \tau_D; \Phi \cdot h(X_{T_n})]. \tag{5}$$

Let $\{D_k\}$ be bounded domains, $D_k \uparrow\uparrow D$. Then $\tau_{D_k} < \tau_D$ and $\Lambda = \{T_n < \tau_{D_k}\} \in \mathcal{F}_{T+}$. Hence if in (5) we replace Φ by $1_\Lambda \Phi$, we see that (5) is true when D there is replaced by D_k. On $\{T < \tau_{D_k}\}$ we have $T_n < \tau_{D_k}$ for all sufficiently large values of n, and $h(X_{T_n})$ converges to $h(X_T)$ boundedly because h is bounded on D_k. Since $(T_n < \tau_{D_k}) \uparrow (T < \tau_{D_k})$, letting $n \to \infty$ we obtain (4) with D replaced by D_k. Finally letting $k \to \infty$ we obtain (4) as stands. □

The following property is the 'strong Markov property' of h-Brownian motion.

Proposition 5.4 *For any optional T, $A \in \mathcal{F}_{T+}$ and any $\mathcal{F}_{\tau_{D-}}$-measurable variable $\Phi \geq 0$, we have*

$$E_h^x[A \cap (T < \tau_D); \Phi(\theta_T)] = E_h^x[A \cap (T < \tau_D);\ E_h^{X_T}(\Phi)].$$

Remark This relation is equivalent to the following:

$$E_h^x[\phi(\theta_T \omega) | \mathcal{F}_T] = E_h^{X_T}[\phi(\omega)] \quad \text{on } \{T < \tau_D\}.$$

Proof Using the monotone class argument, we may suppose that Φ is of the special form $\Phi = \Phi_t 1_{(t<\tau_D)}$, where $\Phi_t \geq 0$ is $\mathcal{F}_t$ measurable.

On $(T < \tau_D)$, we have $\tau_D = T + \tau_D \circ \theta_T$ and $(t < \tau_D) \circ \theta_T = (T + t < \tau_D)$; it follows that

$$\Phi(\theta_T) 1_{A \cap (T < \tau_D)} = \Phi_t(\theta_T) 1_{A \cap (T+t < \tau_D)}.$$

By Proposition 5.3 applied to the optional time $T+t$, and using the strong Markov property of the Brownian motion, we have

$$\begin{aligned} &E_h^x[A \cap (T < \tau_D); \Phi(\theta_T)] \\ &\quad = h(x)^{-1} E^x[A \cap (T + t < \tau_D); \Phi_t(\theta_T) h(X_{T+t})] \\ &\quad = h(x)^{-1} E^x\{A \cap (T < \tau_D); E^{X(T)}[t < \tau_D; \Phi_t h(X_t)]\} \\ &\quad = E_h^x\{A \cap (T < \tau_D); E_h^{X(T)}[\Phi]\}. \end{aligned}$$

□

Proposition 5.5 *For each $x \in D$, P_h^x-almost every path of the h-Brownian motion is continuous in $[0, \tau_D)$.*

Proof Set

$$A = \{X(\cdot) \in \Omega : \ X(\cdot) \text{ is not continuous in } [0, \tau_D)\};$$

and for each rational number $r > 0$,

$$A_r = \{X(\cdot) \in \Omega : \ X(\cdot) \text{ is not continuous in } [0, r], \ r < \tau_D\}.$$

Then

$$A = \bigcup_r A_r.$$

We may suppose that X is separable (see Doob (1953)), then it is easy to see that $A_r \in \mathcal{F}_r$. By Proposition 5.2 and the continuity of Brownian motion, we have for each $r > 0$.

$$P_h^x(A_r) = h(x)^{-1} E^x[A_r; h(X_r)] = 0.$$

Hence we have

$$P_h^x(A) \le \sum_r P_h^x(A_r) = 0.$$

□

The following result is useful.

Proposition 5.6 $\{1_{(t<\tau_D)} h^{-1}(X_t), t > 0\}$ *is a supermartingale on* $(\Omega, \mathcal{F}_t, P_h^x)$.

Proof By the Markov property of h-Brownian motion and Proposition 5.2, we have for $0 < r < t$,

$$\begin{aligned} E_h^x(1_{(t<\tau_D)} h^{-1}(X_t) | \mathcal{F}_r) &= 1_{(r<\tau_D)} E_h^{X_r}(1_{(t-r<\tau_D)} h^{-1}(X_{t-r})) \\ &= 1_{(r<\tau_D)} h^{-1}(X_r) E^{X_r}(1_{(t-r<\tau_D)}) \\ &\le 1_{(r<\tau_D)} h^{-1}(X_r). \end{aligned}$$

□

In contrast to the easy results about τ_D under P^x, obtained in Section 1.5, the behavior of τ_D under P_h^x is a new and rather difficult problem. The following result on the lifetime of h-Brownian motion on the plane is due to Cranston and McConnell (1983).

Theorem 5.7 *If D is a domain in $\mathbb{R}^2$ and $h > 0$ is harmonic in D, then for all $x \in D$:*

$$E_h^x(\tau_D) \le C \, m(D),$$

where C is an absolute constant.

For a simpler proof based on Proposition 5.6, see Chung (1984b). This result is not generally true in $\mathbb{R}^d$ $(d \ge 3)$. A counterexample can be found in Cranston and McConnell (1983). For a Lipschitz domain, see the next section.

5.2 Life in a Lipschitz Domain

We begin this section with a description of a Lipschitz domain. A function ϕ defined on a subset S of $\mathbb{R}^{d-1}$ $(d \geq 2)$ and taking values in $\mathbb{R}^1$ is said to be '*Lipschitzian with modulus C*' iff for any ξ_1 and ξ_2 in S we have

$$|\phi(\xi_1) - \phi(\xi_2)| \leq C|\xi_1 - \xi_2|.$$

When the modulus is not specified ϕ is simply said to be Lipschitzian. This property is preserved under a translation of the origin of the coordinate system (ξ, η) in $\mathbb{R}^d$ but not necessarily under a rotation of the axes, even if the modulus is not fixed.

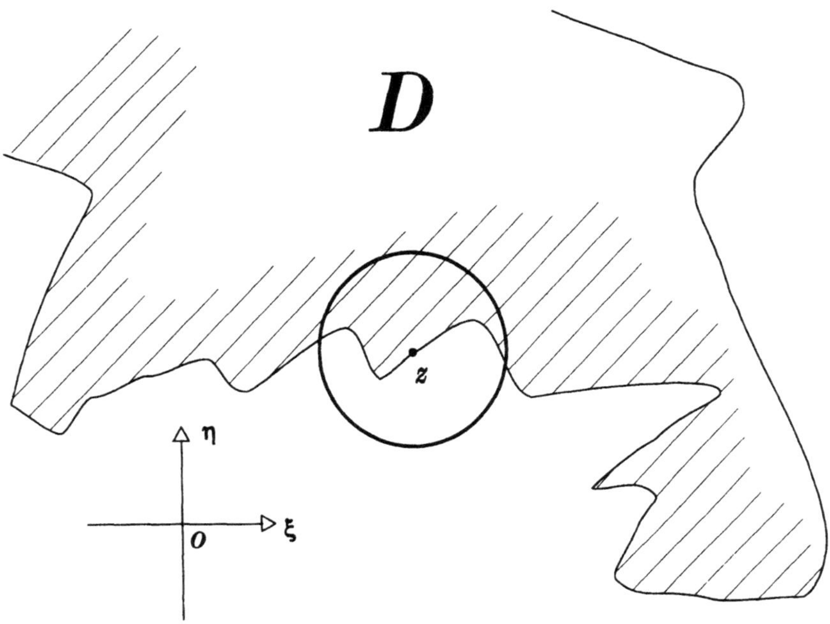

Fig. 5.1.

Let D be a domain in $\mathbb{R}^d$ $(d \geq 2)$, $z \in \partial D$, and $r > 0$. Suppose there exists a coordinate system (ξ, η) as used above, and a Lipschitz function ϕ defined on the projection of the ball $B(z, r)$ on the ξ-hyperplane, such that

$$\begin{aligned} B(z,r) \cap D &= B(z,r) \cap \{(\xi,\eta) : \phi(\xi) < \eta\}; \\ B(z,r) \cap (\partial D) &= B(z,r) \cap \{(\xi,\eta) : \phi(\xi) = \eta\}. \end{aligned} \tag{6}$$

In particular, $z = (\xi_z, \phi(\xi_z))$. Then we say that D is '*locally Lipschitzian at z*'. When this is the case for all $z \in \partial D$, we say that D is a Lipschitz domain. The number r, the coordinate system (ξ, η), and the function ϕ depend in general on

z. But if ∂D is compact, then r and the modulus C may be taken to be the same for all z. To see this,let

$$\partial D \subset \bigcup_{i=1}^{N} B\left(z_i, \frac{r_i}{2}\right),$$

where for $1 \le i \le N < \infty$, $z_i \in \partial D$; here r_i, ϕ_i, C_i and (ξ^i, η^i) will denote the objects associated with z_i in the above description. Let

$$r = \frac{1}{2} \min_{1 \le i \le N} r_i, \; C = \max_{1 \le i \le N} C_i.$$

Then (6) is true for any $z \in \partial D$ for the r defined above and one of the ϕ_i in the corresponding coordinate system (ξ^i, η^i), and all the ϕ_i have the same modulus C. This uniformity will be needed later in Chapter 6.

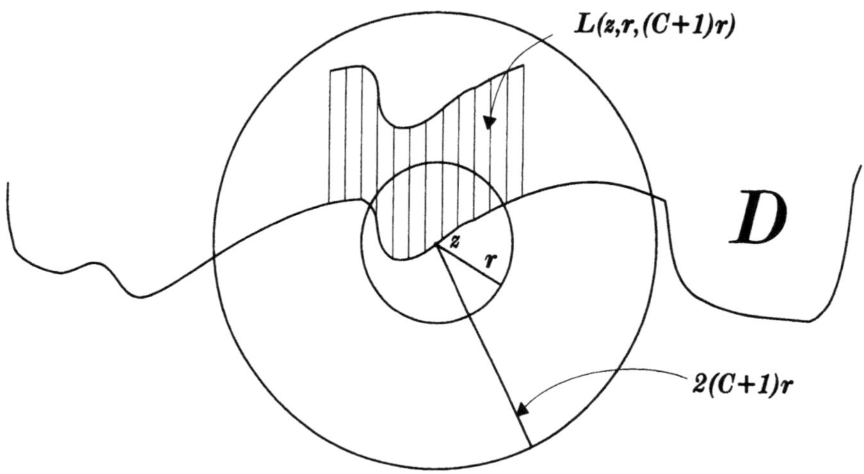

Fig. 5.2.

For later application in Section 8.4, let us define a 'cylinder set' at z, as follows. Suppose $z = (\xi_0, \phi(\xi_0))$ in one of the coordinate systems, and let

$$L(z, r, s) = \{(\xi, \eta) : \; |\xi - \xi_0| < r; \; \phi(\xi) < \eta < \phi(\xi) + s\}$$

where $r > 0$, $s > 0$. Then we can show that

$$D \cap B(z, r) \subset L(z, r, (C+1)r) \subset D \cap B(z, 2(C+1)r)$$

where C is the modulus of ϕ. Next and extremely importantly, $L(z, r, s)$ is a Lipschitz domain! This can be verified by elementary calculus and analytic geometry; however, we must leave the details to the reader. For anyone who considers this obvious, we would propose another exercise, namely: is the set in (6) locally Lipschitzian at z?

It follows from Proposition 1.22 that such a domain is regular. Furthermore, the area measure σ on ∂D is defined.

It is proved in R. Hunt and Wheeden (1968) that for a bounded Lipschitz domain D and a fixed point x_0 in D, there exists a unique function $\{K(x,z) : x \in D,\ z \in \partial D\}$ satisfying the following conditions:

(i) for any fixed $z \in \partial D$, $K(\cdot, z)$ is a strictly positive harmonic function in D;
(ii) for any fixed $x \in D$, $K(x, \cdot)$ is a continuous function in ∂D;
(iii) for any $z \in \partial D$, $w \in \partial D$, $z \neq w$,

$$\lim_{x \to w} K(x,z) = 0;$$

(iv) $K(x_0, z) = 1$ for all $z \in \partial D$.

For convenience, we extend the definition of $K(\cdot, z)$ to $\overline{D}\backslash\{z\}$ by setting $K(w,z) = 0$ for $w \in \partial D$, $w \neq z$. We call $K(\cdot,\cdot)$ the *kernel function* of D. It can be defined by

$$K(x,z) = \left\{ \frac{\mathrm{d}H(x,\cdot)}{\mathrm{d}H(x_0,\cdot)} \right\}(z), \tag{7}$$

where $H(x,\cdot)$ is the harmonic measure on ∂D; see (1.37). It can be shown that the kernel function is exactly Martin's kernel in the general case introduced in Martin (1941), so that we can identify the Martin boundary with the Euclidean boundary ∂D in the Lipschitz case. Using Martin's definition, the kernel function is also given by

$$K(x,z) = \lim_{\substack{y \to z \\ y \in D}} \frac{G(x,y)}{G(x_0,y)} \tag{8}$$

where G is the Green function for D.

As a special case of Martin's general representation theorem, for any strictly positive harmonic function h in D, we have a unique Borel measure μ on ∂D such that

$$h(x) = \int_{\partial D} K(x,z)\mu(\mathrm{d}z), \ x \in D. \tag{9}$$

This result may be used to reduce the study of a general h-Brownian motion to the special case where $h = K(\cdot, z)$, $z \in \partial D$.

Proposition 5.8 *The kernel function $K(\cdot,\cdot)$ of a bounded Lipschitz domain D is jointly continuous on $D \times \partial D$.*

Proof For any $x \in D$ and $z \in \partial D$, take a ball $B \equiv B(x,r) \subset\subset D$. For any $\{x_n\} \subset D$ and $\{z_n\} \subset \partial D$ with $x_n \to x$ and $z_n \to z$, we may assume that $\{x_n\} \subset B(x, r/2) \subset\subset B$. Then we have by condition (i) and (1.25) for each $m \geq 1$, $n \geq 1$,

$$K(x_n, z_m) = \int_{\partial B} K_B(x_n, u) K(u, z_m)\sigma(\mathrm{d}u), \tag{10}$$

where $K_B(\cdot,\cdot)$ is the Poisson kernel for B given in (1.24). Recall for each $u \in \partial B$, $K_B(\cdot, u) \in H_+(B)$; for each $z \in \partial D$, $K(\cdot, z) \in H_+(D)$; and ∂B is compact in D.

Hence by Harnack's inequality applied twice, we see that the function of u in the integrand in (10) is dominated by $CK_B(x_1, u)K(u_0, z_m)$ where C is a constant independent of m and u_0 is a fixed point on ∂B. Since $K_B(x_1, u)$ is bounded in $u \in \partial B$, and $K(u_0, z_m)$ is bounded in m by (ii), as $m \to \infty$ and $n \to \infty$, we can invoke the bounded convergence theorem as well as the continuity of $K_B(\cdot, u)$ and $K(u, \cdot)$ to obtain the limit as

$$\int_{\partial B} K_B(x, u)K(u, z)\sigma(\mathrm{d}u) = K(x, z).$$

□

Remark For any fixed $z \in \partial D$, $K(\cdot, z)$ is unbounded in D. To prove this, suppose that $M \equiv \sup_{x \in D} K(x, z) < \infty$. Take $R > 0$ such that $D \subset B(z, R)$. For each $x \in D$ and $0 < r < |x - z|$, since $K(\cdot, z)$ is harmonic in $D_1 = D \backslash \overline{B(z, r)}$ and continuous on $\overline{D}_1$, we have by Theorem 1.24(b) and condition (iii) above,

$$\begin{aligned} K(x, z) &= E^x[K(X(\tau_{D_1}), z)] \\ &\leq M\, P^x[X(\tau_{D_1}) \in \partial B(z, r)] \\ &\leq M\, P^x[T_{\partial B(z,r)} < T_{\partial B(z,R)}] \\ &= M \frac{g(x) - g(R)}{g(r) - g(R)} \to 0, \quad \text{as } r \to 0, \end{aligned}$$

where g is given in (3.4). For the last equation see e.g. Chung (1982a) pp. 168–169. Hence $K(\cdot, z) \equiv 0$ in D, which is a contradiction to (i).

We note that we have used the same symbol $K(\cdot, \cdot)$ for the Poisson kernel given in Section 1.4, which is a special case of the kernel function introduced here but with a different normalization. In other words, the two differ only by a constant factor.

Now let D be a bounded Lipschitz domain in $\mathbb{R}^d$, $d \geq 2$. For each $z \in \partial D$, the $K(\cdot, z)$-conditioned Brownian motion is defined by virtue of (i) above. It will be called the z-Brownian motion in what follows, and the associated probability and expectation will be denoted by P_z^x and E_z^x. For $d = 2$, its lifetime is almost surely finite by Theorem 5.7, without the Lipschitz assumption. For $d \geq 3$, a recent result due to Cranston (1985) states that (6) is true for a bounded Lipschitz D if the right member is replaced by a constant $C(D)$ depending only on D. We shall give a new proof of this result in Theorem 6.14. It follows that in this case also, the lifetime of the z-Brownian motion is almost surely finite.

Next we investigate the behavior of the z-Brownian motion $X(t)$ as $t \uparrow\uparrow \tau_D$. It is essential to understand that Proposition 5.5 gives no information whatsoever on this score. We do not yet know if the left limit $X(\tau_D-)$ exists. More subtly perhaps, it is by no means obvious that if this limit does exist it should belong to $\mathbb{R}^d \backslash D$, despite the meaning of τ_D. The following result is contained in a more general theorem due to Doob (see Doob (1984)); however, to deduce it from the latter, it would be necessary to prove that in the Lipschitz case the Martin boundary coincides with the Euclidean one. Our proof below is direct and simpler, and may well be new.

Theorem 5.9 *For every $x \in D$ and $z \in \partial D$:*

$$\begin{aligned} P_z^x\{\tau_D < \infty\} &= 1; \\ P_z^x\{\lim_{t \uparrow\uparrow \tau_D} X(t) = z\} &= 1. \end{aligned} \tag{11}$$

Before proceeding to the proof, let us observe that z does not belong to the state space D_∂ of the z-Brownian motion but to the boundary ∂D of D in the usual Euclidean topology of $\mathbb{R}^d$. In Doob's general result cited above, without the Lipschitz assumption, the limit exists in a new topology and belongs to the Martin boundary of D.

Proof of Theorem 5.9 Let $z \in \partial D$, $r_n \downarrow 0$, $B_n = B(z, r_n)$, $D_n = D \backslash \overline{B}_n$ and set

$$T_n = T_{B_n}, \quad R_n = \tau_{B_n \cap D}.$$

We may suppose that $x \in D_n$ for $n \geq 1$. By property (iii) of the kernel K, $K(\cdot, z)$ can be continuously extended onto $\overline{D}_n$ by setting $K(w, z) = 0$ for $w \in \partial D_n \backslash D$. Since $K(\cdot, z)$ is harmonic in D_n, we have by Theorem 1.24 and Proposition 5.3:

$$\begin{aligned} K(x, z) &= E^x\{K(X(\tau_{D_n}), z)\} \\ &= E^x\{T_n < \tau_D;\ K(X(T_n), z)\} = K(x, z) P_z^x\{T_n < \tau_D\}. \end{aligned}$$

It follows that for all $n \geq 1$ we have

$$P_z^x\{T_n < \tau_D\} = 1. \tag{12}$$

This shows that almost every path of the z-Brownian motion hits an arbitrary neighborhood of z before leaving D, but falls short of the conclusion in (11).

The function $K(\cdot, z)$ being bounded on $(\partial B_k) \cap D$ by continuity, let M_k denote a bound. Applying Propositions 5.3 and 5.4 twice, we have the following estimate for all $k < n$:

$$\begin{aligned} & P_z^x\{T_n < \tau_D;\ R_k \circ \theta_{T_n} < \tau_D\} \\ &= E_z^x\{T_n < \tau_D;\ P_z^{X(T_n)}[R_k < \tau_D]\} \\ &= \frac{1}{K(x, z)} E^x\{T_n < \tau_D;\ K(X(T_n), z) P_z^{X(T_n)}[R_k < \tau_D]\} \\ &= \frac{1}{K(x, z)} E^x\{T_n < \tau_D;\ E^{X(T_n)}[R_k < \tau_D;\ K(X(R_k), z)\} \\ &\leq \frac{M_k}{K(x, z)} P^x\{T_n < \tau_D\}. \end{aligned} \tag{13}$$

In the above we need the fact that $\{R_k < \tau_D\} \in \mathcal{F}_{\tau_{D-}}$ in order to apply Proposition 5.4. It is a fundamental property of Brownian motion in $\mathbb{R}^d$, $d \geq 2$, that every singleton is a polar set, i.e. it is almost surely never hit; see Chung (1982a, page 146). Applying this result to the singleton $\{z\}$, recalling the meaning of T_n and using the continuity of the (unconditioned!) Brownian path, we infer that

$$\begin{aligned}\overline{\lim}_{n\to\infty} P^x\{T_n < \tau_D\} &\le P^x\{\lim_{n\to\infty} T_n \le \tau_D\} \\ &\le P^x\{T_{\{z\}} \le \tau_D\} = 0,\end{aligned}$$

because $P^x\{\tau_D < \infty\} = 1$. It then follows from (13) that the left-hand side there converges to zero as $n \to \infty$ for each k. Therefore there exists a subsequence $\{n_j\}$ such that

$$\sum_{j=1}^{\infty} P_z^x\{T_{n_j} < \tau_D;\ R_k \circ \theta_{T_{n_j}} < \tau_D\} < \infty,$$

and consequently by the Borel–Cantelli lemma we have

$$P_z^x\{\limsup_j [T_{n_j} < \tau_D;\ R_k \circ \theta_{T_{n_j}} < \tau_D]\} = 0.$$

Together with (12) this implies that for P_z^x-a.e. ω, there exists an integer $N(\omega) < \infty$ such that

$$X_t(\omega) \in B(z, r_k) \text{ for all } t \in [T_{N(\omega)}(\omega), \tau_D(\omega)).$$

For each k let $N(k)$ be the smallest N for which the above is true. Then $T_{N(k)} \uparrow \tau_D$; otherwise, we would have $X_t = z$ for all $t \in [\lim_{k\to\infty} T_{N(k)}, \tau_D)$, which is impossible because $z \notin D_\partial$. This proves that $X_t \to z$ as $t < \tau_D$, $t \to \tau_D$. □

Next, let D be a regular bounded domain in $\mathbb{R}^d$, $d \ge 2$, and $y \in D$. Then $G(\cdot, y)$ is harmonic in $D\backslash\{y\}$ by the Corollary to Theorem 2.5, and is strictly positive there by Theorem 2.6(i). Hence the $G(\cdot, y)$-conditioned Brownian motion is defined on the state space $(D\backslash\{y\})\cup\{\partial\}$, with lifetime $\tau_{D\backslash\{y\}}$. It will be referred to as the y-Brownian motion in what follows and the associated probability and expectation will be denoted by P_y^x and E_y^x, respectively. The analogue of Theorem 5.9 is true without the Lipschitz assumption, with a supplement on the lifetime as follows.

Theorem 5.10 *Let D be a bounded regular domain in $\mathbb{R}^d$, $d \ge 2$. For each $y \in D$, $x \in D\backslash\{y\}$:*

$$E_y^x\{\tau_{D\backslash\{y\}}\} < \infty \tag{14}$$

$$P_y^x\{\lim_{t\uparrow\uparrow\tau_{D\backslash\{y\}}} X(t) = y\} = 1. \tag{15}$$

Proof The proof of (15) is completely analogous to that of (11) because $G(\cdot, y)$ has the required properties used there for $K(\cdot, z)$; it is therefore left to the reader as an exercise.

The proof of (14) is pedestrian. Writing ζ for $\tau_{D\backslash\{y\}}$ in what follows, we have

$$E_y^x\{\zeta\} = \int_0^\infty P_y^x\{t < \zeta\}\mathrm{d}t = \frac{1}{G(x,y)}\int_0^\infty E^x\{t < \zeta;\ G(X_t, y)\}\mathrm{d}t.$$

But under P^x, $\zeta = \tau_D$ a.s. because $\{y\}$ is a polar set. Hence the last integral above is just

$$E^x\left\{\int_0^{\tau_D} G(X_t, y)\mathrm{d}t\right\} = \int_D G(x,z)G(z,y)\mathrm{d}z \tag{16}$$

by (2.13). It remains to estimate the right member above. In fact, we can show that it is bounded by a constant which depends only on D and $|x-y|$. For $d \geq 3$, let $\rho = \frac{1}{2}|x-y|$ and split the integral $\int_D$ in (16) into $\int_{D_1}$ and $\int_{D_2}$ where $D_1 = B(y,\rho) \cap D$ and $D_2 = D \backslash D_1$. For $z \in D_1$, we have $|x-z| \geq |x-y| - |y-z| \geq \rho$, then by Theorem 2.6,

$$G(x,z) \leq g(x,z) \leq C(\rho)$$

where $C(\rho)$, like $C_1(\rho)$ and $C_2(\rho)$ below, is a constant depending only on ρ. Hence

$$\int_{D_1} \leq C(\rho)\int_{B(y,\rho)} g(z,y)\mathrm{d}z \leq C_1(\rho).$$

For $z \in D_2$, $G(z,y) \leq g(z,y) \leq C_2(\rho)$; hence

$$\int_{D_2} \leq C_2(\rho)\int_D G(x,z)\mathrm{d}z = C_2(\rho)E^x\{\tau_D\}.$$

Therefore by Theorem 1.17, the quantity in (16) is bounded by a constant which depends only on ρ and $m(D)$. For $d = 2$, the proof is similar and uses (2.23). □

By virtue of Theorems 5.9 and 5.10, we can redefine the z- (or y-) conditioned Brownian motion in a more fitting way as follows. The z-Brownian motion is defined on the state space $D \cup \{z\}$, and takes the value z at and after its lifetime τ_D. The process will then have continuous paths on $[0,\infty)$, almost surely. Similarly, the y-Brownian motion is defined on the state space D and takes the value y at and after its lifetime $\tau_{D\backslash\{y\}}$. It has continuous paths in $[0,\infty)$, almost surely. In other words, rather than killing the process at its lifetime, we may let it stop and stay there forever. The same modification is possible for any h-conditioned Brownian motion, by using the representation given in (9), but we need not dwell on this.

The next proposition shows that the z-Brownian motion is simply the Brownian motion conditioned to exit D at z. This interpretation is the key to our applications.

Proposition 5.11 *Let D be a bounded Lipschitz domain in $\mathbb{R}^d$, $d \geq 1$. For any $\mathcal{F}_{\tau_D -}$-measurable function $\Phi \geq 0$ and $x \in D$, we have*

$$E^x[\Phi | X(\tau_D)] = E^x_{X(\tau_D)}[\Phi].$$

Remark The left member above is well defined since $\Phi \geq 0$, but both members in the equation above may equal $+\infty$.

Proof As in the proof of Proposition 5.4, we may assume that $\Phi = 1_{(t<\tau_D)}\Phi_t$, $t > 0$, where Φ_t is $\mathcal{F}_t$-measurable and bounded. Then $E^x_z[\Phi]$ is Borel measurable in z, for fixed x and Φ, by its explicit expression (Proposition 5.2). We need to show that for any $A \in \mathcal{B}(\partial D)$,

$$E^x[X(\tau_D) \in A;\ \Phi] = E^x\{X(\tau_D) \in A;\ E^x_{X(\tau_D)}[\Phi]\}. \tag{17}$$

By Proposition 5.2 and (7), the right-hand side of (17) is equal to

$$\begin{aligned}\int_A E^x_z[\Phi]H(x,\mathrm{d}z) &= \int_A E^x_z[\Phi]K(x,z)H(x_0,\mathrm{d}z)\\ &= \int_A E^x[t<\tau_D;\ \Phi_t K(X_t,z)]H(x_0,\mathrm{d}z)\\ &= E^x[t<\tau_D;\ \Phi_t \int_A K(X_t,z)H(x_0,\mathrm{d}z)]\\ &= E^x[t<\tau_D;\ \Phi_t H(X_t,A)]\\ &= E^x[t<\tau_D;\ \Phi_t E^{X_t}(X(\tau_D)\in A)]\\ &= E^x[X(\tau_D)\in A;\ 1_{(t<\tau_D)}\Phi_t]\end{aligned}$$

which is the left-hand side of (17). □

Proposition 5.11 has the following consequence.

Proposition 5.12 *Let D be a bounded Lipschitz domain and $q \in J_{\mathrm{loc}}$. Then for any $f \in \mathcal{B}_+(\partial D)$, we have*

$$E^x[e_q(\tau_D)f(X(\tau_D))] = \int_{\partial D} f(z)E^x_z[e_q(\tau_D)]H(x,\mathrm{d}z), \tag{18}$$

where H is the harmonic measure defined in (1.37).

Proof Under P^x_z, we have $\tau_D < \infty$, hence $e_q(\tau_D)$ is well defined; see the beginning of Section 4.1. Since

$$\int_0^{\tau_D} q(X_s)\mathrm{d}s = \int_0^\infty 1_{(s<\tau_D)}q(X_s)\mathrm{d}s,$$

it is easy to see that $e_q(\tau_D)$ is $\mathcal{F}_{\tau_D-}$-measurable. Hence Proposition 5.11 is applicable, and we have

$$\begin{aligned}&E^x[e_q(\tau_D)f(X(\tau_D))] = E^x\{E^x[e_q(\tau_D)|X(\tau_D)]f(X(\tau_D))\}\\ &= E^x\{E^x_{X(\tau_D)}[e_q(\tau_D)]f(X_{\tau_D})\} = \int_{\partial D} f(z)E^x_z[e_q(\tau_D)]H(x,\mathrm{d}z).\end{aligned}$$

□

The function $u(x,z) = E^x_z[e_q(\tau_D)]$, $(x,z) \in D \times \partial D$, will be called a 'conditional gauge' for (D,q). It was first introduced by Falkner (1983) for a bounded q. In view of (18), we introduce the following generalization of the harmonic measure:

$$H_q(x, \mathrm{d}z) = u(x, z)H(x, \mathrm{d}z). \tag{19}$$

which we call the q-harmonic measure.

Let us consider the special case where D is a bounded C^2 domain in $\mathbb{R}^d$ $(d \geq 2)$. Let $G(\cdot, \cdot)$ be the Green function for D. For a bounded C^2 domain D, it is known (see e.g. Gilbarg and Trudinger (1977)) that for any $x \in D$ and $z \in \partial D$, $\frac{\partial}{\partial n_z} G(x, z) > 0$ exists and is continuous on $D \times \partial D$, where $\frac{\partial}{\partial n_z}$ denotes the inner normal derivative. By a result due to Widman, Theorem 2.3 in (Widman (1967)), for $d \geq 3$, we have

$$G(x, y) \leq \frac{C\rho(x)\rho(y)}{|x - y|^d}, \ (x, y) \in D \times D, \tag{20}$$

where $\rho(x)$ denotes the distance from x to ∂D and C is a constant depending on D only. For $d = 2$, (20) is still true; see Theorem 6.23 in Chapter 6.

Now let

$$\overline{K}(x, z) = \frac{1}{2} \frac{\partial}{\partial n_z} G(x, z), \quad x \in D, \ z \in \partial D.$$

For any $z \in \partial D$ and $D_0 \subset\subset D$, we take $\{y_n\} \subset D \backslash \overline{D_0}$ such that $y_n \to z$ along the inner normal. Then

$$\overline{K}(x, z) = \frac{1}{2} \lim_{n \to \infty} \frac{G(x, y_n)}{\rho(y_n)}. \tag{21}$$

Since $\{G(\cdot, y_n)/\rho(y_n), \ n \geq 1\}$ is a sequence of harmonic functions uniformly bounded in D_0 by (20), it follows easily from Theorem 1.9 that the limit is harmonic in D_0. Thus $\overline{K}(\cdot, \ z)$ is harmonic in D.

We recall Green's identity:

$$\int_{\partial D} \frac{\partial u}{\partial n_z}(z)\sigma(\mathrm{d}z) = -\int_D \Delta u(y)\mathrm{d}y \tag{22}$$

for each $u \in C^1(\overline{D}) \cap C^2(D)$. By Theorem 2.5, $G(x, \cdot) - g(x, \cdot)$ is harmonic in D, hence we have

$$\int_{\partial D} \frac{\partial}{\partial n_z}[G(x, z) - g(x, z)]\sigma(\mathrm{d}z) = 0. \tag{23}$$

For any ball $B(x, \delta) \subset\subset D$, we have

$$\int_{\partial D} \frac{\partial g}{\partial n_z}(x, z)\sigma(\mathrm{d}z) = \int_{\partial B(x,\delta)} \frac{\partial g}{\partial n_z}(x, z)\sigma(\mathrm{d}z) = 2$$

as in the proof of Proposition 2.8. Collecting these results we obtain for all $x \in D$,

$$\int_{\partial D} \overline{K}(x, z)\sigma(\mathrm{d}z) = 1.$$

By (20) and (21) we have

$$\overline{K}(x,z) \le \frac{C\rho(x)}{|x-z|^d}. \tag{24}$$

Hence, for any $w \in \partial D$, $\delta > 0$, if $x \in D$, $x \to w$, then

$$\int_{\partial D \setminus B(w,\delta)} \overline{K}(x,z)\sigma(\mathrm{d}z) \to 0.$$

Thus the function $\overline{K}(\cdot,\cdot)$ on $D \times \partial D$ satisfies conditions (i)–(iii) of Theorem 1.12 and is therefore the Poisson kernel of D denoted by K there. By (1.37) and (1.38) we have

$$P^x[X(\tau_D) \in \mathrm{d}z] = \overline{K}(x,z)\sigma(\mathrm{d}z). \tag{25}$$

Summing up these results we obtain

Proposition 5.13 *If D is a bounded C^2 domain in $\mathbb{R}^d$, $d \ge 2$, then*

$$\overline{K}(x,z) = \frac{1}{2}\frac{\partial}{\partial n_z}G(x,z)$$

is the Poisson kernel of D, and (25) is true.

We have shown above that $\overline{K}(\cdot,\cdot)$ also satisfies conditions (i)–(iii) in the definition of the kernel function. Therefore, by the uniqueness of the kernel function for a bounded domain with a given point x_0, we have

$$K(x,z) = \frac{\overline{K}(x,z)}{\overline{K}(x_0,z)}, \quad x \in D, \ z \in \partial D. \tag{26}$$

When $D = B(a,r)$ and $x_0 = a$, we have by Theorem 1.13 (where the K is the $\overline{K}$ here):

$$K(x,z) = \frac{r^{d-2}(r^2 - |x-a|^2)}{|x-z|^d} = \sigma_{d-1}(r)\overline{K}(x,z), \tag{27}$$
$$x \in B(a,r), \quad z \in \partial B(a,r),$$

where $\sigma_{d-1}(r) = \sigma(S(a,r)) = (2\pi^{d/2}/\Gamma(d/2))r^{d-1}$ is given in Section 1.1.

5.3 Conditional Gauge for a Small Ball

We begin with a number of estimates of the Green function for a ball. We recall the expression for the Green function for a ball $B = B(a,r)$ given in Section 2.3:

$$G(x,y) = \begin{cases} C_d|x-y|^{-(d-2)} - C_d r^{d-2}|y-a|^{-(d-2)}|x-y^*|^{-(d-2)}, & d \ge 3, \\ \frac{1}{\pi}\ln|x-y|^{-1} - \frac{1}{\pi}\ln r|y-a|^{-1}|x-y^*|^{-1}, & d = 2, \end{cases} \tag{28}$$

where $x \in B$, $y \in B\setminus\{a\}$, and $y^* = a + \frac{r^2}{|y-a|^2}(y-a)$.

We may set $a = 0$; then $\rho(x) = \rho(x, \partial B) = r - |x|$. The mapping $x \to x - a$ does not affect $\rho(x)$ and $x - y$ in the estimates below. Next we set

$$f(x, y) = r^{-1}|y|\,|x - y^*| > 0, \tag{29}$$

and compute the key formula, where $(x, y) = \sum_{i=1}^{d} x_i y_i$:

$$\begin{aligned} f(x,y)^2 &= \frac{|y|^2}{r^2}\left|\frac{r^2 y}{|y|^2} - x\right|^2 = r^2 - 2(x,y) + \frac{|x|^2|y|^2}{r^2} \\ &= |x-y|^2 + \frac{(r^2 - |x|^2)(r^2 - |y|^2)}{r^2}. \end{aligned} \tag{30}$$

For $d \geq 3$, the following estimate is a special case of the Widman's inequality (20) (see Widman (1967)) Here we give a direct and elementary proof for a ball. For a more complete treatment of this case, see Chung (1984b).

Proposition 5.14 *Let G be the Green function for a ball B in $\mathbb{R}^d$ $(d \geq 2)$. We have*

$$G(x, y) \leq C\frac{\rho(x)\rho(y)}{|x - y|^d}, \quad (x, y) \in B \times B, \tag{31}$$

where C is a constant depending only on d.

Proof We may suppose that $x \neq y$, since otherwise (31) is trivial. For $d \geq 3$, we write

$$C_d^{-1} G(x, y) = \frac{f(x,y)^{d-2} - |x-y|^{d-2}}{|x-y|^{d-2} f(x,y)^{d-2}}. \tag{32}$$

Since $f(x, y) > |x - y|$ by (30) and $r^2 - |x|^2 \leq 2r\rho(x)$, the numerator in (32) is less than or equal to

$$\begin{aligned} (d-2)[f(x,y) - |x-y|]f(x,y)^{d-3} &\leq (d-2)[f(x,y)^2 - |x-y|^2]f(x,y)^{d-4} \\ &\leq 4(d-2)\rho(x)\rho(y)f(x,y)^{d-4}. \end{aligned}$$

Substituting this into (32) and using $f(x, y) > |x - y|$ again, we obtain (31).

For $d = 2$, we have by (30)

$$\begin{aligned} \ln f(x,y) &= \frac{1}{2}\ln[|x-y|^2 + r^{-2}(r^2 - |x|^2)(r^2 - |y|^2)] \\ &= \ln|x-y| + \frac{1}{2}\ln\left[1 + \frac{(r^2 - |x|^2)(r^2 - |y|^2)}{r^2|x-y|^2}\right]. \end{aligned} \tag{33}$$

Hence by (28) and (29),

$$G(x, y) \leq \frac{1}{2\pi}\frac{(r^2 - |x|^2)(r^2 - |y|^2)}{r^2|x-y|^2} \leq \frac{2}{\pi}\frac{\rho(x)\rho(y)}{|x-y|^2}. \qquad \square$$

Let K be as in (27) with $a = 0$. Then

$$r^{d-1}\frac{\rho(x)}{|x-z|^d} \le K(x,z) \le 2r^{d-1}\frac{\rho(x)}{|x-z|^d}. \tag{34}$$

Now set

$$G_z(x,y) = \frac{G(x,y)K(y,z)}{K(x,z)}, \quad (x,y) \in B \times B,\ z \in \partial B. \tag{35}$$

The following estimate of G_z is the key to the main results.

Proposition 5.15 *For all* $(x,y) \in B \times B$, $z \in \partial B$,

$$G_z(x,y) \le \begin{cases} A(g(x,y) + g(y,z)) & \text{if } d \ge 3; \\ Ag(x,y) & \text{if } d = 2,\ 0 < r \le 1/3; \end{cases} \tag{36}$$

where A *is a constant depending only on* d.

Proof First let us recall the notation given in Section 2.3:

$$g(x,y) = \begin{cases} C_d|x-y|^{2-d} & \text{if } d \ge 3, \\ \frac{1}{\pi}\ln\left(\frac{1}{|x-y|}\right) & \text{if } d = 2. \end{cases}$$

If $d = 2$, $0 < r \le 1/3$, then by (30),

$$f(x,y)^2 \le (2r)^2 + r^{-2}r^4 = 5r^2 < 1.$$

Hence we have by (28), $G(x,y) \le g(x,y)$. For $d \ge 3$, this is trivial. Hence by (31), we have for $d \ge 2$:

$$G(x,y) \le g(x,y) \wedge \left[\frac{C\rho(x)\rho(y)}{|x-y|^d}\right].$$

If $d \ge 3$,

$$\frac{\rho(x)\rho(y)}{|x-y|^d} = g(x,y)\frac{1}{C_d}\frac{\rho(x)\rho(y)}{|x-y|^2}.$$

If $d = 2$, g is bounded away from 0 in B because then $|x-y| < 2/3$ and $g(x,y) \ge \frac{\ln(3/2)}{\pi}$. Hence, in both cases we have

$$G(x,y) \le Cg(x,y)\left[1 \wedge \frac{\rho(x)\rho(y)}{|x-y|^2}\right]. \tag{37}$$

If $\rho(y) > 2\rho(x)$, then $|x-y| \ge \rho(y) - \rho(x) > \frac{1}{2}\rho(y)$, and so

$$\frac{\rho(x)\rho(y)}{|x-y|^2} \le \frac{4\rho(x)}{\rho(y)};$$

if $\rho(y) \le 2\rho(x)$, then $1 \le \frac{2\rho(x)}{\rho(y)}$. Hence we have

$$1 \wedge \frac{\rho(x)\rho(y)}{|x-y|^2} \le \frac{4\rho(x)}{\rho(y)},$$

and consequently by (37):

$$G(x,y) \leq 4Cg(x,y)\frac{\rho(x)}{\rho(y)}. \tag{38}$$

We have from (27):

$$\frac{K(y,z)}{K(x,z)} \leq \frac{2\rho(y)}{\rho(x)}\left|\frac{x-z}{y-z}\right|^d. \tag{39}$$

It follows from (38) and (39) that

$$G_z(x,y) \leq 8Cg(x,y)\left|\frac{x-z}{y-z}\right|^d. \tag{40}$$

On the other hand, we have by (37):

$$G(x,y) \leq C\ g(x,y)\frac{\rho(x)\rho(y)}{|x-y|^2}. \tag{41}$$

It follows from (39), (41), and the fact that $\rho(y) \leq |y-z|$ that

$$G_z(x,y) \leq 2C\ g(x,y)\left|\frac{y-z}{x-y}\right|^2\left|\frac{x-z}{y-z}\right|^d. \tag{42}$$

Combining (40) and (42), we obtain

$$\begin{aligned} G_z(x,y) &\leq 8C\ g(x,y)\left|\frac{x-z}{y-z}\right|^d\left[1 \wedge \left|\frac{y-z}{x-y}\right|^2\right] \\ &= 8C\ g(x,y)\left|\frac{x-z}{y-z}\right|^{d-2}\left[\left|\frac{x-z}{y-z}\right|^2 \wedge \left|\frac{x-z}{x-y}\right|^2\right]. \end{aligned} \tag{43}$$

Now $|x-z| \leq |x-y| + |y-z|$ implies that

$$\left|\frac{x-z}{y-z}\right| \wedge \left|\frac{x-z}{x-y}\right| \leq 1 + \left[\left|\frac{x-y}{y-z}\right| \wedge \left|\frac{y-z}{x-y}\right|\right] \leq 2.$$

Using this inequality in (43) we obtain, if $d = 2$:

$$G_z(x,y) \leq 32C\ g(x,y);$$

and if $d \geq 3$:

$$\begin{aligned} G_z(x,y) &\leq 32C\left(\frac{|x-z|}{|x-y|\,|y-z|}\right)^{d-2} \\ &\leq 32C\left(\frac{1}{|x-y|} + \frac{1}{|y-z|}\right)^{d-2} \\ &\leq 32C2^{d-2}\left(\frac{1}{|x-y|^{d-2}} + \frac{1}{|y-z|^{d-2}}\right). \end{aligned}$$

□

The next two propositions furnish the crucial inequalities. Of course a ball is a Lipschitz domain, hence we can apply the preceding results to the conditional Brownian motion in a ball.

Proposition 5.16 *For any Borel function $q: \mathbb{R}^d \to \overline{\mathbb{R}}^1$, and any ball $B = B(a,r)$ in $\mathbb{R}^d$ $(d \geq 2)$, we have*

$$\sup_{(x,z)\in B\times\partial B} E_z^x\left\{\int_0^{\tau_B} |q|(X_t)\mathrm{d}t\right\} \leq A \sup_{x\in\overline{B}} \int_B g(x,y)|q|(y)\mathrm{d}y, \tag{44}$$

if $d = 2$, $0 < r \leq 1/3$; or if $d \geq 3$; where A is a constant depending only on d.

Proof By Proposition 5.2,

$$E_z^x\{t < \tau_B,\ |q|(X_t)\} = \frac{1}{K(x,z)} E^x\{t < \tau_B;\ |q|(X_t)K(X_t,z)\}.$$

Integrating over $t \in [0,\infty)$ and using Fubini's theorem, we obtain

$$\begin{aligned} E_z^x\left\{\int_0^{\tau_B} |q|(X_t)\mathrm{d}t\right\} &= \frac{1}{K(x,z)}\int_0^\infty E^x\{t < \tau_B;\ |q|(X_t)K(X_t,z)\}\mathrm{d}t \\ &= \frac{1}{K(x,z)}\int_B G(x,y)|q|(y)K(y,z)\mathrm{d}y \\ &= \int_B G_z(x,y)|q|(y)\mathrm{d}y. \end{aligned} \tag{45}$$

Using the indicated inequalities in (36), and then taking the suprema on both sides of the resulting inequality, we obtain (44) with the constant A at most twice that in (36). □

Theorem 5.17 *Let $q \in J$. Then for each $\varepsilon > 0$, there exists a constant $r(q,\varepsilon) > 0$ such that for any ball B with radius $r \leq r(q,\varepsilon)$, we have*

$$\begin{aligned} \mathrm{e}^{-\varepsilon} &\leq \inf_{(x,z)\in B\times\partial B} E_z^x[\mathrm{e}_q(\tau_B)] \\ &\leq \sup_{(x,z)\in B\times\partial B} E_z^x[\mathrm{e}_q(\tau_B)] \leq \frac{1}{1-\varepsilon}. \end{aligned} \tag{46}$$

Proof The right member of (44) is

$$\sup_{|x-a|\leq r}\int_{|y-a|<r} g(x,y)|q|(y)\mathrm{d}y \leq \sup_{x\in\mathbb{R}^d}\int_{|y-x|<2r} g(x,y)|q|(y)\mathrm{d}y.$$

This converges to zero as $r \to 0$ by (3.5). Hence there exists $r(q,\varepsilon) > 0$ such that for any ball B with radius $r \leq r(q,\varepsilon)$ and all $(x,z) \in B \times \partial B$, we have

$$E_z^x\left\{\int_0^{\tau_B} |q|(X_t)\mathrm{d}t\right\} \leq \varepsilon. \tag{47}$$

Hence by Khas'minskii's Lemma (Lemma 3.7), we have

$$E_z^x[e_q(\tau_B)] \le E_z^x[e_{|q|}(\tau_B)] \le \frac{1}{1-\varepsilon}.$$

Note that the proof of the lemma is valid for the conditional Brownian motion since it uses only the Markov property at $t < \tau$ (= τ_B here). The reader should check this carefully because the conditional Brownian motion is not exactly like the Brownian motion!

Next, by Jensen's inequality, we have

$$\begin{aligned} E_z^x[e_q(\tau_B)] &\ge \exp\left\{E_z^x\left[\int_0^{\tau_B} q(X_t)\mathrm{d}t\right]\right\} \\ &\ge \exp\left\{-E_z^x\left[\int_0^{\tau_B} |q|(X_t)\mathrm{d}t\right]\right\} \ge \mathrm{e}^{-\varepsilon}. \end{aligned}$$

□

5.4 General Gauge Theorem

We can now extend Theorem 4.1 (Harnack inequality for bounded q) and Theorem 4.2 (Gauge Theorem) to the case $q \in J$. Let D be a domain, $q \in J$ and $f \in \mathcal{B}_+(\partial D)$. Set

$$u_f(x) = E^x[e_q(\tau_D)f(X(\tau_D))].$$

Theorem 5.18 *Let D be an arbitrary domain in $\mathbb{R}^d$, $d \ge 1$. If $u_f \not\equiv \infty$ in D, then u_f is bounded in each compact subset C of D. There exists a constant $A = A(D, C, q)$ such that for all $f \in B_+(\partial D)$, we have*

$$\sup_{x\in C} u_f(x) \le A \inf_{x\in C} u_f(x). \tag{48}$$

Proof We may suppose that there exists $x_0 \in C$ such that $u_f(x_0) < \infty$. Then by the strong Markov property, for any $B = B(x_0, r) \subset\subset D$, we have

$$u_f(x_0) = E^{x_0}[e_q(\tau_B)u_f(X(\tau_B))]. \tag{49}$$

We take r to be so small that (46) in Theorem 5.17 holds, say with $\varepsilon = 1/2$, so that for all $(x, z) \in B \times \partial B$:

$$\frac{1}{2} \le E_z^x[e_q(\tau_B)] \le 2. \tag{50}$$

In particular, this applies to the $B = B(x_0, r)$ in (49). We set $u(x, z) = E_z^x[e_q(\tau_B)]$. Then by Proposition 5.12 and (25),

$$u_f(x_0) = \int_{\partial B} u(x_0, z) K(x_0, z) u_f(z) \sigma(\mathrm{d}z), \tag{51}$$

where K is the classical Poisson kernel given in (1.24), namely the $\overline{K}$ in (27). Let $B_1 = B(x_0, \frac{r}{2})$. It is clear from (27) that there are two constants $0 < A_1 < A_2 < \infty$ depending only on d such that for all $x \in B_1$, $z \in \partial B$:

$$r^{1-d} A_1 \leq K(x, z) \leq r^{1-d} A_2.$$

It now follows that

$$u_f(x_0) \geq \frac{1}{2} r^{1-d} A_1 \int_{\partial B} u_f(z) \sigma(\mathrm{d}z).$$

On the other hand, for all $x \in B_1$, we have

$$\begin{aligned} u_f(x) &= \int_{\partial B} u(x, z) K(x, z) u_f(z) \sigma(\mathrm{d}z) \\ &\leq 2r^{1-d} A_2 \int_{\partial B} u_f(z) \sigma(\mathrm{d}z) \leq \frac{4A_2}{A_1} u_f(x_0). \end{aligned}$$

This is the analogue of (4.12). The rest of the proof follows from a chain argument similar to that described in detail in the proof of Theorem 4.1. For this we need (50) for all $B(a, r)$, which is the analogue of (4.5). □

Corollary to Theorem 5.18 *If $f \in \mathcal{B}(\partial D)$, and $u_{|f|} \not\equiv \infty$ in D, then u_f is continuous and q-harmonic in D.*

Proof It follows from the theorem that the present hypothesis implies that $u_{|f|}$ is bounded in $\overline{B}$ for any ball $B \subset\subset D$. Therefore u_f is likewise bounded. We may take B to be small enough that (B, q) is gaugeable (by Proposition 4.11). Then since u_f is bounded on ∂B, we have $u_f \in C(B)$ and u_f is q-harmonic in B by Theorem 4.7(ii). Since q-harmonicity is a local property, this yields the conclusion of the corollary. □

Remark The continuity of u_f also follows from (51), since $u(\cdot, z)$ is bounded and continuous by Theorem 7.11; however, that is a harder result.

Theorem 5.19 *Let D be a domain with $m(D) < \infty$, let $q \in J$, and let $f \in L_+^\infty(\partial D)$. If $u_f \not\equiv \infty$ in D, then it is bounded in $\overline{D}$.*

Proof There exists a sequence of compact sets $K_n \uparrow\uparrow D$ such that for each $n \geq 1$, $E_n \equiv D \backslash K_n$ is either a subdomain if $d \geq 2$, or the union of two disjoint subdomains if $d = 1$. Since $E_n \downarrow \emptyset$, it follows from Theorem 1.17 that

$$\|G_{E_n} 1\|_\infty = \sup_{x \in E_n} E^x(\tau_{E_n}) \leq A_d m(E_n)^{2/d} \to 0$$

as $n \to \infty$. Then by Theorem 4.3(i), for any $0 < \varepsilon \leq 1$ there exists an integer $N \geq 1$, such that

$$\|G_{E_N}(|q|)\|_\infty < \frac{\varepsilon}{2}.$$

Set $E = E_N$, then by Lemma 3.7 we have

$$\sup_{x \in \overline{E}} E^x[e_{|q|}(\tau_E)] \leq \frac{1}{1 - \frac{\varepsilon}{2}} \leq 1 + \varepsilon. \tag{52}$$

Now u_f is bounded on the compact set $K_N = D \backslash E$ by Theorem 5.18. Its boundedness on D follows by exactly the same argument as in Theorem 4.2, using (52) which is the old (4.14). □

Remark The proof of Theorem 5.19 shows that it holds if D is Green-bounded and satisfies the additional assumption that $\inf_K \|G_{D\backslash K}\| = 0$, where K ranges over all compact subsets of D. This seems to be an *ad hoc* assumption but Sturm (1989) showed that (for a Green-bounded D) it is equivalent to either that $\lim_{|x|\to\infty} G_D 1(x) = 0$ or that G_D is a compact operator on $L^\infty(D)$. On the other hand, for the Green-bounded D in Example 3 of Section 4.2, Zhao (1990a) gave an example of bounded q such that the gauge theorem for (D, q) is not true.

The case of Theorem 5.19 for $f \equiv 1$ will be referred to as the *Gauge Theorem*, i.e.: $u_1 \not\equiv \infty$ in D implies u_1 is bounded in $\overline{D}$, or in other words (D, q) is gaugeable. Thus, the theorem is true if $q \in J$ and $m(D) < \infty$, in any dimension. The remarkable feature of this result is the absence of the condition $q \in L^1(D)$, which we know is in general false for an unbounded D; see Examples 1 and 2 in Section 4.2. If we are willing to impose this condition then we have the following theorem.

Theorem 5.20 *Suppose that $q \in J \cap L^1(D)$, then the assertion of Theorem 5.19 is true for any domain D in $\mathbb{R}^d$, $d \geq 3$, or any Green-bounded domain in $\mathbb{R}^2$.*

Proof Let D be any domain in $\mathbb{R}^d$, $d \geq 3$. There exists $\alpha > 0$ such that

$$\sup_x \int_{B(x,\alpha)} g(x,y) 1_D |q|(y) \mathrm{d}y < \frac{1}{4}, \tag{53}$$

since $q \in J$, where g is given in (2.17). Since $1_D q \in L^1(\mathbb{R}^d)$, there exists an open subset E of D such that $K = D \backslash E$ is compact, and

$$\int_E |q(y)| \mathrm{d}y < \frac{\alpha^{d-2}}{4C_d}. \tag{54}$$

Since for $|y - x| \geq \alpha$ we have by Theorem 2.6(ii):

$$G_E(x,y) \leq g(x,y) \leq \frac{C_d}{\alpha^{d-2}}, \tag{55}$$

it follows that

$$\int_{E \cap B(x,\alpha)^c} G_E(x,y) |q(y)| \mathrm{d}y < \frac{1}{4}.$$

Combining the above estimates we obtain

$$\|G_E|q|\| < \frac{1}{2}.$$

Therefore the argument for Theorem 5.19 also applies in this case.

For a Green-bounded domain in $\mathbb{R}^2$, the proof is similar. Instead of (55) we use (2.23):

$$G_E(x,y) \leq \frac{1}{\pi} \ln^+ \frac{1}{|x-y|} + C$$

for $|x-y| \geq \alpha$. □

Observe that the condition that $q \in L^1(D)$ cannot be omitted because it is not true that for a Green-bounded D, there exists a compact subset K such that the Green bound $\|G_{D\setminus K}1\|_\infty$ can be made as small as we wish. One example of this is the D in Example 3 of Section 4.2.

5.5 Continuity of Weak Solutions

In this book we define a q-harmonic function to be a continuous weak solution of the Schrödinger equation; see Section 4.5. In our solution of the Dirichlet boundary value problem (Theorem 4.7), the probabilistic expression u_f in (4.37) is proved to be continuous. From the other direction, the representation of a given q-harmonic function (Theorem 4.15) requires its continuity because its probabilistic expression is continuous as u_f is. Nevertheless, the following result due to Aizenman and Simon (1982) is interesting as it calls for a further probing of our previous methods. The new proof given below combines analytic and probabilistic arguments in a rather delicate manner.

Theorem 5.21 *Let D be an arbitrary domain, and ϕ a weak solution of the Schrödinger equation in D:*

$$\left(\frac{\Delta}{2} + q\right)\phi = 0. \tag{56}$$

Then there exists a continuous function v in D such that $\phi = v$, m-a.e. in D.

Proof Since we are dealing with local properties, we may suppose that D is a ball. As reviewed in Section 3.1, the implicit assumptions for (56) are that

$$\phi \in L^1_{\text{loc}}(D) \text{ and } q\phi \in L^1_{\text{loc}}(D).$$

Therefore by taking a smaller ball, we may in fact suppose that

$$\phi \in L^1(D),\ q\phi \in L^1(D).$$

By a special case of Proposition 2.10, we have

$$\Delta(\phi - G_D(q\phi)) = 0.$$

Hence by Weyl's lemma, there exists a harmonic function h_D in D such that

$$\phi - G_D(q\phi) = h_D, \quad m\text{-a.e. in } D. \tag{57}$$

We do not know if $\phi \in L^1(\partial D, \sigma)$, where σ is the Lebesgue measure on the sphere ∂D; however, there is a smaller concentric ball B for which this is true. Indeed, by Fubini's theorem there exists a ball which satisfies the condition

$$\phi \in L^1(\partial B, \sigma) \tag{58}$$

and (57) holds σ-a.e. on ∂B. Furthermore, we may choose B small enough so that

$$\|G_B|q|\|_\infty < 1, \tag{59}$$

and

$$\sup_{(x,z)\in B\times\partial B} E_z^x\{e_{|q|}(\tau_B)\} \le 2. \tag{60}$$

This is possible by Theorem 4.3(i) and Theorem 5.17, respectively. Note that either (59) or (60) implies that $(B, |q|)$ is gaugeable. Of course (57) holds also when D is replaced by B:

$$\phi - G_B(q\phi) = h_B, \quad m\text{-a.e. in } B, \tag{61}$$

where h_B is a harmonic function in B. We have

$$h_D(x) = E^x\{h_D(X(\tau_B))\}, \ x \in B. \tag{62}$$

This follows from Theorem 1.24(b) because h_D is harmonic in B and continuous in $\overline{B}$. Next we have for a.e. x in B:

$$G_D(q\phi)(x) = G_B(q\phi)(x) + E^x\{G_D[(q\phi)(X(\tau_B))]\}. \tag{63}$$

This follows from the formula below, which is valid for $f \in L^1(D)$, and whenever all terms are finite.

$$\begin{aligned} E^x\left\{\int_0^{\tau_D} f(X_t)\mathrm{d}t\right\} &= E^x\left\{\int_0^{\tau_B} f(X_t)\mathrm{d}t + \int_{\tau_B}^{\tau_D} f(X_t)\mathrm{d}t\right\} \\ &= E^x\left\{\int_0^{\tau_B} f(X_t)\mathrm{d}t\right\} + E^x\left\{E^{X(\tau_B)}\left[\int_0^{\tau_D} f(X_t)\mathrm{d}t\right]\right\}. \end{aligned}$$

Since $q\phi \in L^1(D)$ we have $G_D(q\phi) \in L^1(D)$ by Lemma 2.11, hence all terms in (63) are finite for a.e. x. Substituting (57) and (61) into (63), and using (62), we obtain m-a.e. in B:

$$\begin{aligned} \phi - h_D &= \phi - h_B + E^{\cdot}\{(\phi - h_D)(X(\tau_B))\} \\ &= \phi - h_B + E^{\cdot}\{\phi(X(\tau_B))\} - h_D. \end{aligned}$$

Note that in the first equation above we have used the stipulation mentioned under (58). Therefore we have for m-a.e. in B:

$$h_B(x) = E^x\{\phi(X(\tau_B))\}. \tag{64}$$

Note that this relation is not obvious from (61)!

Now we set

$$\begin{aligned} h_B^*(x) &= E^x\{|\phi|(X(\tau_B))\}, \\ v(x) &= E^x\{e_q(\tau_B)\phi(X(\tau_B))\}, \\ v^*(x) &= E^x\{e_{|q|}(\tau_B)|\phi|(X(\tau_B))\}. \end{aligned}$$

Let x_0 be the center of B. Then $h_B^*(x_0) < \infty$ by (58), hence $h_B^* < \infty$ in B by the Corollary to Theorem 5.18 with B for D, ϕ for f and $q \equiv 0$. Next, we shall prove that $v^*(x_0) < \infty$. This is an essential difficulty which is resolved by recourse to the conditional gauge as in the proof of Theorem 5.19. We have

$$v^*(x) = \int_{\partial B} E_z^x\{e_{|q|}(\tau_B)\}K(x,z)|\phi|(z)\sigma(\mathrm{d}z),$$

where K is the Poisson kernel for B. Then by (60) and (58) we have

$$v^*(x_0) \leq \frac{2}{\sigma(\partial B)} \int_{\partial B} |\phi|(z)\sigma(\mathrm{d}z) < \infty.$$

Hence $v^* < \infty$ in B and v as well as v^* is continuous in B by the Corollary to Theorem 5.18, which does not require ϕ to be bounded.

Observe that v is of the form u_f with $D = B$ and $f = \phi$. But there is an important difference: while the f in (4.37) is bounded, our present ϕ is not. Hence Theorem 5.19 is not applicable. We need the analogue of (4.41), as follows:

$$G_B(|q|v^*) < \infty. \tag{65}$$

Since v^* is not known to be bounded, Theorem 3.2 or Theorem 4.3 cannot be applied. We shall prove (65) by reversing the argument described in detail after (4.41). Let

$$\psi(t,w) = 1_{\{t<\tau_B\}}|q|(X_t)\exp\left[\int_t^{\tau_B} |q|(X_s)\mathrm{d}s\right] |\phi|(X(\tau_B)).$$

Then we have by direct integration:

$$\begin{aligned} E^x\left\{\int_0^\infty \psi(t,w)\mathrm{d}t\right\} &= E^x\{[e_{|q|}(\tau_B) - 1]|\phi|(X(\tau_B))\} \\ &= v^*(x) - h_B^*(x). \end{aligned}$$

This is finite because v^* is. Therefore by the Fubini–Tonelli theorem, the double integral evaluated in reverse order must also by finite. Using the Markov property, we have

$$\begin{aligned}\int_0^\infty E^x\{\psi(t,w)\}dt &= \int_0^\infty E^x\{1_{\{t<\tau_B\}}|q|(X_t)v^*(X_t)\}dt \\ &= E^x\left\{\int_0^{\tau_B}|q|(X_t)v^*(X_t)dt\right\} \\ &= G_B(|q|v^*)(x),\end{aligned}$$

where in the second equation we have again used the Fubini–Tonelli theorem. Thus (65) is true.

Once (65) is proved the calculations in the proof of Theorem 4.7(i) go through as before and yield

$$v = h_B + G_B(qv), \tag{66}$$

where h_B is as in (64).

Comparing (66) with (61), we obtain

$$\phi - v = G_B(q(\phi - v)), \quad m\text{-a.e. in } B. \tag{67}$$

Multiplying both members of the above by q, we obtain by the symmetry of $G_B(\cdot,\cdot)$:

$$\|q(\phi - v)\|_1 \le \|G_B|q|\|_\infty \|q(\phi - v)\|_1.$$

Hence by (59), $\|q(\phi - v)\|_1 = 0$, and so $q(\phi - v) = 0$ m-a.e. in B. Using (67) again, we obtain

$$\phi = v, \quad m\text{-a.e. in } B.$$

This is the desired result. □

5.6 New Approach to the Gauge Theorem

In this section we give a new proof of Theorem 5.19 with $f \equiv 1$. The gauge function u_1 will be denoted by u. The new departure bypasses the Harnack inequality and is totally independent of conditional processes.

We begin with the reminder that u is Borel measurable. For each $x \in D$, we have:

$$\begin{aligned}u(x) &= \lim_{t\downarrow 0} \uparrow E^x\{t < \tau_D;\, e_q(t)u(X_t)\} \\ &= \lim_{t\downarrow 0} \uparrow T_t u(x) \\ &= \lim_{t\downarrow 0} \uparrow \lim_{n\to\infty} \uparrow T_t(u \wedge n)(x),\end{aligned}$$

where the symbol $\uparrow$ indicates increasing convergence, and T_t is as in (3.34). Since T_t has the strong Feller property by Theorem 3.17, $T_t(u \wedge n) \in C_b(D)$. Therefore u is lower semi-continuous in D.

Now set

$$F = \{x \in D : u(x) < \infty\}, \ I = \{x \in D : \ u(x) = \infty\};$$

where both F and I are Borel sets. Suppose F is not empty. Then for any $x \in F$, and any compact subset C of I, we have by the strong Markov property:

$$\infty > u(x) \geq E^x\{T_C < \tau_D;\ e_q(T_C)u(X(T_C))\}. \tag{68}$$

By the definition of T_C, the fact that C is closed, and the continuity of $X(\cdot)$ (right continuity is sufficient), we see that $X(T_C) \in C$ and so $u(X(T_C)) = \infty$. On the other hand, $e_q(T_C) > 0$ as previously noted. Here as elsewhere we have omitted the words 'almost surely' in the assertions. It follows from (68) that

$$\forall x \in F : \ P^x\{T_C < \tau_D\} = 0. \tag{69}$$

Consequently by (2.13), we have

$$\int_C G_D(x,y)\mathrm{d}y = G_D 1_C(x) = E^x\left\{\int_0^{\tau_D} 1_C(X_t)\mathrm{d}t\right\} = 0.$$

Since $G(x,y) > 0$ for all $y \in D$ by Theorem 2.6(i), we have $m(C) = 0$ for any compact subset C of I. This implies by basic measure theory that

$$m(I) = 0. \tag{70}$$

We proceed by setting for each $N \geq 1$:

$$F_N = \{x \in D : \ N < u(x) \leq \infty\}.$$

Since u is lower semi-continuous, F_N is an open set. Since $m(D) < \infty$ and $m(I) = 0$, for any $\delta > 0$ there exists N such that $m(F_N) < \delta$. In the proof of Proposition 4.11, choose δ so that we have $\varepsilon = \frac{1}{2}$, and set $F_N = E$; then we have

$$\sup_{x\in E} E^x\{e_{|q|}(\tau_E)\} \leq 2. \tag{71}$$

Note that E need not be connected, but Proposition 4.11 can be applied to each of its components.

We have now exactly as in (4.15), for all $x \in E$:

$$u(x) \quad = \quad E^x\{\tau_E = \tau_D;\ e_q(\tau_E)\} + E^x\{\tau_E < \tau_D;\ e_q(\tau_E)u(X(\tau_E))\}. \tag{72}$$

Readers should compare this with the similar (4.15) but mark the difference. In the last term of the above, by the definition of τ_E, the fact that $E = F_N$ is open, and the (right) continuity of $X(\cdot)$, we see that $X(\tau_E) \in D\backslash E$ on $\{\tau_E < \tau_D\}$, and so $u(X(\tau_E)) \leq N$. Therefore it follows from (71) and (72) that

$$\sup_{x\in E} u(x) \leq 2 + 2N,$$

and so

$$\sup_{x \in D} u(x) \le 2 + 2N.$$

This is the gauge theorem in view of the introductory remarks of Section 4.3.

The new proof of the gauge theorem is not only short but has the tremendous advantage that it can be generalized to a large class of Markov processes without continuity of paths (right continuity is sufficient as noted above), and to a class of multiplicative functionals other than the Feynman–Kac $e_q(t)$. Let us now indicate how we can retrieve one more property of u. Once we have proved that u is bounded in D, we can deduce its continuity in any ball $B \subset\subset D$ by the proof of Theorem 4.7(ii), hence its continuity in D.

Notes on Chapter 5

The theory of conditional Brownian motion originated with Doob in the study of R.S. Martin's boundary theory for harmonic functions; see Doob (1984). Here in Sections 5.1 and 5.2 we apply the methodology of this theory to a bounded Lipschitz domain. This important particular case does not seem to have received appropriate treatment in the probability literature. Thus we prove Theorems 5.9 and 5.10 directly with no reference to the Martin boundary. Presumably Theorem 5.9 would follow from Doob's general result, provided we also invoke the fact (which is folklore to us) that for a Lipschitz domain the Martin boundary is equivalent to the Euclidean one. Theorem 5.7 is of recent origin and goes back to an open problem in a paper by Charles Lamb, see Cranston and McConnell (1983).

It may be intimated here that the rather abstruse notion of a general h-conditioned process tends to obscure the workaday significance of the conditioning as verified in Proposition 5.11. When $h = K(\cdot, z)$, this is nothing but the old-fashioned conditioning under '$X(\tau_D) = z$'. Nevertheless the basic sample function properties of the conditional processes are sufficiently different from those of the unconditional process (i.e. with $h \equiv 1$) to require attention, particularly in respect of the behavior at the lifetime. Otherwise why should the 'intuitively obvious' Theorem 5.9 be so hard to prove? Apropos, Proposition 5.5 ought to be proved in the same manner as Theorem 1.4, by first constructing the paths on a countable dense set such as the V there, then verifying its uniform continuity on $V \cap [0, n]$, etc. The recourse to separability is merely expedient, but it was Doob's original approach.

For a recent exposition on Lipschitz domains with some discussion of probabilistic formulations, see Jerison and Kenig (1982). For further work on the lifetime of h-processes extending some of the results mentioned in the text, see Bañuelos (1992) in which other references are listed.

Proposition 5.15 was given in Brossard (1985); the proof given here is taken from Chung (1987a). Theorem 5.17 was proved by Zhao (1983), who used it to deduce both Theorem 5.18 and 5.19. Theorem 5.18 was proved in Aizenman and Simon (1982) by an ingenious time-reversal from τ_B. This is an instructive example in which a trick may replace an idea, or vice versa, and the reader may decide for himself which is which.

For certain topics treated in this book, we have initially considered only the case of a bounded domain. Even the case where D is unbounded but $m(D) < \infty$ presents difficulties and is not always discussed in other treatments. The eventual extensions to more general cases arose as afterthoughts, as may be apparent from some passages in the text. There are other possibilities for generalization, such as the splitting of q into q^+ and q^-, which are not mentioned here. It is somewhat disconcerting to discover that if we are willing to assume $q \in L^1(D)$ for an unbounded D, then easy generalizations are possible as in Theorem 5.20. Why should such a reasonable assumption hamper applications? However, a mathematical physicist consulted on this question demurred with the Coulomb potential. For the extension of the function q, i.e. the measure $q \cdot m$ to a more general measure μ, see e.g. Boukricha, Hansen and Hueber (1987) and Sturm (1989).

Theorem 5.21 was proved for $d \geq 3$ in Aizenman and Simon (1982) using a complex operator-analytic method. To test the power of our equipment developed in these chapters, we sought a new proof along the same lines, which resulted in Pop-Stojanovic and Rao (1990). The latter still uses L^p estimates as well as Theorem 8.9, and contains a misleading gap (the validity of (64)). The present treatment thoroughly reworks their main argument but employs only the results in our previous text. We have nothing against L^p estimates but we are content to do without them. We note the unusual use of the Fubini–Tonelli theorem in the verification of (65). This indispensable tool of analysis is used, sometimes unmentioned and often without reference to the second name, countless times in the course of the book as a whole.

Section 5.6 is based on Chung and Rao (1988) with a further simplification so that the use of general methodology is minimized. That paper contains a number of generalizations of the main topics of this book to a larger class of Markov processes and multiplicative functionals, including the equivalence theorem (Theorem 5.19), the super-gauge theorem (Theorem 8.9), and generalized Green potentials. One challenging open problem is the search for an operator which would take the place of the Schrödinger operator in the general context.

In a somewhat different direction, a comprehensive study of the ideas relating to the gauge has been undertaken by Sturm (1989 and 1991). Another approach from the standpoint of harmonic spaces is given in Boukricha, Hansen and Hueber (1987). Further references to other relevant works may be found in these papers.

6. Green Functions

6.1 Basic Properties of the q-Green Function

Let D be a bounded and regular domain in $\mathbb{R}^d$, $d \geq 2$, and $q \in J_{\text{loc}}$. We assume that (D, q) is gaugeable.

We recall from (3.44) that:

$$V(x, y) = \int_0^\infty u(t; x, y)\mathrm{d}t, \ (x, y) \in D \times D, \tag{1}$$

where $u(t; \cdot, \cdot)$ is the density for the stopped Feynman–Kac semigroup T_t; see Theorems 3.10 and 3.17. Thus for $f \in L^1(D)$:

$$E^x[t < \tau_D; \ e_q(t)f(X_t)] = \int_D u(t; x, y)f(y)\mathrm{d}y. \tag{2}$$

We recall that we called $V(\cdot, \cdot)$ the q-Green function. In Chapter 3 we studied the associated operator V. Here we deal with the finer properties of the function $V(\cdot, \cdot)$.

We begin with an inequality for $u(\cdot\, ; \cdot\, , \cdot)$, due to Simon (1982).

Lemma 6.1 *If (D, q) is gaugeable, then there exist strictly positive constants t_0, α and C depending only on D and q such that*

$$\begin{aligned} u(t; x, y) &\leq Ct^{-d/2}\exp(-|x-y|^2/4t) \ \textit{if} \ 0 < t \leq t_0, &(3)\\ u(t; x, y) &\leq C\exp(-\alpha t) \ \textit{if} \ t > t_0. &(4)\end{aligned}$$

Proof Since $q \in J_{\text{loc}}$, by Proposition 3.8, there exists $t_0 > 0$ such that for any $0 < t \leq t_0$,

$$\sup_{x \in D} E^x\{e_{4q}(t)\} \leq 2.$$

First let $f \in L^2(D)$ and set

$$S_t f(x) = E^x[t < \tau_D; \ e_{2q}(t)f(X_t)].$$

Then, as in Step 4 of the proof of Theorem 3.10, for $x \in D$ and $0 < t \leq t_0$ we have:

$$\begin{aligned}|S_tf(x)|^2 &\le E^x[t<\tau_D;\ e_{4q}(t)]E^x[t<\tau_D;\ f(X_t)^2]\\ &\le 2(2\pi t)^{-d/2}\|f\|_2^2.\end{aligned}$$

It follows that

$$\|S_t\|_{2,\infty}\le C\,t^{-d/4},$$

and consequently by Step 5 in the proof of Theorem 3.10,

$$\|S_t\|_{1,\infty}\le\|S_{t/2}\|_{1,2}\|S_{t/2}\|_{2,\infty}\le\|S_{t/2}\|_{2,\infty}^2\le Ct^{-d/2}. \tag{5}$$

Here and hereafter C will denote a constant whose value may change.

Let B be a Borel set in D. We have by (5),

$$\begin{aligned}(T_t1_B(x))^2 &\le E^x[t<\tau_D;\ e_{2q}(t)1_B(X_t)]E^x[t<\tau_D;\ 1_B(X_t)]\\ &= [S_t1_B(x)][P_t^D1_B(x)]\\ &\le \|S_t\|_{1,\infty}m(B)(2\pi t)^{-d/2}\mathrm{e}^{-\rho^2/2t}m(B)\\ &\le Ct^{-d}\mathrm{e}^{-\rho^2/2t}m(B)^2,\end{aligned}$$

where $\rho=\rho(x,B)$.

Hence by (2) with $f=1_B$:

$$\int_B u(t;x,y)\mathrm{d}y\le Ct^{-d/2}\mathrm{e}^{-\rho^2/4t}m(B). \tag{6}$$

For x and y in D, taking $B=B(y,\delta)$ and letting $\delta\downarrow 0$, we obtain (3) from (6) by the continuity of $u(t;x,\cdot)$.

Next, since (D,q) is gaugeable, by (3.35), (4.66) and (4.67), there exist C and $\alpha>0$ such that

$$\|T_t\|_1\le C\mathrm{e}^{-\alpha t},\quad t>0. \tag{7}$$

Hence for all $t>t_0$ it follows as before from the continuity of $u(t;x,\cdot)$ and (7) that

$$\begin{aligned}u(t;x,y) &\le \|T_t\|_{1,\infty}\le\|T_{t-t_0}\|_1\,\|T_{t_0}\|_{1,\infty}\\ &\le \|T_{t_0}\|_{1,\infty}C\mathrm{e}^{-\alpha(t-t_0)}.\end{aligned}$$

This reduces to (4) with a different constant C. □

Theorem 6.2 *If (D,q) is gaugeable, then the q-Green function $V(\cdot\,,\cdot)$ has the following properties.*

(a) *$V(\cdot\,,\cdot)$ is finite, symmetric and continuous in $(x,y)\in D\times D$, $x\ne y$.*

(b) *For any $x\in D$,*

$$\lim_{y\to\partial D}V(x,y)=0.$$

(c) *There exists $C>0$ such that*

$$V(x,y) \leq C\, g(x-y),\ (x,y) \in D \times D,$$

where $g(\cdot)$ is as in (3.4) for $d \geq 3$, but for $d = 2$ it is redefined as follows:

$$g(x) = \left(\ln \frac{1}{|x|}\right) \vee 1. \tag{8}$$

Proof (a) Firstly, it follows from Lemma 6.1 that $V(x,y) < \infty$ for $x \neq y$. For each $t > 0$, $u(t;\cdot,\cdot)$ is symmetric and continuous by Theorem 3.17. Hence the symmetry of $V(\cdot\,,\cdot)$ is obvious. Its continuity for $x \neq y$ follows by dominated convergence from Lemma 6.1.

(b) Since D is regular, we have $u(t;\cdot,\cdot) \in C_0(D \times D)$ by Theorem 3.17. Hence (b) follows by dominated convergence from Lemma 6.1.

(c) We have by Lemma 6.1:

$$V(x,y) \leq C \int_0^{t_0} p\left(t; \frac{x}{2}, \frac{y}{2}\right) \mathrm{d}t + \frac{C}{\alpha}.$$

Hence by Lemma 3.4,

$$V(x,y) \leq C\, g\left(\frac{x-y}{2}\right) + \frac{C}{\alpha}. \tag{9}$$

If $d \geq 3$, this reduces to $C\, g(x,y)$ for a different constant C because D is bounded so that $|x-y|$ is bounded. If $d = 2$, we may choose $t_0 \leq \frac{1}{2}$ in Lemma 6.1 so that Lemma 3.4 is applicable for $|x-y| \leq t_0$. Again using the boundedness of $|x-y|$ and the revised definition (8), we see, after a tedious checking of cases, that the right member of (9) is bounded by the $C\, g(x-y)$. □

The following theorem is a strengthened version of Theorem 3.20.

Theorem 6.3 *For all $(x,y) \in D \times D$, $x \neq y$, we have*

$$V(x,y) = G(x,y) + \int_D V(x,u) q(u) G(u,y) \mathrm{d}u, \tag{10}$$

and

$$V(x,y) = G(x,y) + \int_D G(x,u) q(u) V(u,y) \mathrm{d}u. \tag{11}$$

Proof If $|x-y| > \delta > 0$, then either $|x-u| > \frac{\delta}{2}$ or $|u-y| > \frac{\delta}{2}$. Hence using Theorem 6.2(c) and Theorem 2.6(ii) (for the latter we can also use Theorem 6.2(c) with $q \equiv 0$), we have

$$V(x,u)|q(u)|G(u,y) \leq C\, g\left(\frac{\delta}{2}\right) |q(u)|\ [g(u-y) + g(x-u)]. \tag{12}$$

Since D is bounded, it follows from the last assertion of the Corollary to Proposition 3.1 and (12) that the set of functions

$$\{V(x,\cdot)q(\cdot)G(\cdot\,,y) : \ (x,y) \in D \times D, \ |x-y| > \delta\}$$

is uniformly integrable over D. On the other hand, for each $u \in D$, the function

$$(x,y) \to V(x,u)q(u)G(u,y)$$

is continuous except possibly at $x = u$ or $y = u$. Therefore the integral on the right-hand side of (10) is continuous in $(x,y) \in D \times D$, $|x-y| > \delta$. Since δ is arbitrary, both members of (10) are continuous in $(x,y) \in D \times D$, $x \neq y$. Since D is bounded and (D,q) is gaugeable, $V1 \in L^\infty(D)$ by Theorem 4.19. Hence by Theorem 3.20(ii) with $\lambda = 0$, equation (10) holds for almost all $(x,y) \in D \times D$. Therefore it holds as stated by continuity. The proof of (11) is exactly the same. □

The next result is an extension of the Corollary to Theorem 2.5.

Theorem 6.4 *For any fixed $x \in D$, the function $V(x,\cdot)$ is a weak solution of the Schrödinger equation*

$$\left(\frac{\Delta}{2} + q\right) u = 0$$

on $D \backslash \{x\}$.

Proof Let $D_x = D\backslash\{x\}$. For any $\phi \in C_c^\infty(D_x)$, let S_ϕ be the support of ϕ and $\delta = \rho(x, S_\phi) > 0$. It follows from Theorem 6.2(c) and (10) that $V(x,\cdot)$ and $\int_D V(x,u)q(u)G(u,\cdot)\mathrm{d}u$ are bounded on S_ϕ. Hence we have by Theorem 6.3 and Proposition 2.8,

$$\begin{aligned}
&\int_{D_x} V(x,y)\Delta\phi(y)\mathrm{d}y \\
&= \int_D G(x,y)\Delta\phi(y)\mathrm{d}y + \int_D \left[\int_D V(x,u)q(u)G(u,y)\mathrm{d}u\right] \Delta\phi(y)\mathrm{d}y \\
&= -2\phi(x) + \int_D V(x,u)q(u)\left[\int_D G(u,y)\Delta\phi(y)\mathrm{d}y\right]\mathrm{d}u \\
&= -2\int_{D_x} V(x,u)q(u)\phi(u)\mathrm{d}u,
\end{aligned}$$

since $\phi(x) = 0$. By definition this means that

$$\left(\frac{\Delta}{2} + q\right) V(x,\cdot) = 0 \text{ on } D_x.$$

□

6.2 Inequalites for Green Functions in a Lipschitz Domain in $\mathbb{R}^d$ $(d \geq 3)$

We shall now give an inequality which is crucial in proving the conditional gauge theorem and studying the Green function for the Schrödinger operator in a bounded Lipschitz domain in $\mathbb{R}^d$, $d \geq 3$. This result was proved recently by Cranston, Fabes and Zhao (1988) and will be referred to as the '3G Theorem'.

Theorem 6.5 *Let D be a bounded Lipschitz domain in $\mathbb{R}^d$, $d \geq 3$, and let $G(\cdot,\cdot)$ denote its Green function (i.e. the $G_D(\cdot,\cdot)$ in (2.14)). There exists a constant $C > 0$ depending only on D such that for all x, y and z in D, we have:*

$$\frac{G(x,y)G(y,z)}{G(x,z)} \leq C \frac{|x-z|^{d-2}}{|x-y|^{d-2}|y-z|^{d-2}}, \tag{13}$$

and

$$\frac{G(x,y)G(y,z)}{G(x,z)} \leq C(|x-y|^{2-d} + |y-z|^{2-d}). \tag{14}$$

Note that the right member of (13) may be written as

$$C\frac{g(x,y)g(y,z)}{g(x,z)},$$

where g is given in (2.17).

To prove the theorem, we require a number of properties of a bounded Lipschitz domain, which we shall prove or quote as lemmas. The symbols C, C_0, C_1, ..., will be used to denote strictly positive constants which depend only on D and are not necessarily the same in different instances. We recall that $\rho(x)$ is the distance from x to ∂D.

The first two properties we need below are the conditions for 'non-tangentially accessible' (abbreviated as NTA) domains defined by Jerison and Kenig in (1982, Section 3). Although Jerison and Kenig described the following lemmas as 'easy to see', we shall spell out the details below to give the reader an idea of the geometry of a Lipschitzian boundary.

Lemma 6.6 *Any bounded Lipschitz domain D in $\mathbb{R}^d$ $(d \geq 2)$ is a NTA domain; i.e. it satisfies the following conditions.*

(i) *The 'corkscrew condition' for D. There exist constants $C_0 \geq 1$ and $\alpha > 0$ such that for any $z \in \partial D$ and $0 < r \leq \alpha$, we can find a point $A = A_r(z)$ in D satisfying the conditions:*

$$|A - z| \leq C_0 r, \text{ and } \rho(A) \geq r.$$

(ii) *The corkscrew condition above also holds when D is replaced by its complement D^c.*

(iii) *The Harnack chain condition. For any given $C_1 > 0$ and any $x \in D$, $y \in D$ such that $|x - y| \leq C_1[\rho(x) \wedge \rho(y)]$, we can find a finite number of balls $B(a_i, r_i)$*

$(0 \leq i \leq n)$ in D such that $a_0 = x$, $a_n = y$, and $B\left(a_i, \frac{r_i}{2}\right) \cap B\left(a_{i+1}, \frac{r_{i+1}}{2}\right) \neq \emptyset$ for $0 \leq i \leq n-1$. Here the number n may depend on the constant C_1 and the domain D, but not on (x, y).

Proof By the definition of a Lipschitz domain given in Section 5.2, to each $z \in \partial D$ there corresponds a local coordinate system (ξ, η), $\xi \in \mathbb{R}^{d-1}$, $\eta \in \mathbb{R}^1$, a function ϕ and constants $C > 0$ and $s_0 > 0$ such that

$$|\phi(\xi_1) - \phi(\xi_2)| \leq C|\xi_1 - \xi_2|, \tag{15}$$

and

$$D \cap B(z, s_0) = \{(\xi, \eta) \in B(z, s_0) : \phi(\xi) < \eta\}. \tag{16}$$

Since ∂D is compact, the two strictly positive constants C and s_0 can be chosen to depend on D only. It follows from (16) and the continuity of ϕ that

$$(\partial D) \cap B(z, s_0) = \{(\xi, \eta) \in B(z, s_0) : \phi(\xi) = \eta\}. \tag{17}$$

For simplicity, for each $z \in \partial D$ we shall place the origin of the corresponding coordinates at z, i.e. $z = (0^{d-1}, 0)$. Here we use 0^{d-1} to denote the 0-vector in $\mathbb{R}^{d-1}$. Similarly if $y = (\xi, \eta)$ we shall denote ξ by y^{d-1} later. Consequently by (17) we have

$$\phi(0^{d-1}) = 0. \tag{18}$$

(i) Let $C_0 = 2C$ and $\alpha = \frac{s_0}{4C}$. Without loss of generality we may suppose that $C \geq 1$ in what follows. For any $z \in \partial D$ and $0 < r \leq \alpha$, let $A = (0^{d-1}, C_0 r)$ in the coordinate system corresponding to z. We then have

$$|A - z| = C_0 r \leq C_0 \alpha = \frac{s_0}{2}.$$

Hence by (16), $A \in D \cap B(z, s_0)$.

For any $w \in \partial D$, if $|w - z| \geq s_0$, then $|w - A| \geq s_0 - |A - z| \geq 4Cr - 2Cr \geq r$; if $|w - z| < s_0$, then $w = (\xi, \phi(\xi))$. Thus for $|\xi| \geq r$, we have $|w - A| \geq |\xi - 0^{d-1}| \geq r$; and for $|\xi| < r$, we have $|\phi(\xi)| \leq C|\xi| < Cr$ by (15), and so $|w - A| \geq 2Cr - |\phi(\xi)| \geq Cr \geq r$. Hence in all cases we have $|w - A| \geq r$. Therefore $\rho(A) \geq r$.

(ii) This can be verified in a similar way by setting $A = (0^{d-1}, -C_0 r)$.

(iii) For any $C_1 > 0$ and any $x \in D$, $y \in D$ satisfying

$$|x - y| \leq C_1[\rho(x) \wedge \rho(y)], \tag{19}$$

we need to construct a finite number of balls satisfying the condition in (iii). First let $s_1 > 0$ be a small constant which may depend on C_1 and D, to be determined later. There exists a domain $D_1 \subset\subset D$ such that D_1 contains the compact set $\{x \in D : \rho(x) \geq s_1\}$; see the Appendix to Chapter 1. Let $s = \rho(\partial D_1, \partial D)$. Then $0 < s \leq s_1$, and there exist finitely many balls $B_i = B(w_i, \frac{s}{2})$, $1 \leq i \leq N$, whose union covers $\overline{D}_1$, and such that $B(w_i, s) \subset D$, $1 \leq i \leq N$. We consider two cases.

Case (a): both x and y belong to $D_1 \subset \cup_{i=1}^N B_i$. Then there exists a set $\{B_{i_k}, 1 \leq k \leq l\}$ which is a subset of $\{B_i, 1 \leq i \leq N\}$ such that each B_{i_k} intersects $B_{i_{k+1}}$ for $1 \leq k \leq l-1$, and $x \in B_{i_1}$, $y \in B_{i_l}$. For a proof of this assertion, see Chung (1982a, Exercise 3(b), page 205). Let $a_0 = x$, $a_k = w_{i_k}$, $1 \leq k \leq l$, $a_{l+1} = y$, and $r_k = s$, $0 \leq k \leq l+1$. Then the balls $B(a_k, r_k)$, satisfy the condition in (iii) and the number $l+1 \leq N+1$ does not depend on (x, y).

Case (b): at least one of x and y does not belong to D_1. Then we must have

$$\rho(x) \wedge \rho(y) < s_1, \tag{20}$$

for otherwise both x and y belong to $\{x \in D: \rho(x) \geq s_1\} \subset D_1$. In the remainder of this chapter, for any $x \in D$, we shall denote by x^* a point on ∂D such that $\rho(x) = |x - x^*|$. We may assume that $\rho(x) \leq \rho(y)$. Thus if we take $s_1 \leq \frac{s_0}{C_1+1}$, then by (19) and (20), $|x - x^*| < s_0$ and $|y - x^*| \leq |y - x| + |x - x^*| < (C_1 + 1)s_1 \leq s_0$, i.e. both x and y belong to $B(x^*, s_0)$. We shall construct the required balls in $D \cap B(x^*, s_0)$ using a coordinate system (ξ, η) corresponding to $x^* = (0^{d-1}, 0)$. We recall that a point (ξ, η) in $B(x^*, s_0)$ belongs to D if and only if $\phi(\xi) < \eta$. Hence for $x = (x^{d-1}, x_d)$ and $y = (y^{d-1}, y_d)$ in $D \cap B(x^*, s_0)$, we have

$$\phi(x^{d-1}) < x_d \text{ and } \phi(y^{d-1}) < y_d. \tag{21}$$

Note that $|x| = \rho(x)$ and by (19),

$$|y| \leq |y - x| + |x| \leq (C_1 + 1)\rho(x). \tag{22}$$

We now choose n points a_i, $0 \leq i \leq n$, where n is to be determined later, as follows (these will be the centers of the required balls):

$$a_i = (a_i^{d-1}, a_{i,d}),$$

where

$$a_i^{d-1} = \frac{n-i}{n} x^{d-1} + \frac{i}{n} y^{d-1}$$

and

$$a_{i,d} = \phi(a_i^{d-1}) + \frac{n-i}{n}[x_d - \phi(x^{d-1})] + \frac{i}{n}[y_d - \phi(y^{d-1})].$$

Obviously we have $a_0 = x$ and $a_n = y$. By (21) we have

$$a_{i,d} > \phi(a_i^{d-1}),$$

hence by (16) we need only verify that $a_i \in B(x^*, s_0)$ to ensure that $a_i \in D$. By (15), (19), and (22), we have

$$\begin{aligned}
|a_i - x^*| &= |a_i| \le |a_i^{d-1}| + |a_{i,d}| \\
&\le |a_i^{d-1}| + |\phi(a_i^{d-1})| + |x_d| + |\phi(x^{d-1})| + |y_d| + |\phi(y^{d-1})| \\
&\le (C+1)(|a_i^{d-1}| + |x| + |y|) \\
&\le 2(C+1)(|x| + |y|) \\
&\le 2(C+1)(C_1+2)\rho(x) \\
&< 2(C+1)(C_1+2)s_1.
\end{aligned}$$

Thus if we take

$$s_1 = \frac{s_0}{4(C+1)(C_1+2)},$$

then $s_1 \le \frac{s_0}{C_1+1}$ as required before, and we have

$$|a_i - x^*| < \frac{s_0}{2}, \tag{23}$$

which is more than we need.

Next, we set for $0 \le i \le n$,

$$r_i = \frac{1}{C+1}\rho(x). \tag{24}$$

To ensure that $B(a_i, r_i) \subset D$ we must verify that $\rho(a_i) \ge \frac{1}{C+1}\rho(x)$; in other words, for each $z \in \partial D$,

$$|a_i - z| \ge \frac{1}{C+1}\rho(x). \tag{25}$$

If $z \notin B(x^*, s_0)$, then by (23),

$$|a_i - z| \ge |x^* - z| - |a_i - x^*| \ge s_0 - \frac{s_0}{2} \ge \frac{1}{C+1}\rho(x);$$

if $z \in B(x^*, s_0)$, then $z = (z^{d-1}, \phi(z^{d-1}))$ and

$$\begin{aligned}
|a_i - z| \ \ge\ & |a_i^{d-1} - z^{d-1}| \vee |\phi(a_i^{d-1}) - \phi(z^{d-1}) \\
& + \frac{n-i}{n}[x_d - \phi(x^{d-1})] + \frac{i}{n}[y_d - \phi(y^{d-1})]|.
\end{aligned} \tag{26}$$

Hence if $|a_i^{d-1} - z^{d-1}| \ge \frac{1}{C+1}\rho(x)$, then (25) holds by (26), while if $|a_i^{d-1} - z^{d-1}| < \frac{1}{C+1}\rho(x)$, then by (15),

$$|\phi(a_i^{d-1}) - \phi(z^{d-1})| < \frac{C}{C+1}\rho(x), \tag{27}$$

and since $x_d - \phi(x^{d-1}) \ge \rho(x)$ and $y_d - \phi(y^{d-1}) \ge \rho(y) \ge \rho(x)$ we have

$$\frac{n-i}{n}[x_d - \phi(x^{d-1})] + \frac{i}{n}[y_d - \phi(y^{d-1})] \ge \rho(x). \tag{28}$$

It follows from (26)–(28) that in this case we have

$$|a_i - z| \geq \rho(x) - \frac{C}{C+1}\rho(x) = \frac{1}{C+1}\rho(x).$$

Thus (25) holds in all cases.

Our final step is to determine the number n so that $B(a_i, \frac{r_i}{2})$ intersects $B(a_{i+1}, \frac{r_{i+1}}{2})$, for $0 \leq i \leq n-1$. By (15) and (22),

$$\begin{aligned} |a_{i+1} - a_i| &\leq \frac{1}{n}(2|x| + 2|y| + |\phi(x^{d-1})| + |\phi(y^{d-1})|) \\ &\quad + |\phi(a_{i+1}^{d-1}) - \phi(a_i^{d-1})| \\ &\leq \frac{C+2}{n}(|x| + |y|) + C|a_{i+1}^{d-1} - a_i^{d-1}| \\ &\leq \frac{2(C+1)}{n}(|x| + |y|) \\ &\leq \frac{2(C+1)(C_1+2)}{n}\rho(x). \end{aligned}$$

Thus if we take $n = [2(C+1)^2(C_1+2)] + 1$, then we have

$$|a_{i+1} - a_i| < \frac{1}{C+1}\rho(x) = \frac{r_{i+1} + r_i}{2},$$

which implies that $B(a_{i+1}, \frac{r_{i+1}}{2}) \cap B(a_i, \frac{r_i}{2}) \neq \emptyset$.

The balls $B(a_i, r_i)$, $1 \leq i \leq n$, constructed above satisfy the condition (iii), since n does not depend on (x, y). (The fact that for each x, these balls are all in $B(x^*, s_0)$ for a suitably chosen s_0 shows the local character of the construction in Case (b), which relies on the Lipschitzian condition for the boundary.) □

Lemma 6.7 *Let D be a bounded Lipschitz domain in $\mathbb{R}^d$ ($d \geq 2$). Suppose that for a constant $C_1 > 0$ we have*

$$|x - z| \leq C_1[\rho(x) \wedge \rho(z)], \tag{29}$$

then there exists a constant $C_2 > 0$ such that

$$G(x, z) \geq C_2|x - z|^{2-d}. \tag{30}$$

Proof We may assume that $\rho(x) \leq \rho(z)$ and that the Lipschitz constant C in (15) is greater than 3.

If the given constant $C_1 < 1$, then we consider the ball $B = B(x, \rho(x)) \subset D$. We have $z \in B$ because $|z - x| < \rho(x)$. Since $D \supset B$, $G = G_D \geq G_B$ by (2.14). By the formulae (2.27) and (2.28) for the Green function of a ball, we have in the case $d \geq 3$:

$$\begin{aligned} G(x, z) &\geq G_B(x, z) = C_d[|x - z|^{2-d} - \rho(x)^{2-d}] \\ &\geq C_d[1 - C_1^{d-2}]|x - z|^{2-d}; \end{aligned}$$

and in the case $d = 2$:

$$G(x,z) \geq G_B(x,z) = \frac{1}{\pi} \ln \frac{\rho(x)}{|x-z|} \geq \frac{1}{\pi} \ln \frac{1}{C_1}.$$

Thus (30) holds provided $C_1 < 1$.

Suppose $C_1 \geq 1$. Since $\frac{2}{C+1} < 1$, we need only consider the case where $\frac{2}{C+1}\rho(x) < |x-z| \leq C_1\rho(x)$. By Lemma 6.6(iii), there is a Harnack chain which joins x and z: $B_i = B(a_i, r_i) \subset D$, $0 \leq i \leq n$, $a_0 = x$, $a_n = z$ and $B(a_i, \frac{r_i}{2}) \cap B(a_{i+1}, \frac{r_{i+1}}{2}) \neq \emptyset$, $0 \leq i \leq n-1$. Here $n \leq N$ and N is a constant depending only on C_1 and D. Set

$$k = \min\{j : \ x \notin B_i \text{ for } j \leq i \leq n\}.$$

By (24), $r_n = \frac{1}{C+1}\rho(x) < |x-z|$, thus $x \notin B_n$, and so $0 < k \leq n$. Then set

$$w = a_k.$$

Since $x \in B_{k-1}$ we have by (24),

$$\begin{aligned} |x-w| &\leq |x-a_{k-1}| + |a_{k-1}-a_k| \\ &\leq r_{k-1} + \frac{r_{k-1}+r_k}{2} \leq \frac{2}{C+1}\rho(x). \end{aligned} \tag{31}$$

Since $\rho(x) \leq \rho(w) + |x-w| \leq \rho(w) + \frac{2}{C+1}\rho(x)$, we have

$$\rho(x) \leq \frac{C+1}{C-1}\rho(w),$$

and consequently,

$$|x-w| \leq \frac{2}{C-1}[\rho(x) \wedge \rho(w)].$$

Since $\frac{2}{C-1} < 1$ because $C > 3$, we can apply the previous case (for $C_1 < 1$) to obtain

$$G(x,w) \geq C_2|x-w|^{2-d}. \tag{32}$$

The function $G(x,\cdot)$ is harmonic in $\cup_{k \leq i \leq n} B_i$; and $\{B_i : k \leq i \leq n\}$ is a Harnack chain which joins w and z. Hence by (1.26) in the proof of Harnack's inequality (Theorem 1.14),

$$G(x,z) \geq C'G(x,w),$$

where $C' = \left(3^{d-1}4\right)^{-N}$.

It follows by (31), (32) and the fact that $\frac{2}{C+1}\rho(x) < |x-z|$ that

$$\begin{aligned} G(x,z) &\geq C'C_2|x-w|^{2-d} \\ &\geq C'C_2\left(\frac{2}{C+1}\right)^{2-d}\rho(x)^{2-d} \\ &\geq C'C_2|x-z|^{2-d}. \end{aligned}$$

This completes the proof of Lemma 6.7. □

The next lemma is a property of a bounded Lipschitz domain, or more generally a bounded NTA domain, known as the boundary Harnack principle. For any $z \in \partial D$ and $r > 0$, we denote by $H_0^+(z,r)$ the set of strictly positive harmonic functions h in $D \cap B(z,2r)$ such that $h(x) \to 0$ as $x \in D \cap B(z,2r)$, $x \to (\partial D) \cap B(z,2r)$.

Lemma 6.8 (Boundary Harnack Principle) *Let D be a bounded Lipschitz domain in $\mathbb{R}^d$, $d \geq 2$. Then there exist constants $r_0 > 0$ and $C > 0$ depending only on D with the following property. For any $z \in \partial D$, $0 < r \leq r_0$, and any u and v in $H_0^+(z,r)$, we have*

$$\frac{u(x)}{v(x)} \leq C \frac{u(y)}{v(y)} \tag{33}$$

for all x and y in $D \cap \overline{B}(z,r)$.

For the proof we refer to Jerison and Kenig (1982, Theorem 5.1) which implies Lemma 6.8. Earlier forms of the boundary Harnack principle were given by Dahlberg (1977), Wu (1978), and Ancona (1978).

Lemma 6.9 (Carleson's Estimate) *Let D and r_0 be as in Lemma 6.8, and $u \in H_0^+(z,r)$, where $0 < r \leq r_0$. Then for any $0 < C_1 < 1$ any $x \in D \cap \overline{B}(z,r)$ and any $y \in D \cap \overline{B}(z,r)$ with $\rho(y) \geq C_1 r$, we have*

$$u(x) \leq C\, u(y), \tag{34}$$

where C is a constant which depends only on C_1 and D.

For the proof, see Jerison and Kenig (1982, Lemma 4.11).

The constant C in Lemma 6.8 and 6.9 does not depend on the choice of r in $(0, r_0]$. This independence is essential in later applications.

From now on in this section D is a bounded Lipschitz domain in $\mathbb{R}^d$, and $d \geq 3$. The case $d = 2$ will be treated later in Section 6.3.

Let $C_0 \geq 1$, $\alpha > 0$ and $A_r(\cdot)$, $0 < r \leq \alpha$ be as given in Lemma 6.6. For $0 < r \leq \alpha$ and $x \in D$, set $x_r = A_r(x^*)$ where x^* is defined in the proof of Lemma 6.6 if $\rho(x) < r$, while $x_r = x$ if $\rho(x) \geq r$. Then by Lemma 6.6(i), we have

$$|x_r - x| \leq |x_r - x^*| + |x^* - x| \leq (C_0 + 1)r. \tag{35}$$

We shall use the following notation:

$$Q(x,y,z) = \frac{G(x,y)G(y,z)}{G(x,z)}.$$

In what follows we shall say that '(13) holds for (x', y', z')' to mean '(13) holds when the triple (x,y,z) there is replaced by (x', y', z')'. In fact, we shall establish (13) using such transformations of the variables.

Lemma 6.10 *Suppose that* $0 < r \le \alpha$, $|x-y| \ge (2C_0+1)r$ *and* $|x-z| \ge (2C_0+1)r$. *If (13) holds for* (x_r, y, z), *then it holds for* (x, y, z). *A similar result is true if we interchange* x *and* z *in the preceding statement.*

Proof Obviously we may suppose that $\rho(x) < r$. Since

$$|y - x^*| \ge |y - x| - \rho(x) \ge (2C_0 + 1)r - r = 2C_0 r$$

and

$$|z - x^*| \ge |z - x| - \rho(x) \ge (2C_0 + 1)r - r = 2C_0 r,$$

we have both $G(\cdot, y)$ and $G(\cdot, z) \in H_0^+(x^*, C_0 r)$ by the Corollary to Theorem 2.5; moreover, since

$$|x_r - x^*| \le C_0 r, \quad |x - x^*| = \rho(x) < r \le C_0 r,$$

we have by Lemma 6.8 with $u = G(\cdot, y)$, $v = G(\cdot, z)$ and r replaced by $C_0 r$,

$$\frac{G(x, y)}{G(x, z)} \le C \frac{G(x_r, y)}{G(x_r, z)},$$

and consequently

$$Q(x, y, z) \le C\, Q(x_r, y, z). \tag{36}$$

It follows from (35) and the inequalities in the assumption that

$$\begin{aligned} |x_r - z| &\le |x - z| + (C_0 + 1)r \le |x - z| + \frac{C_0 + 1}{2C_0 + 1}|x - z| \\ &= \frac{3C_0 + 2}{2C_0 + 1}|x - z| \end{aligned} \tag{37}$$

and

$$\begin{aligned} |x_r - y| &\ge |x - y| - (C_0 + 1)r \ge |x - y| - \frac{C_0 + 1}{2C_0 + 1}|x - y| \\ &= \frac{C_0}{2C_0 + 1}|x - y|. \end{aligned} \tag{38}$$

By hypothesis we have

$$Q(x_r, y, z) \le C \frac{|x_r - z|^{d-2}}{|x_r - y|^{d-2}|y - z|^{d-2}}. \tag{39}$$

Using (36) to (38) in (39) we obtain

$$Q(x, y, z) \le C \frac{|x - z|^{d-2}}{|x - y|^{d-2}|y - z|^{d-2}}.$$

This is (13). □

We continue with the proof of (13). Since (13) is symmetric in x and z, we may assume that

$$\rho(x) \le \rho(z). \tag{40}$$

Let $d(D)$ denote the diameter of D and set

$$C_1 = C_0 \vee \frac{d(D)}{3\alpha}, \tag{41}$$

where C_0 and α are given in Lemma 6.6.

The proof of (13) will be carried out in several mutually overlapping cases.

Lemma 6.11 *If* $|x - z| \leq (14C_1 + 11)\rho(x)$, *then (13) holds.*

Proof Under the hypothesis and (40), we have by Lemma 6.7,

$$G(x, z) \geq C|x - z|^{2-d}.$$

By Theorem 2.6(ii), we have

$$G(x, y) \leq C_d|x - y|^{2-d},$$

and

$$G(y, x) \leq C_d|y - z|^{2-d}.$$

Clearly (13) follows from the above inequalities. □

Lemma 6.12 *If* $|x - y| \leq (7C_1 + 4)\rho(x)$, *then (13) holds provided* $\rho(x) \leq \rho(z)$.

Proof By virtue of Lemma 6.11, we may assume that

$$|x - z| > (14C_1 + 11)\rho(x). \tag{42}$$

Since

$$|y - x^*| \leq |y - x| + \rho(x) \leq (7C_1 + 5)\rho(x), \quad y \in D \cap \overline{B}(x^*, (7C_1 + 5)\rho(x));$$

and since

$$|z - x^*| \geq |z - x| - \rho(x) > (14C_1 + 10)\rho(x),$$

we have $G(\cdot, z) \in H_0^+(x^*, (7C_1 + 5)\rho(x))$ by the Corollary to Theorem 2.5. Using Carleson's lemma (Lemma 6.9) with $r = (7C_1 + 5)\rho(x)$ and x and y interchanged, noting that $\rho(x) = (7C_1 + 5)^{-1}r$, we have

$$G(y, z) \leq C\, G(x, z).$$

It follows from this that

$$Q(x, y, z) \leq C\, G(x, y) \leq C|x - y|^{2-d}. \tag{43}$$

By the hypothesis of the lemma and (42), we have

$$|y - z| \leq |x - y| + |x - z| \leq 2|x - z|.$$

Hence (13) follows from (43) and the above inequality. □

We shall prove Theorem 6.5 by using Lemma 6.10 repeatedly to reduce the general case to the two special cases treated in Lemmas 6.11 and 6.12.

Proof of Theorem 6.5 Since

$$|x-z|^{d-2} \le 2^{d-2}(|x-y|^{d-2} + |y-z|^{d-2}),$$

(13) implies (14). Hence we need only prove (13).

We assume first that

$$|y-z| \ge 2|x-z|. \tag{44}$$

Let

$$r = \frac{|x-z|}{3C_1+2}.$$

Then by (41),

$$\begin{aligned} r &< \frac{d(D)}{3C_1} \le \alpha, \\ |z-x| &= (3C_1+2)r > (2C_0+1)r \end{aligned} \tag{45}$$

and by (44)

$$|z-y| \ge 2|x-z| > (2C_0+1)r.$$

Hence by Lemma 6.10, it suffices to prove (13) for (x, y, z_r).

By (44) and (45), we have

$$|x-y| \ge |y-z| - |x-z| \ge |x-z| > (2C_0+1)r.$$

By (35) with x replaced by z, we have

$$\begin{aligned} |x-z_r| &\ge |x-z| - |z-z_r| \\ &\ge (3C_1+2)r - (C_0+1)r \\ &\ge (2C_0+1)r. \end{aligned}$$

Using Lemma 6.10 again, we need only prove (13) for (x_r, y, z_r). By (35) we have

$$\begin{aligned} |x_r - z_r| &\le |x-z| + |x_r - x| + |z_r - z| \\ &\le |x-z| + 2(C_0+1)r \\ &\le (3C_1+2)r + 2(C_1+1)r \\ &\le (5C_1+4)[\rho(x_r) \wedge \rho(z_r)] \end{aligned}$$

since for any $x \in D$, $\rho(x_r) \ge r$ by definition of x_r. Thus (13) for (x_r, y, z_r) follows from Lemma 6.11. Therefore we have proved (13) for (x, y, z) under the assumption (44).

We now consider the alternative case:

$$|y - z| < 2|x - z|. \tag{46}$$

This time we set

$$r = \frac{|x - y|}{6C_1 + 3}.$$

Then $r \leq \alpha$ as before, and

$$|x - y| = (6C_1 + 3)r > (2C_0 + 1)r.$$

By (46),

$$|x - z| \geq |x - y| - |y - z| \geq |x - y| - 2|x - z|,$$

so that

$$|x - z| \geq \frac{1}{3}|x - y| = (2C_1 + 1)r > (2C_0 + 1)r.$$

Thus by Lemma 6.10, it suffices to prove (13) for (x_r, y, z).

We have by (35) and the definition of x_r,

$$\begin{aligned} |x_r - y| &\leq |x - y| + |x - x_r| \leq (6C_1 + 3)r + (C_0 + 1)r \\ &\leq (7C_1 + 4)r \leq (7C_1 + 4)\rho(x_r). \end{aligned} \tag{47}$$

If $\rho(x_r) \leq \rho(z)$, then (13) holds for (x_r, y, z) by Lemma 6.12 and (47).

If $\rho(z) < \rho(x_r)$, then we consider (13) for (z, y, x_r), which is equivalent to (13) for (x_r, y, z). If $|y - x_r| \geq 2|z - x_r|$, then (z, y, x_r) satisfies the assumption (44) with x replaced by z and z replaced by x_r there, hence (13) holds for (z, y, x_r) in this case. It remains to consider the case

$$|y - x_r| < 2|z - x_r|. \tag{48}$$

But (48) is the analogue of (46) with x replaced by z and z replaced by x_r. By exactly the same argument as before, we choose a point z_s with $s = \frac{|z-y|}{6C_1+3}$ and reduce (13) for (z, y, x_r) to (13) for (z_s, y, x_r). Thus we obtain the inequality similar to (47) with x replaced by z and r replaced by s:

$$|z_s - y| \leq (7C_1 + 4)\rho(z_s). \tag{49}$$

Thus if $\rho(x_r) \leq \rho(z_s)$, then since (13) for (z_s, y, x_r) is equivalent to (13) for (x_r, y, z_s), the latter follows from (47) and Lemma 6.12; if $\rho(z_s) < \rho(x_r)$, then (13) for (z_s, y, x_r) follows from (49) and Lemma 6.12. □

The following corollary to Theorem 6.5 is a limiting form of the latter. It is essential for the application to the conditional Brownian motion in Chapter 7.

Corollary 6.13 *For a bounded Lipschitz domain D in $\mathbb{R}^d$, $d \geq 3$, there exists a constant $C = C(D) > 0$ such that for all x and y in D, and z on ∂D:*

$$\frac{G(x,y)K(y,z)}{K(x,z)} \leq C(|x-y|^{2-d} + |y-z|^{2-d}). \tag{50}$$

Proof Let $\{z_n\}$ be a sequence of points in D converging to z as $n \to \infty$. By Theorem 6.5 we have for all $n \geq 1$,

$$\frac{G(x,y)G(y,z_n)}{G(x,z_n)} \leq C(|x-y|^{2-d} + |y-z_n|^{2-d}). \tag{51}$$

Since by (5.8),

$$\frac{K(y,z)}{K(x,z)} = \lim_{n\to\infty} \frac{G(y,z_n)}{G(x,z_n)}, \tag{52}$$

(50) follows from (51) and (52). □

The following theorem is an extension of Theorem 5.7 to the higher dimensional case; we have already used it in Theorem 5.9 to obtain the first relation in (5.11).

Theorem 6.14 *For a bounded Lipschitz domain D in $\mathbb{R}^d$, $d \geq 3$, there exists a constant $C(D) > 0$ such that for all $x \in D$, $z \in \partial D$:*

$$E_z^x(\tau_D) \leq C(D).$$

Proof We have by Proposition 5.2 for all $(x,z) \in D \times \partial D$,

$$\begin{aligned} E_z^x(\tau_D) &= E_z^x\left(\int_0^\infty 1_{(t<\tau_D)}\mathrm{d}t\right) \\ &= \frac{1}{K(x,z)}\int_0^\infty E^x[t<\tau_D;\ K(X_t,z)]\mathrm{d}t \\ &= \int_D \frac{G(x,y)K(y,z)}{K(x,z)}\mathrm{d}y \\ &\leq 2\,C \sup_{x\in\overline{D}} \int_D |x-y|^{2-d}\mathrm{d}y, \end{aligned}$$

where the inequality follows from Corollary 6.13 by combining two integrals. It is clear that if r is taken such that $m(D) = m(B(0,r))$, then the last integral in the above is not greater than

$$2C\int_{B(0,r)} |y|^{2-d}\mathrm{d}y = C_1 m(D)^{2/d}.$$

□

6.3 Inequalities for Green Functions in a Jordan Domain in $\mathbb{R}^2$

In this section, we consider inequalities for Green functions in the 2-dimensional case. We shall prove the analogues of the results in Section 6.2 for a general class of domains in $\mathbb{R}^2$ called the Jordan domains. Let us start by defining Jordan domains. Let

$$B = \{x \in \mathbb{R}^2 : |x| < 1\}, \quad B^* = \{x \in \mathbb{R}^2 : |x| > 1\}$$

and

$$S = \{x \in \mathbb{R}^2 : |x| = 1\}.$$

A set $\Gamma \subset \mathbb{R}^2$ is called a closed Jordan curve iff there exists a homeomorphism (1–1 continuous mapping) from S onto Γ. In other words, there exists a continuous function f from $[0, 1]$ into $\mathbb{R}^2$ with $f(0) = f(1)$, and $f(t_1) \neq f(t_2)$ for any $0 \leq t_1 < t_2 < 1$, such that $\Gamma = \{f(t) : 0 \leq t < 1\}$. A domain D in $\mathbb{R}^2$ is called a Jordan domain iff D is bounded and ∂D consists of finitely many disjoint closed Jordan curves.

If D is a bounded Lipschitz domain in $\mathbb{R}^2$, then D is a Jordan domain. To see this, we first notice that ∂D consists of finitely many connected paths since otherwise there must be an accumulation point $z \in \partial D$, which contradicts the local representation (17) of ∂D at z. Then for each component Γ of ∂D, there are finitely many balls $B(z_i, s_i)$, $z_i \in \partial D$, $s_i > 0$, $1 \leq i \leq k$ such that $\Gamma \subset \cup_{i=1}^k B(z_i, s_i)$ and the corresponding local representation (17) holds for each $1 \leq i \leq k$:

$$(\partial D) \cap B(z_i, s_i) = \{(\xi, \eta) \in B(z_i, s_i) : \phi_i(\xi) = \eta\}, \tag{53}$$

where ϕ_i is a continuous function. Thus it is easily seen that Γ is a closed Jordan curve by joining the finitely many continuous curves given by (53).

Let D be a Jordan domain in $\mathbb{R}^2$ and G its Green function. The main result is as follows.

Theorem 6.15 (**3G Theorem for** $d = 2$) *Let D be a Jordan domain in $\mathbb{R}^2$. There exists a constant $C = C(D) > 0$ such that for all x, y, and z in D we have*

$$\frac{G(x, y)G(y, z)}{G(x, z)} \leq C[g(x - y) + g(y - z)], \tag{54}$$

where $g(u)$ is given in (8).

To prove Theorem 6.15 we need the following preparations. A closed Jordan curve Γ has an interior and an exterior to be denoted by Int Γ and Ext Γ respectively; these are nonempty open sets with common boundary Γ. If D_1 and D_2 are two domains in $\mathbb{R}^2$ the ϕ is said to be an *extended conformal mapping* from D_1 onto D_2 iff ϕ is a 1–1 conformal mapping from D_1 onto D_2, and also a homeomorphism from $\overline{D}_1$ onto $\overline{D}_2$. We quote the following fundamental theorem (see e.g. Curtiss (1978, Theorem 13.7.1)).

Proposition 6.16 (Extended Riemann Mapping Theorem) *Let Γ be a closed Jordan curve. Then there exists an extended conformal mapping Ψ from Int Γ onto B, and an extended conformal mapping Ψ^* from Ext Γ onto B^*.*

For the Jordan domain D, we have

$$\partial D = \bigcup_{i=0}^{m} \Gamma_i,$$

where Γ_i $(0 \leq i \leq m)$ are disjoint closed Jordan curves and $D \subset$ Int Γ_0.

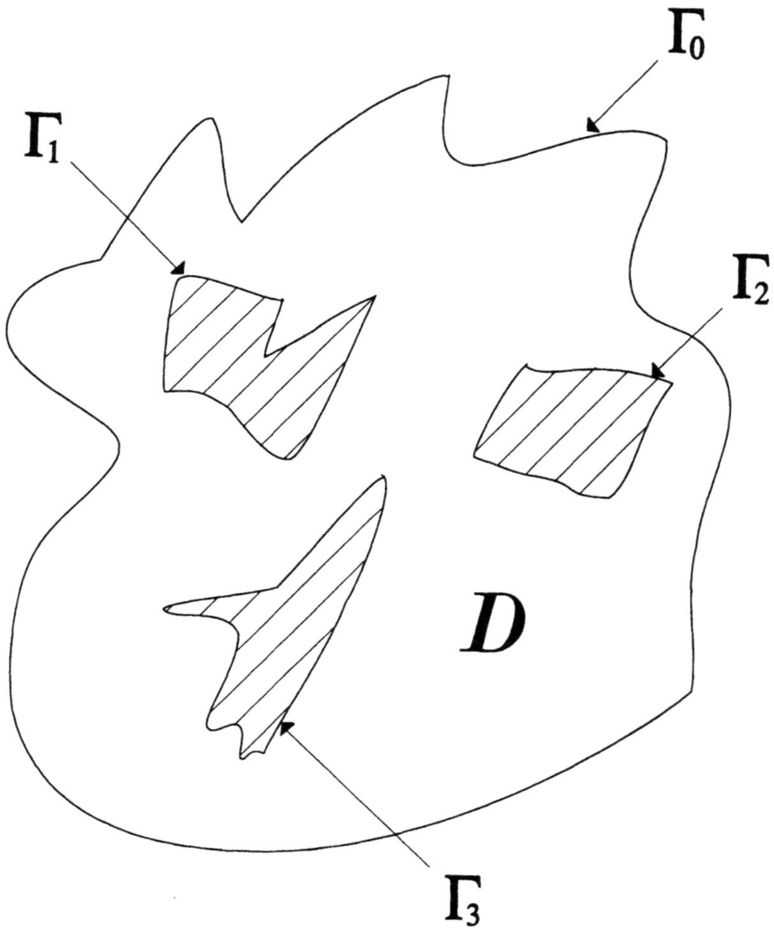

Fig. 6.1.

Lemma 6.17 *For a Jordan domain D, there exists an extended conformal mapping ϕ from D onto a bounded C^∞ domain.*

Proof Since $D = (\text{Int}\,\Gamma_0) \cap (\text{Ext}\,\Gamma_1) \cap \cdots \cap (\text{Ext}\,\Gamma_m)$, we can use Proposition 6.16 $m + 1$ times to obtain the desired mapping ϕ. In the following, we use $\Psi[\Gamma]$ and

$\Psi^*[\Gamma]$ to denote the extended conformal mappings given in Proposition 6.16 for the curve Γ.

We set

$$\Psi_1 = \Psi^*[\Gamma_1],$$

and for $1 < i \le m$ inductively set

$$\Psi_i = \Psi^*[\Psi_{i-1} \cdots \Psi_1(\Gamma_i)],$$

and

$$\Psi_0 = \Psi[\Psi_m \cdots \Psi_1(\Gamma_0)].$$

Thus the desired mapping ϕ can be obtained by restricting $\Psi_0\Psi_m \cdots \Psi_1$ on $\overline{D}$. We note that in the above procedure each curve Γ_i has been mapped into the unit circle once, then the following conformal mappings ensure that it remains a C^∞ curve. □

Lemma 6.18 *Let ϕ be an extended conformal mapping from the Jordan domain D onto E, and let G and G_E be the Green functions of D and E respectively. Then we have for all x and y in D:*

$$G(x, y) = G_E(\phi(x), \phi(y)).$$

Proof We recall that for a bounded regular domain D, $G(\cdot, \cdot)$ is characterized by the following conditions (see the Corollary to Theorem 2.6). For any $y \in D$, we have

(i) $G(\cdot, y)$ is positive and harmonic in $D \backslash \{y\}$;

(ii) $\lim_{x \to \partial D} G(x, y) = 0$;

(iii) $\frac{1}{\pi} \ln \frac{1}{|x-y|} - G(x, y)$ is bounded for $x \in D$.

Hence we need only verify these conditions for the function $F(x, y) \equiv G_E(\phi(x), \phi(y))$.

Since a harmonic function remains harmonic after any analytic transformation of variables, $F(\cdot, y)$ is harmonic in $D \backslash \{y\}$. Since ϕ is a homeomorphism from $\overline{D}$ onto $\overline{E}$, $\lim_{x \to \partial D} F(x, y) = 0$ for $y \in D$. Finally we have

$$\begin{aligned} &\frac{1}{\pi} \ln \frac{1}{|x-y|} - F(x, y) \\ &= \frac{1}{\pi} \ln \frac{|\phi(x) - \phi(y)|}{|x-y|} + \left[\frac{1}{\pi} \ln \frac{1}{|\phi(x) - \phi(y)|} - G_E(\phi(x), \phi(y)) \right]. \quad (55) \end{aligned}$$

Since condition (iii) holds when G is replaced by G_E, the second term on the right-hand side of (55) is bounded for $x \in D$. The first term is bounded because $\phi' \neq 0$ in D for any conformal mapping ϕ. □

Lemma 6.19 *For any ball $B = B(a, r)$, $a \in \mathbb{R}^2$, $r > 0$, and $B^* = \mathbb{R}^2 \backslash \overline{B}$, we have for all x and y in B:*

$$\begin{aligned} \frac{1}{2\pi} \ln\left[1 + \frac{\rho_B(x)\rho_B(y)}{|x-y|^2}\right] &\leq G_B(x, y) \\ &\leq \frac{1}{2\pi} \ln\left[1 + 4\frac{\rho_B(x)\rho_B(y)}{|x-y|^2}\right]. \end{aligned} \tag{56}$$

Suppose that $R > r$. For all x and y in $B^* \cap B(a, R)$, we have

$$\begin{aligned} \frac{1}{2\pi} \ln\left[1 + \frac{\rho_{B^*}(x)\rho_{B^*}(y)}{|x-y|^2}\right] &\leq G_{B^*}(x, y) \\ &\leq \frac{1}{2\pi} \ln\left[1 + 4\frac{R^2}{r^2}\frac{\rho_{B^*}(x)\rho_{B^*}(y)}{|x-y|^2}\right]. \end{aligned} \tag{57}$$

Proof By a change of variables, we need only prove these inequalities for $B = B(0, 1)$ as discussed at the beginning of this section. Using the formula for the Green function of B given in (5.28), we have

$$G_B(x, y) = \frac{1}{\pi} \ln \frac{f(x, y)}{|x-y|} = \frac{1}{2\pi} \ln \frac{f(x, y)^2}{|x-y|^2}, \tag{58}$$

where $f(x, y) = |y|\ |x - y^*|$ and $y^* = \frac{y}{|y|^2}$. We have by (5.30),

$$\begin{aligned} f(x, y)^2 &= |x-y|^2 + (1 - |x|^2)(1 - |y|^2) \\ &= |x-y|^2 + \rho_B(x)\rho_B(y)(1 + |x|)(1 + |y|). \end{aligned} \tag{59}$$

Thus (56) follows from (58) and (59).

For x and y in $B^* \cap B(0, R)$, we have x^* and y^* in B and

$$\begin{aligned} \rho_B(x^*) &= 1 - \frac{1}{|x|} = \frac{\rho_{B^*}(x)}{|x|}; \\ |x^* - y^*| &= \frac{|x-y|}{|x|\ |y|}. \end{aligned}$$

Hence

$$\frac{\rho_B(x^*)\rho_B(y^*)}{|x^* - y^*|^2} = |x|\ |y| \frac{\rho_{B^*}(x)\rho_{B^*}(y)}{|x-y|^2}. \tag{60}$$

Thus (57) follows from (56), (60) and the equality

$$G_{B^*}(x, y) = G_B(x^*, y^*).$$

□

Lemma 6.20 (a) *For any constant $\beta > 0$, there exists a constant $C = C(\beta) > 0$ such that for all $a \in (0, \infty)$ we have*

$$\frac{1}{C} \ln(1 + a) \leq \ln(1 + \beta a) \leq C \ln(1 + a).$$

(b) *For any* $0 < b \leq 4$, *we have*

$$\frac{1}{4}b \leq \ln(1+b) \leq b.$$

These inequalities are proved by elementary calculus.

We shall reduce the proof of Theorem 6.15 to the 'smooth' case when D is a C^∞ domain, using Lemmas 6.17 and 6.18. It turns out that we can give the proof even for a C^2 domain, by using certain general inequalities for its Green function. These inequalities are of independent interest and will be given in Theorem 6.23. We begin with a basic geometric property of a bounded C^2 domain, as follows. There exists a number $r_0 > 0$ depending only on D such that for any $z \in \partial D$ and $r \in (0, r_0]$, we can find two balls B_1^z and B_2^z both with radius r, and lying respectively in the interior and exterior of D, i.e.:

$$B_1^z \subset D, \quad B_2^z \subset (\overline{D})^c. \tag{61}$$

To see this, we note that for a bounded C^2 domain D, by the compactness of ∂D, there exist constants $s_0 > 0$ and $M > 0$ depending only on D such that for each $z \in \partial D$, there exists a local coordinate (ξ, η) and a C^2 function ϕ with $|\phi''(\xi)| \leq M$ such that

$$D \cap B(z, s_0) = \{(\xi, \eta) \in B(z, s_0) : \ \phi(\xi) < \eta\}. \tag{62}$$

We may place the origin of the coordinates at z and use the tangent line of ∂D at z as the x-axis. Thus we have $\phi(0) = \phi'(0) = 0$, hence by the Taylor's expansion we have

$$-M\xi^2 \leq \phi(\xi) \leq M\xi^2. \tag{63}$$

Now we take $r_0 = \frac{s_0}{2} \wedge \frac{1}{2M}$. For any $r \in (0, r_0]$, we set $B_1^z = B((0,r), r)$ and $B_2^z = B((0,-r), r)$. It is easy to verify by (63) that

$$B_1^z \subset \{(\xi, \eta) \in B(z, s_0) : \ \phi(\xi) < \eta\},$$

and

$$B_2^z \subset \{(\xi, \eta) \in B(z, s_0) : \ \phi(\xi) > \eta\}.$$

Thus (61) follows from (62).

Lemma 6.21 *Let D be a bounded C^2 domain in $\mathbb{R}^2$, $z \in \partial D$ and $u \in H_0^+(z, \alpha)$ for some $\alpha > 0$, where H_0^+ is as defined before Lemma 6.8. Then for any sequence $\{x_n\}$ in D with $x_n \to z$ we have*

$$\underline{\lim}_n \frac{u(x_n)}{\rho(x_n)} > 0; \quad \overline{\lim}_n \frac{u(x_n)}{\rho(x_n)} < \infty. \tag{64}$$

Proof Set

$$r = \frac{\alpha}{4} \wedge r_0.$$

Since $x_n \to z$ we may assume that for all $n \geq 1$,

$$|x_n - z| < r. \tag{65}$$

For each $n \geq 1$ there is z_n on ∂D with $|x_n - z_n| = \rho(x_n)$. Let $B_1^z = B(a, r)$ and $B_1^{z_n} = B(a_n, r)$, where B_1^z and $B_1^{z_n}$ are as defined in (61). We have $a_n = z_n + r n_{z_n}$ and $a = z + r n_z$, where n_w is the unit inner normal vector at $w \in \partial D$. Note that n_w is continuous in w on ∂D for a C^2 domain D. Since

$$|z_n - z| \leq |z_n - x_n| + |x_n - z| = \rho(x_n) + |x_n - z| \leq 2|x_n - z|,$$

we have

$$a_n \to a. \tag{66}$$

For $x \in B(a_n, r)$, (65) implies

$$|x - z| \leq |x - z_n| + |z_n - z| < 4r \leq \alpha,$$

thus for $n \geq 1$ we have

$$B(a_n, r) \subset B(z, \alpha).$$

Using Poisson's formula (Theorem 1.13) we have for each $n \geq 1$:

$$\begin{aligned} u(x_n) &= \frac{r^2 - |x_n - a_n|^2}{2\pi r} \int_{\partial B(a_n, r)} \frac{u(w)}{|w - x_n|^2} \sigma(\mathrm{d}w) \\ &= \frac{\rho(x_n)(r + |x_n - a_n|)}{2\pi r} \int_{\partial B(0,r)} \frac{u(w + a_n)}{|w + a_n - x_n|^2} \sigma(\mathrm{d}w). \end{aligned}$$

It follow from (66) and Fatou's lemma that

$$\begin{aligned} \underline{\lim}_n \frac{u(x_n)}{\rho(x_n)} &= \frac{1}{\pi} \underline{\lim}_n \int_{\partial B(0,r)} \frac{u(w + a_n)}{|w + a_n - x_n|^2} \sigma(\mathrm{d}w) \\ &\geq \frac{1}{\pi} \int_{\partial B(0,r)} \frac{u(w + a)}{|w + a - z|^2} \sigma(\mathrm{d}w) > 0, \end{aligned}$$

proving the first inequality in (64).

For each $n \geq 1$, let $B_2^{z_n} = B(b_n, r)$, where $B_2^{z_n}$ is defined in (61), and set

$$A_n = \{x : \ r < |x - b_n| < 2r\}.$$

For each $x \in \overline{A}_n$, we have

$$|x - z| \leq |z - b_n| + |b_n - x_n| + |x_n - z| \leq 5r < 2\alpha,$$

hence

$$\overline{A}_n \subset \overline{B}(z, 5r) \subset B(z, 2\alpha).$$

Since $u \in H_+^0(z, \alpha)$, u vanishes continuously on $(\partial D) \cap B(z, 2\alpha)$, and

$$M = \sup_{x \in D \cap \overline{B}(z,5r)} u(x) < \infty.$$

Consider the Dirichlet boundary value problem:

$$\begin{array}{ll} \Delta h = 0 & x \in A_n; \\ h(x) = M & \text{if } |x - b_n| = 2r; \\ h(x) = 0 & \text{if } |x - b_n| = r. \end{array}$$

Its unique solution h_n can be shown to be

$$h_n(x) = \frac{M}{\ln 2} \ln \frac{|x - b_n|}{r}, \quad x \in A_n.$$

Since $\partial(D \cup A_n) = [\partial A_n \cap D] \cup [\partial D \cap \overline{A}_n]$ and u is bounded by M and zero on $\partial A_n \cap D$ and $\partial D \cap \overline{A}_n$, respectively, u is bounded above by h_n in $D \cap A_n$ by the maximum principle (Proposition 1.10). Since $|x_n - b_n| = \rho(x_n) + r$, we have $x_n \in D \cap A_n$. Hence for each $n \geq 1$,

$$\begin{aligned} u(x_n) \leq h(x_n) &= \frac{M}{\ln 2} \ln \frac{|x_n - b_n|}{r} \\ &= \frac{M}{\ln 2} \ln \left(1 + \frac{\rho(x_n)}{r}\right) \\ &\leq \frac{M}{r \ln 2} \rho(x_n). \end{aligned}$$

This implies the second inequality in (64). □

Lemma 6.22 *For a bounded C^2 domain D in $\mathbb{R}^2$, if x and y satisfy*

$$\frac{1}{C'}[\rho(x) \vee \rho(y)] \leq |x - y| \leq C'[\rho(x) \wedge \rho(y)]$$

for some constant $C' > 1$, then there is another constant $C'' > 1$ depending only on D and C' such that

$$\frac{1}{C''} \leq G(x, y) \leq C''. \tag{67}$$

Proof Since $|x - y| \leq C'[\rho(x) \wedge \rho(y)]$, the lower bound in (67) follows from Lemma 6.7 with $d = 2$.

Let $z \in \partial D$ such that $|x - z| = \rho(x)$, and let B denote the B_2^z defined in (61) and r_0 its radius. Then $D \subset B^*$, $\rho_{B^*}(x) = \rho(x)$ and

$$\rho_{B^*}(y) \leq \rho_{B^*}(x) + |x - y| \leq (C' + 1)|x - y|.$$

Set

$$R = \sup_{x,y \in \overline{D}} |x - y|. \tag{68}$$

Then $D \subset B(z, R)$. Thus we have by Lemma 6.19,

$$\begin{aligned} G(x,y) \leq G_{B^*}(x,y) &\leq \frac{1}{2\pi} \ln\left[1 + \frac{4R^2}{r_0^2} \frac{\rho_{B^*}(x)\rho_{B^*}(y)}{|x-y|^2}\right] \\ &\leq \frac{1}{2\pi} \ln\left[1 + \frac{4R^2}{r_0^2} C'(C'+1)\right], \end{aligned}$$

which establishes the upper bound in (67). □

Theorem 6.23 *For a bounded C^2 domain D in $\mathbb{R}^2$, there exists a constant $C = C(D) > 0$ such that for all x and y in D,*

$$\frac{1}{C} \ln\left(1 + \frac{\rho(x)\rho(y)}{|x-y|^2}\right) \leq G(x,y) \leq C \ln\left(1 + \frac{\rho(x)\rho(y)}{|x-y|^2}\right). \tag{69}$$

Proof By the symmetry of (69) in x and y, we may assume

$$\rho(x) \geq \rho(y). \tag{70}$$

Consider first the case: $\rho(x) \geq 2|x-y|$. Set $B = B(x, \rho(x)) \subset D$. Since $|y-x| < \rho(x)$, $y \in B$. We also have $\rho_B(x) = \rho(x)$ and

$$\rho_B(y) = \rho(x) - |x-y| \geq \frac{1}{2}\rho(x) \geq \frac{1}{2}\rho(y).$$

Hence by Lemma 6.19 we obtain

$$\begin{aligned} G(x,y) &\geq G_B(x,y) \geq \frac{1}{2\pi} \ln\left(1 + \frac{\rho_B(x)\rho_B(y)}{|x-y|^2}\right) \\ &\geq \frac{1}{2\pi} \ln\left(1 + \frac{\rho(x)\rho(y)}{2|x-y|^2}\right). \end{aligned} \tag{71}$$

Let $z \in \partial D$ be such that $|x-z| = \rho(x)$ and let $B = B_2^z$ as in the preceding proof. Then $D \subset B^*$, $\rho_{B^*}(x) = \rho(x)$ and

$$\rho_{B^*}(y) \leq \rho_{B^*}(x) + |x-y| = \rho(x) + |x-y|. \tag{72}$$

Since

$$\rho(x) \leq \rho(y) + |x-y| \leq \rho(y) + \frac{1}{2}\rho(x),$$

we have

$$\rho(x) \leq 2\rho(y) \text{ and } |x-y| \leq \frac{1}{2}\rho(x) \leq \rho(y). \tag{73}$$

It follows from (72) and (73) that

$$\rho_{B^*}(y) \leq 3\rho(y).$$

Using Lemma 6.19, we obtain

$$G(x,y) \le G_{B^*}(x,y) \le \frac{1}{2\pi}\ln\left(1+\frac{4R^2}{r_0^2}\frac{\rho_{B^*}(x)\rho_{B^*}(y)}{|x-y|^2}\right)$$
$$\le \frac{1}{2\pi}\ln\left(1+\frac{12R^2}{r_0^2}\frac{\rho(x)\rho(y)}{|x-y|^2}\right), \tag{74}$$

where R is defined as in (68). Thus (69) follows from (71), (74) and Lemma 6.20(a).

From now on we may assume that

$$\rho(x) < 2|x-y|. \tag{75}$$

Then we have by (70) and (75),

$$\frac{\rho(x)\rho(y)}{|x-y|^2} \le 4. \tag{76}$$

Therefore by Lemma 6.20(b), the inequalities in (69) are equivalent to

$$\frac{1}{C}\frac{\rho(x)\rho(y)}{|x-y|^2} \le G(x,y) \le C\frac{\rho(x)\rho(y)}{|x-y|^2}. \tag{77}$$

Consider first the case $|x-y| < r_0$. Let x^* and y^* be in ∂D such that $|x-x^*| = \rho(x)$ as in the proof of Lemma 6.6.

If $\rho(y) > \frac{1}{8}|x-y|$, then we have

$$\frac{1}{2}[\rho(x)\vee\rho(y)] \le |x-y| < 8[\rho(x)\wedge\rho(y)]. \tag{78}$$

It then follows from Lemma 6.22 that there exists $C > 1$ such that

$$\frac{1}{C} \le G(x,y) \le C; \tag{79}$$

hence (77) follows from (78) and (79).

If $\rho(y) \le \frac{1}{8}|x-y| < \rho(x)$, then for $i = 1,2$ let

$$y_i = y^* + \frac{i|x-y|}{8}n_{y^*}.$$

We have

$$\rho(y_i) = \frac{i}{8}|x-y| \text{ and } |y_1-y_2| = \frac{1}{8}|x-y|,$$

hence by Lemma 6.22,

$$\frac{1}{C} \le G(y_1,y_2) \le C. \tag{80}$$

Since

$$\frac{1}{8}|x-y| < \rho(x) < 2|x-y|, \tag{81}$$

and

$$|y - y_1| \le \frac{1}{8}|x - y|,$$

we have

$$\frac{7}{16}[\rho(x) \vee \rho(y_1)] \le |x - y_1| \le 9[\rho(x) \wedge \rho(y_1)];$$

hence it follows from Lemma 6.22 that

$$\frac{1}{C} \le G(x, y_1) \le C. \tag{82}$$

Noting that

$$|x - y^*| \ge |x - y| - \rho(y) \ge \frac{|x - y|}{4}, \quad \text{and} \quad |y_2 - y^*| = \frac{|x - y|}{4},$$

we see that both $G(x, \cdot)$ and $G(y_2, \cdot)$ belong to $H_0^+(y^*, \frac{|x-y|}{8})$. Hence we have by Lemma 6.8:

$$\frac{1}{C}\frac{G(y_2, y)}{G(y_2, y_1)} \le \frac{G(x, y)}{G(x, y_1)} \le C\frac{G(y_2, y)}{G(y_2, y_1)}. \tag{83}$$

It then follows from (80)–(83) that

$$\frac{1}{C}G(y_2, y) \le G(x, y) \le C\, G(y_2, y). \tag{84}$$

Let $B_1 = B_1^{y^*}$ and $B_2 = B_2^{y^*}$ be as defined in (61). Noting that

$$\frac{1}{4}|x - y| \le |y - y_2| \le \frac{1}{2}|x - y|,$$

we have by Lemma 6.19 and (75),

$$\begin{aligned} G(y_2, y) &\ge G_{B_1}(y_2, y) \ge \frac{1}{2\pi}\ln\left(1 + \frac{\rho_{B_1}(y_2)\rho_{B_1}(y)}{|y - y_2|^2}\right) \\ &\ge C'\frac{\rho(y)}{|x - y|}, \end{aligned} \tag{85}$$

and

$$\begin{aligned} G(y_2, y) &\le G_{B_2^*}(y_2, y) \le \frac{1}{2\pi}\ln\left(1 + \frac{4R^2}{r_0^2}\frac{\rho_{B_2^*}(y_2)\rho_{B_2^*}(y)}{|y - y_2|^2}\right) \\ &\le C''\frac{\rho(y)}{|x - y|}. \end{aligned} \tag{86}$$

Thus we obtain (77) from (81) and (84)–(86).

If $\rho(x) \le \frac{1}{8}|x - y|$, then for $i = 1, 2$ we also set

$$x_i = x^* + \frac{i|x - y|}{8}n_{x^*}.$$

As in (80) we have,

$$\frac{1}{C} \leq G(x_1, x_2) \leq C. \tag{87}$$

Noting that

$$\begin{aligned} |x_1 - y_1| &\leq |x - y| + \rho(x) + \rho(y) + |x_1 - x^*| + |y_1 - y^*| \\ &\leq \frac{3}{2}|x - y| = 12[\rho(x_1) \wedge \rho(y_1)], \end{aligned}$$

and

$$\begin{aligned} |x_1 - y_1| &\geq |x - y| - \rho(x) - \rho(y) - |x_1 - x^*| - |y_1 - y^*| \\ &\geq \frac{1}{2}|x - y| = 4[\rho(x_1) \vee \rho(y_1)], \end{aligned}$$

we have by Lemma 6.22,

$$\frac{1}{C} \leq G(x_1, y_1) \leq C. \tag{88}$$

Since

$$|y_1 - x^*| \geq |x - y| - |y_1 - y| - |x - x^*| \geq \frac{|x - y|}{4},$$

and

$$|x_2 - x^*| = \frac{|x - y|}{4},$$

both $G(\cdot, y_1)$ and $G(\cdot, x_2)$ belong to $H_0^+(x^*, \frac{|x-y|}{8})$. Hence we have by Lemma 6.8:

$$\frac{1}{C}\frac{G(x, x_2)}{G(x_1, x_2)} \leq \frac{G(x, y_1)}{G(x_1, y_1)} \leq C\frac{G(x, x_2)}{G(x_1, x_2)}. \tag{89}$$

As before in (85) and (86), we now have

$$C'\frac{\rho(x)}{|x - y|} \leq G(x, x_2) \leq C''\frac{\rho(x)}{|x - y|}. \tag{90}$$

Then using the equation

$$G(x, y) = \frac{G(x, y)}{G(x, y_1)}\frac{G(x, y_1)}{G(x_1, y_1)}G(x_1, y_1),$$

we obtain (77) from (83)–(90).

It remains to consider the case $|x - y| \geq r_0$. If (77) is not true then we can find two sequences $\{x_n\}$ and $\{y_n\}$ in D with $|x_n - y_n| \geq r_0$ such that as $n \to \infty$,

$$\frac{G(x_n, y_n)}{\rho(x_n)\rho(y_n)} \to \infty \text{ or } 0. \tag{91}$$

We may assume that $x_n \to x_0$, $y_n \to y_0$, and

$$|x_n - x_0| < \frac{r_0}{4}, \quad |y_n - y_0| < \frac{r_0}{4},$$

by taking subsequences. Then $|x_0 - y_0| \geq r_0$. If both x_0 and y_0 are in D, then

$$\frac{G(x_n, y_n)}{\rho(x_n)\rho(y_n)} \to \frac{G(x_0, y_0)}{\rho(x_0)\rho(y_0)}, \tag{92}$$

which contradicts (91).

If $x_0 \in D$ and $y_0 \in \partial D$, then since $|x_n - y_0| \geq \frac{r_0}{2}$, $G(x_n, \cdot) \in H_0^+(y_0, \frac{r_0}{4})$ and $y_n \in D \cap B(y_0, \frac{r_0}{4})$ for all $n \geq 1$. Using Lemma 6.8 we have

$$G(x_n, y_n) = \frac{G(x_n, y_n)}{G(x_1, y_n)} G(x_1, y_n) \leq C \frac{G(x_n, y_1)}{G(x_1, y_1)} G(x_1, y_n). \tag{93}$$

Hence by Lemma 6.21, we obtain

$$\overline{\lim}_n \frac{G(x_n, y_n)}{\rho(x_n)\rho(y_n)} \leq C \frac{G(x_0, y_1)}{\rho(x_0) G(x_1, y_1)} \overline{\lim}_n \frac{G(x_1, y_n)}{\rho(y_n)} < \infty.$$

Similarly

$$G(x_n, y_n) \geq \frac{1}{C} \frac{G(x_n, y_1)}{G(x_1, y_1)} G(x_1, y_n) \tag{94}$$

and

$$\underline{\lim}_n \frac{G(x_n, y_n)}{\rho(x_n)\rho(y_n)} \geq \frac{1}{C} \frac{G(x_0, y_1)}{\rho(x_0) G(x_1, y_1)} \underline{\lim}_n \frac{G(x_1, y_n)}{\rho(y_n)} > 0.$$

These relations contradict (91). The case in which $x_0 \in \partial D$ and $y_0 \in D$ leads to a contradiction by symmetry.

Finally if $x_0 \in \partial D$ and $y_0 \in \partial D$, then by (93), (94), and Lemma 6.21 we have,

$$\overline{\lim}_n \frac{G(x_n, y_n)}{\rho(x_n)\rho(y_n)} \leq \frac{C}{G(x_1, y_1)} \overline{\lim}_n \frac{G(x_n, y_1)}{\rho(x_n)} \overline{\lim}_n \frac{G(x_1, y_n)}{\rho(y_n)} < \infty,$$

and

$$\underline{\lim}_n \frac{G(x_n, y_n)}{\rho(x_n)\rho(y_n)} \geq \frac{1}{C G(x_1, y_1)} \underline{\lim}_n \frac{G(x_n, y_1)}{\rho(x_n)} \underline{\lim}_n \frac{G(x_1, y_n)}{\rho(y_n)} > 0.$$

These relations contradict (91).

We have therefore proved (77) in all cases. □

We are now ready for a preliminary statement of our final result. Again, and with apologies to the impatient reader, the proof revolves around the careful inspection of a number of subcases.

Theorem 6.24 *For a Jordan domain D in $\mathbb{R}^2$, there exists a constant $C = C(D) > 0$ such that for all x, y and $z \in D$, we have*

$$\frac{G(x, y) G(y, z)}{G(x, z)} \leq C[G(x, y) + G(y, z) + 1]. \tag{95}$$

Proof We first prove (95) for a bounded C^2 domain in $\mathbb{R}^2$. By Theorem 6.23, we need only verify (95) when $G(x,y)$ is replaced by

$$F(x,y) \equiv \ln\left(1 + \frac{\rho(x)\rho(y)}{|x-y|^2}\right).$$

Let

$$f(x,y) = \frac{\rho(x)\rho(y)}{|x-y|^2}.$$

By the symmetry of (95) in x and z, we may assume that

$$f(x,y) \ge f(y,z). \tag{96}$$

If $f(x,y) \ge 4$, then $\rho(x) \vee \rho(y) \ge 2|x-y|$. Since

$$\begin{aligned} \rho(x) \vee \rho(y) &\le |x-y| + [\rho(x) \wedge \rho(y)] \\ &\le \frac{1}{2}[\rho(x) \vee \rho(y)] + [\rho(x) \wedge \rho(y)], \end{aligned}$$

we have

$$\rho(x) \wedge \rho(y) \ge \frac{1}{2}[\rho(x) \vee \rho(y)]. \tag{97}$$

If $f(x,y) < 4$, then $\rho(x) \wedge \rho(y) < 2|x-y|$, hence

$$\rho(x) \vee \rho(y) \le |x-y| + [\rho(x) \wedge \rho(y)] < 3|x-y|. \tag{98}$$

The proof of the theorem will be divided into four overlapping cases which exhaust the possibilities.

Case 1: $f(y,z) \ge 4$. We have by (96) and (97),

$$\rho(x)\rho(z) \ge \frac{1}{2}[\rho(x)\rho(y) \vee \rho(y)\rho(z)].$$

This together with the inequality $|x-z| \le 2(|x-y| \vee |y-z|)$ implies that

$$f(x,z) \ge \frac{1}{8}[f(x,y) \wedge f(y,z)],$$

and consequently,

$$\frac{F(x,y)F(y,z)}{F(x,z)} \le 8[F(x,y) \vee F(y,z)]. \tag{99}$$

Case 2: $f(y,z) < 4$ and $f(x,z) \ge 4$. Since

$$F(y,z) < \ln 5 \le F(x,z),$$

we obtain

$$\frac{F(x,y)F(y,z)}{F(x,z)} \leq F(x,y). \tag{100}$$

Case 3: $f(y,z) < 4 \leq f(x,y)$ and $f(x,z) < 4$. We have by Lemma 6.20(b),

$$\begin{aligned}\frac{F(x,y)F(y,z)}{F(x,z)} &\leq 4\, F(x,y)\frac{f(y,z)}{f(x,z)} \\ &= 4\, F(x,y)\frac{\rho(y)|x-z|^2}{\rho(x)|y-z|^2}.\end{aligned} \tag{101}$$

We have by (97),

$$\frac{1}{2}\rho(x) \leq \rho(y) \leq 2\rho(x), \tag{102}$$

and by (98) with x and y replaced by y and z, respectively:

$$\begin{aligned}|x-y| &\leq \frac{1}{2}[\rho(x) \vee \rho(y)] \leq \rho(y) \\ &\leq \rho(y) \vee \rho(z) < 3|y-z|.\end{aligned} \tag{103}$$

By (101)–(103) and the inequality

$$|x-z|^2 \leq 2(|x-y|^2 + |y-z|^2)$$

we obtain

$$\frac{F(x,y)F(y,z)}{F(x,z)} \leq 160\, F(x,y). \tag{104}$$

Case 4: $f(x,y) < 4$, $f(y,z) < 4$, and $f(x,z) < 4$. By Lemma 6.20(b) and (98) for (x,y) and (y,z) respectively, we have

$$\begin{aligned}\frac{F(x,y)F(y,z)}{F(x,z)} &\leq 4\frac{f(x,y)f(y,z)}{f(x,z)} \\ &= 4\frac{\rho(y)^2|x-z|^2}{|y-z|^2|x-y|^2} \\ &\leq 8\left[\frac{\rho(y)^2}{|x-y|^2} + \frac{\rho(y)^2}{|y-z|^2}\right] \\ &\leq 144.\end{aligned} \tag{105}$$

Thus it follows from (99), (100), (104), and (105) that for all x, y, and z in $\mathbb{R}^2$:

$$\frac{F(x,y)F(y,z)}{F(x,z)} \leq 160[F(x,y) + F(y,z) + 1], \tag{106}$$

hence we obtain (95) for a bounded C^2 domain.

Now for a Jordan domain D in $\mathbb{R}^2$, by Lemma 6.17 and Lemma 6.18, there exists an extended conformal mapping ϕ from D to a bounded C^∞ domain E with

$$G(x,y) = G_E(\phi(x), \phi(y)). \tag{107}$$

Since (95) holds when G is replaced by G_E, and x, y, and z are replaced by $\phi(x)$, $\phi(y)$, and $\phi(z)$, respectively, it holds as stands by substituting from(107). □

Theorem 6.15 is an easy consequence of Theorem 6.24.

Proof of Theorem 6.15 By Theorem 2.6(ii) and mindful of (8), we have for all x and y in D:

$$G(x,y) \le C\ g(x-y).$$

Hence the right member of (95) does not exceed

$$C_1\left[\left(\ln\frac{1}{|x-y|}\vee 1\right)+\left(\ln\frac{1}{|y-z|}\vee 1\right)+1\right] \le 2C_1[g(x-y)+g(y-z)].$$

□

Corollary 6.25 *For a bounded Lipschitz domain D in $\mathbb{R}^2$, there exists a constant $C = C(D) > 0$ such that*

$$\frac{G(x,y)K(y,z)}{K(x,z)} \le C[g(x-y)+g(y-z)], \tag{108}$$

where $K(\cdot,\cdot)$ is the kernel function of D and $g(x) = \ln\frac{1}{|x|}\vee 1$.

This follows from Theorem 6.24 in the same way that Corollary 6.13 follows from Theorem 6.5.

Corollary 6.25 is only stated for the special case of a bounded Lipschitz domain because this is the case for which the kernel function was defined in Section 5.2. However, the proof required only the equation (5.8):

$$K(x,z) = \lim_n \frac{G(x,y_n)}{G(x_0,y_n)}$$

as $y_n \to z$. For a general domain, this equation defines the Martin kernel K. Hence the validity of the corollary may be extended to the case of a Jordan domain provided K is interpreted in this way.

We end this chapter with another application of Theorem 6.23 and the conformal mapping method. The result will be used in the proof of Theorem 7.27.

Proposition 6.26 *Let D be a Jordan domain in $\mathbb{R}^2$. Then for any $(\xi, w) \in \overline{D}\times D$ with $\xi \neq w$, we have*

$$\lim_{D\times D\ni(x,y)\to(\xi,\xi)} \frac{G(x,w)G(w,y)}{G(x,y)} = 0. \tag{109}$$

Proof By Lemma 6.17 there exists an extended conformal mapping ϕ from D to a bounded C^∞ domain E. Let $\tilde{x} = \phi(x)$, $\tilde{y} = \phi(y)$, $\tilde{w} = \phi(w)$, $\alpha = \rho_E(\tilde{x})$, $\beta = \rho_E(\tilde{y})$ and $a = \rho_E(\tilde{w})$. Then it follows from Lemma 6.18 and Theorem 6.23 that

$$
\begin{aligned}
&\frac{G(x,w)G(w,y)}{G(x,y)} \\
&\le C \ln\left(1+\frac{\alpha a}{|\tilde{x}-\tilde{w}|^2}\right)\ln\left(1+\frac{\beta a}{|\tilde{y}-\tilde{w}|^2}\right)\left[\ln\left(1+\frac{\alpha\beta}{|\tilde{x}-\tilde{y}|^2}\right)\right]^{-1} \\
&\le \frac{Ca^2}{|\tilde{x}-\tilde{w}|^2|\tilde{y}-\tilde{w}|^2}\cdot\left\{\alpha\beta\left[\ln\left(1+\frac{\alpha\beta}{|\tilde{x}-\tilde{y}|^2}\right)\right]^{-1}\right\}. \qquad (110)
\end{aligned}
$$

Since

$$
\lim_{x\to\xi}|\tilde{x}-\tilde{w}| = \lim_{y\to\xi}|\tilde{y}-\tilde{w}| = |\phi(\xi)-\phi(w)| > 0,
$$

the first factor in the last member of (110) converges to a finite limit as $(x,y)\to(\xi,\xi)$. Since $\alpha\beta$ is bounded and $|\tilde{x}-\tilde{y}|\to|\phi(\xi)-\phi(\xi)|=0$ as $(x,y)\to(\xi,\xi)$, the second factor converges to zero by elementary calculus. □

Notes on Chapter 6

A recent probabilistic proof of the boundary Harnack principle (Lemma 6.8), see Bass and Burdzy (1989), may be still further probabilisable.

The $3G$ Theorem for the case $d \ge 3$ is given in Cranston, Fabes and Zhao (1988). The exposition has been radically improved and expanded to make it more accessible to the reluctant reader. We are indebted to Eugene Fabes for looking over this part of the manuscript and sending us his comments. In the case $d = 2$, the result is partially given in Zhao (1987). G. Sweers has pointed out a mistake in the latter which is corrected here. The inequality (95) has been proved recently in McConnell (1990) for simply connected Greenian domains in $\mathbb{R}^2$, where C may be taken to be an absolute constant.

The exact order bound given in Theorem 6.23 for a C^2-domain in $\mathbb{R}^2$ seems new; the analogous result in $\mathbb{R}^d$, $d \ge 3$ is known but not needed here.

7. Conditional Gauge and q-Green Function

7.1 Conditional Gauge Theorem

In Section 5.4, we proved the Gauge Theorem for a bounded domain D: if the gauge $u(x) = E^x\{e_q(\tau_D)\} \not\equiv \infty$ in D, then it is bounded in $\overline{D}$. In this section, we shall prove the Conditional Gauge Theorem for a bounded Lipschitz domain D: if the conditional gauge $u(x,z) = E^x_z\{e_q(\tau_D)\} \not\equiv \infty$ in $D \times \partial D$, then it is bounded in $D \times \partial D$. For the gauge theorem, no assumption about the boundary is imposed, not even its regularity in the Dirichlet sense. By contrast, the conditional gauge theorem requires a certain smoothness of the boundary. Ad hoc assumptions on D and q may be and have been considered, but we shall settle the case in which D is a bounded Lipschitz domain in $\mathbb{R}^d$, $d \geq 2$ and $q \in J_{\text{loc}}$. For the case $d = 1$, see Theorem 9.9 and the Appendix to Section 9.2.

Let $G(\cdot,\cdot)$ denote the Green function and $K(\cdot,\cdot)$ the kernel function for D (see Section 5.2). The key to the conditional gauge theorem is the following lemma which uses the 3G Theorem and its corollary in Sections 6.2 and 6.3.

Lemma 7.1 *Given $\varepsilon > 0$, there exists a constant $\delta(\varepsilon) = \delta(\varepsilon, D, q) > 0$ such that for any open subset U of D with $m(U) \leq \delta(\varepsilon)$, we have*

$$\sup_{(x,z)\in D\times\partial D} E^x_z\left\{\int_0^{\tau_U} |q(X_t)|\mathrm{d}t\right\} < \varepsilon. \tag{1}$$

Proof Recalling the definition of p_h^D with $h = K(\cdot, z)$ from Section 5.1 and using Fubini's Theorem a number of times, we have for any $(x,z) \in D \times \partial D$:

$$\begin{aligned}
E^x_z\left\{\int_0^{\tau_U} |q(X_t)|\mathrm{d}t\right\} &\leq E^x_z\left\{\int_0^{\tau_D} 1_U(X_t)|q(X_t)|\mathrm{d}t\right\} \\
&= \int_0^\infty \int_U p^D_{K(\cdot,z)}(t;x,y)|q(y)|\mathrm{d}y\,\mathrm{d}t \\
&= \frac{1}{K(x,z)} \int_0^\infty \int_U p^D(t;x,y)|q(y)|K(y,z)\mathrm{d}y\,\mathrm{d}t \\
&= \frac{1}{K(x,z)} \int_U G(x,y)|q(y)|K(y,z)\mathrm{d}y.
\end{aligned}$$

Now by (6.50) for $d \geq 3$, and (6.108) for $d = 2$, the last integral in the above is bounded by

$$C \int_U [g(x-y) + g(y-z)]|q(y)|\mathrm{d}y,$$

where C is a constant depending only on D and q. We recall that for $d \geq 3$, $g(x) = g(|x|) = |x|^{2-d}$; while for $d = 2$,

$$g(x) = g(|x|) = \left(\ln \frac{1}{|x|}\right) \vee 1, \tag{2}$$

as in Theorem 6.2. Hence we need only prove that

$$\sup_{x \in D} \int_U g(x-y)|q(y)|\mathrm{d}y \to 0 \text{ as } m(U) \to 0. \tag{3}$$

This is just a consequence of the last assertion of the Corollary to Proposition 3.1. □

The next result is a simple topological lemma.

Lemma 7.2 *Let D be a bounded domain. Given any $\delta > 0$, there exist two subdomains D_0 and U of D such that $D_0 \subset\subset D$, $m(U) < \delta$ and $D = D_0 \cup U$.*

Proof Define for $a > 0$:

$$U(a) = \{x \in D : d(x, \partial D) < a\}. \tag{4}$$

Then $U(a)$ is open and $m(U(a)) \to 0$ as $a \to 0$. Choose a so that $m(U(a)) < \frac{\delta}{2}$. Since $D \backslash U(a)$ is a compact set in D, there exist finite balls $B_i \subset\subset D$, $i = 1, \ldots, m$ such that $D \backslash U(a) \subset \cup_{i=1}^m B_i$. Obviously, the union of the two open sets $\cup_{i=1}^m B_i$ and $U(a)$ is D. To make these sets connected, we proceed as follows. For any curve Γ in D and $r > 0$, we define a 'tube' around Γ:

$$\beta(r) = \{x : d(x, \Gamma) < r\}.$$

Then $m(\beta(r)) \to 0$ as $r \to 0$ and for sufficiently small $r > 0$, $\overline{\beta(r)} \subset D$. Since D is connected, any two balls B_i and B_j can be joined by a curve in D, hence by a tube β_{ij} with $\overline{\beta_{ij}} \subset D$. Let

$$D_0 = \left[\cup_{i=1}^m B_i\right] \bigcup \left[\cup_{1 \leq i < j \leq m} \beta_{ij}\right].$$

If the connected components of $U(a)$ are $\{U_n\}$, then for each n there is a tube β_n connecting U_n and U_{n+1} with $m(\beta_n) < \frac{\delta}{2^{n+2}}$. Let $U = \cup_n (U_n \cup \beta_n)$. Then D_0 and U satisfy the conditions of the lemma. □

We take $\varepsilon = \frac{1}{2}$ in Lemma 7.1 and use the corresponding $\delta(\frac{1}{2})$ for the δ in Lemma 7.2 to obtain D_0 and U. Observe that $\partial D_0 \subseteq U$ and $\partial D \subseteq \partial U$. We can now state the next important step.

Lemma 7.3 *There exist two constants $C_1 > 0$ and $C_2 > 0$ such that for all $(y, z) \in \partial D_0 \times \partial D$:*

$$C_1 \leq E_z^y\{\tau_D = \tau_U, e_q(\tau_D)\} \leq C_2. \tag{5}$$

Proof Our first fundamental observation is that the conditional Brownian motion is defined as a Markov process in Section 5.1. Hence, Khas'minskii's lemma (Lemma 3.7) is applicable to it and yields, by (1) with $\varepsilon = \frac{1}{2}$:

$$E_z^x\{e_{|q|}(\tau_U)\} \leq 2 \tag{6}$$

for $(x, z) \in D \times \partial D$. This is a stronger result than the right-hand inequality in (5).

To prove the left-hand inequality in (5), we first prove the crucial inequality below:

$$\inf_{(y,z)\in\partial D_0\times\partial D} P_z^y\{\tau_D = \tau_U\} > 0. \tag{7}$$

By Proposition 5.3, we have

$$P_z^y\{\tau_U < \tau_D\} = \frac{E^y\{\tau_U < \tau_D; K(X(\tau_U), z)\}}{K(y, z)}. \tag{8}$$

The function of y in the numerator above is harmonic in U; indeed it is the solution of the Dirichlet boundary value problem in U with boundary value $K(\cdot, z)$ on $\partial U \backslash \partial D$, and boundary value 0 on ∂D. Its harmonicity follows from Theorem 1.23. On the other hand, $K(\cdot, z)$ is harmonic even in D by definition. We state the following simple lemma, which is a consequence of the maximum principle (Proposition 1.10), for ready reference.

Lemma 7.4 *If h_1 and h_2 are two harmonic functions in a domain U such that $h_1 \leq h_2$ in U, but $h_1 \not\equiv h_2$ in U. Then $h_1 < h_2$ in U.*

We apply this lemma to the two functions of y in the right member of (8) for fixed z. Since the numerator is bounded while the denominator is unbounded in D because z is a 'pole' (see the remark after Proposition 5.8), we infer that the function of (y, z) given in (8) is strictly less than one in $U \times \partial D$. We shall now show that it is jointly continuous in $U \times \partial D$. For this purpose we need only consider the numerator there:

$$\phi(y, z) = E^y\{\tau_U < \tau_D; K(X(\tau_U), z)\}.$$

For each $z \in \partial D$ we know from the above that $\phi(\cdot, z)$ is continuous in U; hence it is sufficient to show that the set of functions $\{\phi(y, \cdot), y \in U\}$ is equi-continuous on ∂D. Let $z \in \partial D$ and $z' \in \partial D$, then we have

$$\begin{aligned} |\phi(y, z) - \phi(y, z')| &\leq \int_{\partial U \backslash \partial D} H(y, \mathrm{d}w)|K(w, z) - K(w, z')| \\ &\leq \sup_{w \in \partial U \backslash \partial D} |K(w, z) - K(w, z')|, \end{aligned}$$

where H is the harmonic measure for U defined in (1.37). Since $K(\cdot, \cdot)$ is jointly continuous in $D \times \partial D$, the above supremum converges to zero as $|z - z'| \to 0$. This is the desired result.

Therefore the function of (y,z) in (8) is bounded away from one on the compact set $\partial D_0 \times \partial D$, which is equivalent to the inequality (7).

Denoting the infimum in (7) by C, and using Jensen's inequality for the conditional expectation (see Chung (1974, Theorem 9.1.4)) we obtain, by a lucky computation (lucky because all the inequalities go in the right direction!):

$$\begin{aligned}
&E_z^y\{\tau_D = \tau_U;\ e_q(\tau_D)\} \\
&\geq P_z^y(\tau_D = \tau_U)\exp\left\{E_z^y\left[\int_0^{\tau_U} q(X_t)\mathrm{d}t \,\middle|\, \tau_D = \tau_U\right]\right\} \\
&\geq C\exp\left\{E_z^y\left[-\int_0^{\tau_U} |q(X_t)|\mathrm{d}t \,\middle|\, \tau_D = \tau_U\right]\right\} \\
&\geq C\exp\left\{-\frac{1}{C}E_z^y\left[\int_0^{\tau_U} |q(X_t)|\mathrm{d}t\right]\right\} \geq C\exp\left(-\frac{1}{2C}\right).
\end{aligned}$$

In the last step above we have used (1) again with $\varepsilon = 1/2$. □

In what follows we shall use the alternative notation:

$$\begin{aligned}
u(x) &= E^x\{e_q(\tau_D)\}, \\
u(x,z) &= E_z^x\{e_q(\tau_D)\}.
\end{aligned}$$

Theorem 7.5 (Conditional Gauge Theorem) *Let D be a bounded Lipschitz domain in $\mathbb{R}^d$, $d \geq 2$; and $q \in J_{\text{loc}}$. If $u(x,z) \not\equiv \infty$ in $D \times \partial D$, then $u(x,z)$ is bounded in $D \times \partial D$. This is the case if and only if (D,q) is gaugeable.*

Proof With D_0 and U as specified above for Lemma 7.3, we define a sequence of hitting times as follows:

$$\begin{aligned}
T_0 = 0,\ T_{2n-1} &= T_{2n-2} + \tau_{D_0} \circ \theta_{T_{2n-2}} \\
T_{2n} &= T_{2n-1} + \tau_U \circ \theta_{T_{2n-1}};\ \ n \geq 1.
\end{aligned}$$

By Theorems 5.7 and 6.14, under any P_z^x we have $\tau_D < \infty$ a.s., and consequently all $T_n \leq \tau_D < \infty$ a.s. Observe that if $T_{2n} = \tau_D$, then since $X(t) = \partial$ for all $t \geq \tau_D$, all $T_k = \tau_D$ for $k \geq 2n$. Since T_n increases with n, let $T = \lim_n T_n \leq \tau_D$. Suppose that $T < \tau_D$, then there are infinitely many oscillations of the z-path between ∂D_0 and $\partial U \backslash \partial D$ in the time interval (T_{n_0}, T) for all n_0, contradicting the continuity of the z-path at T (Proposition 5.5), therefore $T = \tau_D$. Now by Theorem 5.9, $X(t) \to z$ as $t \uparrow\uparrow \tau_D$. This is impossible if $T_n < \tau_D$ for all n, since $X(T_{2n-1}) \in \partial D_0$ and $d(z, \partial D_0) > 0$. Therefore $T_n = \tau_D$ for some n, and that can only happen for an even value of n.

As a consequence, and using the strong Markov property of the conditional process (Proposition 5.4), we can write

$$\begin{aligned}
E_z^x\{e_q(\tau_D)\} &= \sum_{n=1}^{\infty} E_z^x\{T_{2n-2} < \tau_D,\ \tau_D = T_{2n};\ e_q(\tau_D)\} \\
&= \sum_{n=1}^{\infty} E_z^x\left\{T_{2n-2} < \tau_D;\ e_q(T_{2n-1})E_z^{X(T_{2n-1})}[\tau_D = \tau_U;\ e_q(\tau_D)]\right\}.
\end{aligned}$$

Now on $\{T_{2n-2} < \tau_D\}$, we have $X(T_{2n-1}) \in \partial D_0$; hence by Lemma 7.3, the nth term in the above, say u_n, satisfies the inequalities

$$C_1 E_z^x\{T_{2n-2} < \tau_D;\ e_q(T_{2n-1})\} \leq u_n \leq C_2 E_z^x\{T_{2n-2} < \tau_D;\ e_q(T_{2n-1})\}.$$

By Proposition 5.3, the above conditional expectation is equal to

$$\frac{1}{K(x,z)} E^x\{T_{2n-2} < \tau_D;\ K(X(T_{2n-1}), z) e_q(T_{2n-1})\}.$$

There exist constants $C_3 > 0$ and $C_4 > 0$ such that for all $x \in \overline{D}_0$, $y \in \partial D_0$ and $z \in \partial D$, we have

$$C_3 \leq \frac{K(y,z)}{K(x,z)} \leq C_4, \tag{9}$$

because $K(\cdot,\cdot)$ is continuous in $D \times \partial D$ and $\overline{D}_0 \times \partial D$ is a compact subset of $D \times \partial D$. It follows that

$$\begin{aligned} C_1 C_3 E^x\{T_{2n-2} < \tau_D;\ e_q(T_{2n-1})\} &\leq u_n \\ &\leq C_2 C_4 E^x\{T_{2n-2} < \tau_D;\ e_q(T_{2n-1})\}, \end{aligned}$$

and so

$$\begin{aligned} C_1 C_3 \sum_{n=1}^{\infty} E^x\{T_{2n-2} < \tau_D;\ e_q(T_{2n-1})\} &\leq E_z^x\{e_q(\tau_D)\} \\ \leq C_2 C_4 \sum_{n=1}^{\infty} E^x\{T_{2n-2} < \tau_D;\ e_q(T_{2n-1})\}. \end{aligned}$$

To simplify the formulae, we introduce the symbol $\approx$ as follows:

$$f_1(x,z) \approx f_2(x,z) \text{ for } (x,z) \in A$$

if there exist constants $C > 0$ and $C' > 0$ not depending on (x,z) such that for all $(x,z) \in A$,

$$C f_2(x,z) \leq f_1(x,z) \leq C' f_2(x,z).$$

As a particular case, f_1 or f_2 may be a function of x alone. Using this abbreviation we may record our final result as follows:

$$u(x,z) \approx \sum_{n=1}^{\infty} E^x\{T_{2n-2} < \tau_D;\ e_q(T_{2n-2})\} \text{ for } (x,z) \in \overline{D}_0 \times \partial D. \tag{10}$$

The point is that the function of (x,z) is reduced, apart from a numerical factor, to a function of x alone. It follows that there exists a constant $C > 0$ independent of x such that for all $x \in \overline{D}_0$:

$$\sup_{z \in \partial D} u(x,z) \leq C \inf_{z \in \partial D} u(x,z). \tag{11}$$

By (5.18) with $f \equiv 1$, we have for all $x \in D$:

$$u(x) = \int_{\partial D} u(x,z) H(x, \mathrm{d}z). \tag{12}$$

Hence

$$\inf_{z \in \partial D} u(x,z) \le u(x) \le \sup_{z \in \partial D} u(x,z). \tag{13}$$

To prove the first assertion of the theorem, suppose that $u(x_0, z_0) < \infty$ for some $x_0 \in D$, $z_0 \in \partial D$. We may suppose that $x_0 \in D_0$ by enlarging D_0, if necessary. Then by (11),

$$\sup_{z \in \partial D} u(x_0, z) \le C u(x_0, z_0) < \infty,$$

and consequently $u(x_0) < \infty$ by (13). Therefore by the gauge theorem (Theorem 5.19),

$$\sup_{x \in \overline{D}} u(x) = M < \infty. \tag{14}$$

This implies by (13) that

$$\sup_{x \in \overline{D}} \inf_{z \in \partial D} u(x,z) \le M,$$

and so by (11) that

$$\sup_{x \in \overline{D}_0} \sup_{z \in \partial D} u(x,z) \le CM. \tag{15}$$

We have thus proved that $u(x,z)$ is bounded in $\overline{D}_0 \times \partial D$.

For $x \in D \backslash \overline{D}_0$ and $z \in \partial D$, we repeat the argument in Theorem 4.2 as follows.

$$\begin{aligned} u(x,z) &= E_z^x\{\tau_U = \tau_D;\ e_q(\tau_D)\} + E_z^x\{\tau_U < \tau_D;\ e_q(\tau_U) u(X(\tau_U), z)\} \\ &\le E_z^x\{e_q(\tau_U)\} \left(1 + \sup_{(y,z) \in \overline{D}_0 \times \partial D} u(y,z)\right) \\ &\le E_z^x\{e_q(\tau_U)\}(1 + CM) \le 2(1 + CM) \end{aligned}$$

by (15) and (6). This establishes the boundedness of $u(\cdot, \cdot)$ in $D \times \partial D$.

It remains to prove the second assertion of the theorem. If $u(x_0, z_0) < \infty$ for some $(x_0, z_0) \in D \times \partial D$, then we have already proved in (14) that (D, q) is gaugeable. Conversely, if $u(x_0) < \infty$ for some $x_0 \in D$, then by (13), $u(x_0, z_0) < \infty$ for some $z_0 \in \partial D$. □

The next property of the conditional gauge is easier to prove but almost as important.

Theorem 7.6 *Let D and q be as in Theorem 7.5. Then*

$$\inf_{(x,z) \in D \times \partial D} u(x,z) > 0. \tag{16}$$

Proof Inspection of the proof of Lemma 7.1 shows that it contains the result

$$\sup_{(x,z)\in D\times\partial D} E_z^x\left\{\int_0^{\tau_D} |q(X_t)|\mathrm{d}t\right\} = C < \infty. \tag{17}$$

Hence by Jensen's inequality

$$\begin{aligned} E_z^x\left\{\exp\left[\int_0^{\tau_D} q(X_t)\mathrm{d}t\right]\right\} &\geq \exp\left\{-E_z^x\int_0^{\tau_D} |q(X_t)|\mathrm{d}t\right\} \\ &\geq \mathrm{e}^{-C} > 0. \end{aligned}$$

Observe that this result is independent of Theorem 7.5 and includes the case where $u(x,z) \equiv \infty$ in $D\times\partial D$. □

Using the new symbol $\approx$ introduced in the proof of Theorem 7.5, we can combine the last two theorems as follows.

Theorem 7.7 *Let D and q be as in Theorem 7.5. If (D,q) is gaugeable, then*

$$u(x,z) \approx 1 \textit{ in } D\times\partial D. \tag{18}$$

In terms of the q-harmonic measure introduced in (5.19), and with an obvious extension of the usage of the new symbol, this may be expressed as follows:

$$H_q(\cdot,\cdot) \approx H_0(\cdot,\cdot) \text{ in } D\times\mathcal{B}(\partial D).$$

7.2 Approximation and Continuity of the Conditional Gauge

In this section D is a bounded Lipschitz domain, $q \in J_{\mathrm{loc}}$ and (D,q) is gaugeable. We proved in Theorem 7.5 that in this case the conditional gauge $u(\cdot,\cdot)$ is bounded on $D\times\partial D$. In this section we prove that it is continuous there by approximating D with subdomains.

Lemma 7.8 *There exists a constant $C_0 > 0$ such that for any domain $G \subseteq D$, we have for all $(x,z)\in G\times\partial D$:*

$$E_z^x\{e_q(\tau_G)\} \leq C_0. \tag{19}$$

Proof By Theorem 7.7 there exist constants $C_1 > 0$ and $C_2 > 0$ such that

$$C_1 \leq E_z^x\{e_q(\tau_D)\} \leq C_2 \tag{20}$$

for all $(x,z)\in D\times\partial D$. It follows trivially that

$$E_z^x\{\tau_G = \tau_D;\ e_q(\tau_G)\} \leq C_2.$$

On the other hand, we have

$$\begin{aligned} C_2 &\geq E_z^x\{\tau_G < \tau_D;\ e_q(\tau_D)\} \\ &= E_z^x\{\tau_G < \tau_D;\ e_q(\tau_G) E_z^{X(\tau_G)}[e_q(\tau_D)]\} \\ &\geq C_1 E_z^x\{\tau_G < \tau_D;\ e_q(\tau_G)\}. \end{aligned}$$

Therefore, combining the above two inequalities, we obtain

$$E_z^x\{e_q(\tau_G)\} \leq C_2 + \frac{C_2}{C_1} = C_0.$$

□

Remark It is trivial from (17) that

$$E_z^x\{e_q(\tau_G)\} \geq \mathrm{e}^{-C}.$$

Actually, the constant C_2 in (20) must be at least one. For there exists a regular boundary point z (see Chung (1982a, page 186)), and by Theorem 4.7(iii), $u(x) \to 1$ as $x \to z$, $x \in D$. Thus for any $\varepsilon > 0$ there exists $x_0 \in D$ with $u(x_0) > 1 - \varepsilon$. Hence by (13), there exists $z_0 \in \partial D$ with $u(x_0, z_0) > 1 - \varepsilon$. Since ε is arbitrary, we conclude that $\sup_{(x,z) \in D \times \partial D} u(x,z) \geq 1$.

Theorem 7.9 *Let D_n $(n \geq 1)$ be Lipschitz domains such that $D_n \uparrow\uparrow D$. Then*

$$\lim_n E_z^x\{|e_q(\tau_{D_n}) - e_q(\tau_D)|\} = 0$$

uniformly in $D_1 \times \partial D$.

Proof For $0 < \varepsilon < 1$, let $\delta = \delta(\varepsilon/2)$ be as in Lemma 7.1, and for this δ, let D_0 and U be as in Lemma 7.2. Then we have as in (6), for $(x,z) \in U \times \partial D$:

$$E_z^x\{e_{|q|}(\tau_U)\} \leq 1 + \varepsilon.$$

Using this and the inequality

$$|e_q(t) - 1| \leq e_{|q|}(t) - 1,$$

we obtain

$$E_z^x\{|e_q(\tau_U) - 1|\} \leq \varepsilon. \tag{21}$$

We shall write $\tau_{D_n} = T_n$, $\tau_D = T$ and $e_q(\cdot) = e(\cdot)$ in what follows. Then by the strong Markov property,

$$E_z^x\{|e(T_n) - e(T)|\} \leq E_z^x\{e(T_n) E_z^{X(T_n)}[|1 - e(T)|]\}.$$

Since $D_n \uparrow\uparrow D$, we may suppose that $X(T_n) \in U$ for all $n \geq 1$. Let

$$\begin{aligned} E_1 &= E_z^x\{e(T_n) E_z^{X(T_n)}[\tau_U = T;\ |1 - e(T)|\ ]\}; \\ E_2 &= E_z^x\{e(T_n) E_z^{X(T_n)}[\tau_U < T;\ |1 - e(T)|\ ]\}. \end{aligned}$$

It follows from (19) with $G = D_n$ and (21) that

$$E_1 \le \varepsilon E_z^x\{e(T_n)\} \le C_0\varepsilon. \tag{22}$$

On the other hand, it follows from (19) with $G = D$ that for $(y,z) \in D \times \partial D$,

$$E_z^y[|1 - e(T)|] \le 1 + E_z^y[e(T)] \le 1 + C_0,$$

and so

$$\begin{aligned} E_2 &\le E_z^x\{e(T_n)E_z^{X(T_n)}[\tau_U < T;\ E_z^{X(\tau_U)}[|1 - e(T)|]]\} \\ &\le (1 + C_0)E_z^x\{e(T_n)P_z^{X(T_n)}[\tau_U < T]\}. \end{aligned}$$

The last E_z^x in the above is given explicitly by

$$\frac{1}{K(x,z)}E^x\{e(T_n)E^{X(T_n)}[\tau_U < T;\ K(X(\tau_U), z)]\}.$$

As in (9) above, there is a constant C_3 such that

$$\frac{K(y,z)}{K(x,z)} \le C_3$$

for all $x \in \overline{D}_1$, $y \in \overline{D}_1$, $z \in \partial D$. It follows by (4.31) and (4.32) that

$$\begin{aligned} E_2 &\le (1 + C_0)C_3E^x\{e(T_n)P^{X(T_n)}[\tau_U < T]\} \\ &\le (1 + C_0)C_3C \sup_{y \in \partial D_n} P^y[\tau_U < \tau_D]. \end{aligned}$$

Now the function

$$y \to \phi(y) = P^y\{\tau_U < \tau_D\}$$

is harmonic in U with $\phi(y) = 1$ on $\partial U \backslash \partial D$ and $\phi(y) = 0$ on ∂D. Since D being Lipschitzian is regular, it follows from Theorem 1.23 that

$$\lim_n [\sup_{y \in \partial D_n} \phi(y)] = 0.$$

Thus $\lim_n E_2 = 0$ uniformly in $(x,z) \in D_1 \times \partial D$. Together with (22) this establishes the theorem. □

Theorem 7.10 *For any domain $G \subset\subset D$, the function*

$$(x,z) \to E_z^x\{e_q(\tau_G)\}$$

is continuous in $G \times \partial D$.

Proof Since

$$E_z^x\{e_q(\tau_G)\} = \frac{1}{K(x,z)}E^x\{e_q(\tau_G)K(X(\tau_G), z)\}$$

and $K(\cdot,\cdot)$ is continuous and > 0 in $G \times \partial D$, it is sufficient to show that the function

$$(x, z) \to \phi(x, z) = E^x\{e_q(\tau_G)K(X(\tau_G), z)\}$$

is continuous in $G \times \partial D$. Since (D, q) is gaugeable, it follows from Proposition 4.4 that there exists a constant C such that for all $G \subset D$:

$$\sup_{x \in G} E^x\{e_q(\tau_G)\} \le C.$$

Let $z_n \in \partial D$, $z \in \partial D$, and $z_n \to z$. We have

$$\sup_{x \in G} |\phi(x, z_n) - \phi(x, z)| \le C \sup_{y \in \partial G} |K(y, z_n) - K(y, z)| \to 0$$

by the joint continuity of $K(\cdot,\cdot)$ on $G \times \partial D$. Thus the family of functions $\{\phi(x, \cdot),$ $x \in G\}$ is equi-continuous on ∂D. For fixed $z \in \partial D$, $K(\cdot, z)$ is a bounded continuous function on ∂G. Hence $\phi(\cdot, z)$ is of the form u_f in (4.37) with $D = G$, $f = K(\cdot, z)$, and consequently by Theorem 4.7(ii) it is continuous in G.

We have therefore proved that $\phi(x, z)$ is continuous in $x \in G$ for $z \in \partial D$, and equi-continuous in $z \in \partial D$ for all x in G. This implies that it is jointly continuous in $G \times \partial D$. □

Combining Theorems 7.9 and 7.10, we state the main result in this section as follows.

Theorem 7.11 *Let D be a bounded Lipschitz domain, $q \in J_{\text{loc}}$ and let (D, q) be gaugeable. Then the conditional gauge $u(x, z)$ is bounded, continuous and strictly positive in $D \times \partial D$.*

Proof Only the continuity requires consideration. For any subdomain $G \subset\subset D$, there exists a sequence of C^2 domains $\{D_n\}$ with $D_n \uparrow\uparrow D$ and $G \subseteq D_1$ (see the remark at the end of the Appendix to Chapter 1). By Theorems 7.9 and 7.10, $u(x, z)$ is a uniform limit of continuous functions $E_z^x\{e_q(\tau_{D_n})\}$ in $G \times \partial D$, so it is continuous there. Since G is arbitrary, this implies the continuity of $D \times \partial D$. □

A further elaboration of this theorem will be given in Section 7.4.

7.3 Extended Conditional Gauge Theorem

In this section we extend Theorem 7.5 to include interior points. As before, D is a bounded Lipschitz domain and $q \in J_{\text{loc}}$.

In Section 5.2 we defined both the y-conditioned and the z-conditioned Brownian motion, for $y \in D$ and $z \in \partial D$. We can define the conditional gauge using the same notation in both cases, namely for $x \in D$ and $y \in \overline{D}$:

$$u(x, y) = E_y^x\{e_q(\zeta)\},$$

where ζ is the lifetime of the process. We recall that if $y \in \partial D$, then $\zeta = \tau_D$; while if $y \in D$, then $\zeta = \tau_{D\backslash\{y\}}$. By Theorems 5.9 and 5.10, we have $\zeta < \infty$ a.s. in both cases. We stress that 'a.s.' here means P_y^x-a.s. for a fixed $y \in \overline{D}$ and for all $x \in D\backslash\{y\}$.

The main result can be stated exactly as Theorem 7.5, by simply changing ∂D there into $\overline{D}$.

Theorem 7.12 (Extended Conditional Gauge Theorem) *Let D and q be as in Theorem 7.5. If $u(x_0, y_0) < \infty$ for some $(x_0, y_0) \in D \times \overline{D}$, $x_0 \neq y_0$, then $u(\cdot, \cdot)$ is bounded in $D \times \overline{D}$. This is the case if and only if (D, q) is gaugeable.*

Proof The basic ideas of the proof are similar to those of Theorem 7.5, but some new inequalities are required. For clarity, we shall state and prove a number of lemmas. We begin by strengthening Lemma 7.1.

Extended Lemma 7.1 *The conclusion of Lemma 7.1 can be strengthened as follows: if $m(U) < \delta(\varepsilon)$, then*

$$\sup_{\substack{(x,y)\in D\times\overline{D} \\ x\neq y}} E_y^x\left\{\int_0^{\tau_U} |q(X_t)|\mathrm{d}t\right\} < \varepsilon. \tag{23}$$

Proof First we consider y in D only. Proceeding as in the proof of Lemma 7.1 but replacing $K(\cdot, z)$ by $G(\cdot, y)$ everywhere and using Theorems 6.5 and 6.15 we obtain (23) with $\overline{D}$ replaced by D. Together with Lemma 7.1, this implies (23) as shown. □

In what follows we take $\varepsilon = 1/2$ in the Extended Lemma 7.1 and use the corresponding $\delta = \delta(1/2)$ in Lemma 7.2 to obtain domains D_0 and U so that the analogue of (6) holds:

$$E_y^z\{e_{|q|}(\tau_U)\} \le 2. \tag{24}$$

We construct two more domains D_1 and D_2 like D_0 in Lemma 7.2 such that

$$x_0 \in D_0 \subset\subset D_1 \subset\subset D_2 \subset\subset D,$$

and choose $r > 0$ to be sufficiently small that

$$\begin{aligned} 2r &< |x_0 - y_0|, \\ 5r &< d(\overline{D}_0, \partial D_1) \wedge d(\overline{D}_1, \partial D_2) \wedge d(\overline{D}_2, \partial D), \end{aligned} \tag{25}$$

and

$$v_d(5r) = m(B(x, 5r)) < \delta\left(\frac{1}{2}\right).$$

We define a function F on $D \times \overline{D}$ as follows:

$$F(x,y) = \begin{cases} \frac{G(x,y)}{G(x_0,y)} & \text{if } (x,y) \in D \times D \\ K(x,y) & \text{if } (x,y) \in D \times \partial D, \end{cases}$$

where x_0 is the reference point in the definition of $K(x,z)$ in (5.8).

Lemma 7.13 *$F(\cdot,\cdot)$ is strictly positive and jointly continuous on $\overline{D}_1 \times (\overline{D} \backslash D_2)$.*

Proof From Theorem 2.6(i) and (iii), (5.8), and the conditions (i) and (ii) for $K(\cdot,\cdot)$ in Section 5.2, we see that $F(\cdot,\cdot) > 0$ on $D \times \overline{D}$ and that for each $x \in D_2$, $F(x,\cdot)$ is continuous in $\overline{D} \backslash D_2$. Note that for each $y \in \overline{D} \backslash D_2$, $F(\cdot,y)$ is harmonic in D_2 by the Corollary to Theorem 2.5 and the condition (i) for $K(\cdot,\cdot)$ in Section 5.2, and $\overline{D}_1$ is a compact subset of D_2. Hence it follows from Harnack's convergence theorem (see Chung, (1982a, page 161, Exercise 1) that if $y_n \to y_0$ in $\overline{D} \backslash D_2$ as $n \to \infty$, then $F(x,y_n)$ converges to $F(x,y_0)$ uniformly for $x \in \overline{D}_1$, and so the family $\{F(x,\cdot)\colon x \in \overline{D}_1\}$ is equi-continuous on $\overline{D} \backslash D_2$. Therefore $F(\cdot,\cdot)$ is jointly continuous on $\overline{D_1} \times (\overline{D} \backslash D_2)$. □

The next step is an analogue of Lemma 7.3. We recall the notation $\approx$ introduced in Section 7.1.

Lemma 7.14 *We have*

$$E^x_y\{\zeta = \tau_{U \backslash \{y\}};\ e_q(\zeta)\} \approx 1 \text{ in } \partial D_1 \times (\overline{D} \backslash D_2). \tag{26}$$

Proof To prove this we need the analogue of (7):

$$\inf_{(x,y) \in \partial D_1 \times (\overline{D} \backslash D_2)} P^x_y\{\zeta = \tau_{U \backslash \{y\}}\} > 0. \tag{27}$$

For $(x,y) \in U \times (\overline{D} \backslash D_2)$, $x \neq y$,

$$P^x_y\{\tau_{U \backslash \{y\}} < \zeta\} = \frac{P^x\{\tau_U < \tau_D;\ F(X(\tau_U),y)\}}{F(x,y)}. \tag{28}$$

Note that if $y \in D \backslash D_2$, then we may replace F in (28) by G, after cancelling the factor $G(x_0,y)$. We have also replaced ζ by τ_D and $U \backslash \{y\}$ by U on the right-hand side of (28) because $P^x\{T_{\{y\}} < \infty\} = 0$, a fact used before (see the proof of Theorem 5.10).

We can now obtain (27) from (28) by exactly the same argument used to obtain (7) from (8) in Lemma 7.3. Namely, we replace $K(\cdot,\cdot)$ there by $F(\cdot,\cdot)$, D_0 by D_1 and ∂D by $\overline{D} \backslash D_2$, and use the property of $F(\cdot,\cdot)$ given in Lemma 7.13. Once (27) is proved, the rest of the proof of (26) is also the same as in the proof of Lemma 7.3. □

We now extend the first major step (11) in the proof of Theorem 7.5. As in Lemma 7.14, the boundary ∂D is replaced by a closed region $\overline{D} \backslash D_2$.

Lemma 7.15 *There exists a constant C_1 independent of x such that for all $x \in D_1$, we have*

$$\sup_{y \in \overline{D} \backslash D_2} u(x,y) \leq C_1 \inf_{y \in \overline{D} \backslash D_2} u(x,y). \tag{29}$$

Proof Let $T_0 = 0$, $T_{2n-1} = T_{2n-2} + \tau_{D_1} \circ \theta_{T_{2n-2}}$

$$T_{2n} = T_{2n-1} + \tau_{U\setminus\{y\}} \circ \theta_{T_{2n-1}}, \quad n \geq 1.$$

Then $\lim_n T_n = \zeta$; in fact $T_{2n} = \zeta$ for some n. This is argued out as in the proof of Theorem 7.5 using the two facts in Theorem 5.10: $\zeta < \infty$ and $X(t) \to y$ as $t \uparrow\uparrow \zeta$ a.s. Thus we have by Lemma 7.14:

$$\begin{aligned} u(x,y) &= \sum_{n=1}^{\infty} E_y^x \left\{T_{2n-2} < \zeta; e_q(T_{2n-1}) E_y^{X(T_{2n-1})}[\zeta = \tau_{U\setminus\{y\}};\, e_q(\zeta)]\right\} \\ &\approx \sum_{n=1}^{\infty} E_y^x\{T_{2n-2} < \zeta;\, e_q(T_{2n-1})\}. \end{aligned} \tag{30}$$

So far the above result holds for $(x, y) \in D \times (\overline{D}\setminus D_2)$. Now the general term in the last series is equal to

$$\frac{1}{F(x,y)} E^x\{T_{2n-2} < \tau_D;\; F(X(T_{2n-1}), y) e_q(T_{2n-1})\}. \tag{31}$$

By Lemma 7.13 and the compactness of the set, we have

$$F(\cdot,\cdot) \approx 1 \text{ in } \overline{D}_1 \times (\overline{D}\setminus D_2),$$

which implies

$$\frac{F(w,y)}{F(x,y)} \approx 1 \text{ for } (w,x,y) \text{ in } \overline{D}_1 \times \overline{D}_1 \times (\overline{D}\setminus D_2). \tag{32}$$

Using (32) in (31) and then in (30), we obtain

$$u(x,y) \approx \sum_{n=1}^{\infty} E^x\{T_{2n-2} < \tau_D;\; e_q(T_{2n-1})\} \text{ in } \overline{D}_1 \times (\overline{D}\setminus D_2),$$

where the right-hand side no longer depends on y. This relation clearly implies the conclusion of the lemma. □

The next step should be a complement to (29) with $\overline{D}\setminus D_2$ replaced by D_2, but we need to restrict y to be at a distance from x. The result is Lemma 7.17 which is preceded by a local version, as follows.

Lemma 7.16 *There exists C_2 not depending on x or w such that for all $x \in D_1$ and $w \in \overline{D}_2$ with $|w - x| > r$, we have*

$$\sup_{y \in B(w,r/4)} u(x,y) \leq C_2 \inf_{y \in B(w,r/4)} u(x,y). \tag{33}$$

Proof We need a new estimate similar to (5) and (26) above, as follows:

$$E_y^x\{\zeta = \tau_{B(w,r)\setminus\{y\}};\; e_q(\zeta)\} \approx 1 \text{ for } (x,y,w) \text{ in } \Lambda, \tag{34}$$

where

$$\Lambda = \{(x,y,w):\ w \in \overline{D}_2,\ |y-w| \le \frac{r}{4},\ |x-w| = \frac{r}{2}\}.$$

Just as in the proof of Lemma 7.3, this will follow from

$$\inf_{(x,y,w)\in\Lambda} P_y^x\{\tau_{B(w,r)\setminus\{y\}} < \zeta\} > 0. \tag{35}$$

For $(x,y,w) \in \Lambda$, by (5.4) and Poisson's formula (Theorem 1.13), we have

$$\begin{aligned} P_y^x\{\tau_{B(w,r)\setminus\{y\}} < \zeta\} &= \frac{E^x\{G(X(\tau_{B(w,r)}),y)\}}{G(x,y)} \\ &= \frac{\Gamma(d/2)(r^2-|x-w|^2)}{2\pi^{d/2} r G(x,y)} \int_{\partial B(0,r)} \frac{G(z+w,y)}{|z+w-x|^d}\sigma(\mathrm{d}z). \end{aligned}$$

From the right-hand side above it is easy to see that $(x,y,w) \to P_y^x\{\tau_{B(w,r)\setminus\{y\}} < \zeta\}$ is continuous on Λ. Next, by comparing two harmonic functions in $B(w,r)\setminus\{y\}$ as we did in Lemmas 7.3 and 7.14, we deduce that (35) is true.

Now for any fixed $x \in \overline{D}_1$, $w \in \overline{D}_2$ with $|w-x| > r$, and $y \in B(w, r/4)$, we define a sequence of hitting times S_n:

$$\begin{aligned} &S_0 = 0,\ S_{2n-1} = S_{2n-2} + T_{B(w,r/2)} \circ \theta_{S_{2n-2}} \\ &S_{2n} = S_{2n-1} + \tau_{B(w,r)\setminus\{y\}} \circ \theta_{S_{2n-1}},\quad n \ge 1. \end{aligned}$$

Noting that on the set $(S_{2n-2} < \zeta)$, we have $(X(S_{2n-1}), y, w) \in \Lambda$; then using Theorem 5.10 as before, followed by (34), we obtain the following.

$$\begin{aligned} u(x,y) &= \sum_{n=1}^{\infty} E_y^x\{S_{2n-2} < \zeta,\ \zeta = S_{2n};\ e_q(\zeta)\} \\ &= \sum_{n=1}^{\infty} E_y^x\{S_{2n-2} < \zeta;\ e_q(S_{2n-1}) E_y^{X(S_{2n-1})}[\zeta = \tau_{B(w,r)\setminus\{y\}}; e_q(\zeta)]\} \\ &\approx \sum_{n=1}^{\infty} E_y^x\{S_{2n-2} < \zeta;\ e_q(S_{2n-1})\} \\ &= \frac{1}{G(x,y)} \sum_{n=1}^{\infty} E^x\{S_{2n-2} < \tau_D;\ e_q(S_{2n-1}) G(X(S_{2n-1}),y)\}. \end{aligned} \tag{36}$$

Note that under the conditions of the lemma we have

$$|x-y| \ge |x-w| - |y-w| > \frac{r}{4},$$

and

$$|X(S_{2n-1}) - y| \ge |X(S_{2n-1}) - w| - |y-w| \ge \frac{r}{4} \text{ on } \{S_{2n-2} < \zeta\}.$$

Since $G(\cdot,\cdot)$ is continuous and > 0 on the compact set $\{(x,y)\colon d(x,\overline{D}_2) \le r,$ $d(y,\overline{D}_2) \le r$ and $|x-y| \ge r/4\}$, it follows that

$$\frac{G(X(S_{2n-1}),y)}{G(x,y)} \approx 1 \text{ for } (x,y) \in \overline{D}_1 \times B\left(w, \frac{r}{4}\right).$$

Using this in (36) we obtain

$$u(x,y) \approx \sum_{n=1}^{\infty} E^x\{S_{2n-2} < \tau_D;\ e_q(S_{2n-1})\}.$$

The right-hand side of the above is independent of y, which implies (33). □

To lead from Lemma 7.16 to Lemma 7.17 we need a 'chain argument' which is somewhat more complicated than the usual kind, and will therefore be meticulously spelled out here. For each $x \in D_1$ we define

$$D_2(x) = D_2 \backslash \overline{B}(x, 2r).$$

Since $2r < d(D_1, \partial D_2)$, we have $\overline{B}(x,2r) \subseteq D_2$, so each $D_2(x)$ is also a domain by a geometric argument. Since $\overline{D}_2$ is compact, it can be covered as shown below:

$$\overline{D}_2 \subseteq \bigcup_{i=1}^{N} B\left(w_i, \frac{r}{4}\right), \tag{37}$$

where $w_i \in \overline{D}_2$, $i = 1, \cdots, N$. Of course, each $\overline{D}_2(x)$ is also covered as in (37), but we want to cover it using only those balls with $|w_i - x| > r$, as stipulated in Lemma 7.16. Suppose that $|w_i - x| \le r$, then $B(w_i, r/4) \subsetneq B(x, 2r)$ and so $B(w_i, r/4)$ does not intersect $\overline{D}_2(x)$. Therefore we can omit such balls in covering $\overline{D}_2(x)$, i.e.:

$$\overline{D}_2(x) \subseteq \bigcup_{\substack{1 \le i \le N \\ |w_i - x| > r}} B\left(w_i, \frac{r}{4}\right). \tag{38}$$

Thus by the Lemma in the proof of Theorem 4.1, for any two points y and y' in each $D_2(x)$, there exists a subset of balls in (38): $\{B(w_{i_j}, r/4)\colon 1 \le j \le l\}$ such that $y \in B(w_{i_1}, r/4)$, $B(w_{i_j}, r/4) \cap B(w_{i_{j+1}}, r/4) \ne \emptyset$, $1 \le j < l$ and $y' \in B(w_{i_l}, r/4)$. Now if we pick $y_j \in B(w_{i_j}, r/4) \cap B(w_{i_{j+1}}, r/4)$, $1 \le j < l$ and apply (33) of Lemma 7.16 successively to the sequence of points $y, y_1, \cdots, y_{l-1}, y'$, then we obtain

$$u(x,y) \le C_2 u(x,y_1) \le C_2^2\, u(x,y_2) \cdots \le C_2^l u(x,y'),$$

where $1 \le l \le N$. We have thus proved the next result.

Lemma 7.17 *There exists C_3 (not depending on x) such that for all $x \in D_1$, we have*

$$\sup_{y \in D_2, |y-x| > 2r} u(x,y) \le C_3 \inf_{y \in D_2, |y-x| > 2r} u(x,y). \tag{39}$$

Proof In fact, we may take $C_3 = C_2^N$ where C_2 is as in (33) and N is as in (37). Note that it is essential that N does not depend on x, and that is the point of the detailed linking argument given above. □

Combining Lemmas 7.15 and 7.17, we have for all $x \in D_1$:

$$\sup_{y\in\overline{D},|y-x|>2r} u(x,y) \le (C_1 \vee C_3) \inf_{y\in\overline{D},|y-x|>2r} u(x,y). \tag{40}$$

Noting that $\partial D \subseteq \{y \in \overline{D}\colon |y-x| > 2r\}$ for all $x \in D_1$; since $x_0 \in D_1$ and $|x_0 - y_0| > 2r$ we obtain by (13) and (40):

$$u(x_0) \le \sup_{y\in\partial D} u(x_0,y) \le (C_1 \vee C_3) u(x_0,y_0) < \infty.$$

Therefore by the gauge theorem (Theorem 5.19),

$$\sup_{x\in\overline{D}} u(x) = M < \infty.$$

Thus using (40) and (13) again we obtain for all $x \in D_1$:

$$\begin{aligned}\sup_{y\in\overline{D},|y-x|>2r} u(x,y) &\le (C_1 \vee C_3) \inf_{y\in\partial D} u(x,y)\\ &\le (C_1 \vee C_3) u(x)\\ &\le (C_1 \vee C_3) M = C_4. \end{aligned}\tag{41}$$

Now we shall deal with a pair (x,y) at short distance, but again we must make a concession on the domain, retreating from D_1 to $\overline{D}_0$.

Lemma 7.18 *For all $x \in \overline{D}_0$, we have*

$$\sup_{y\in\overline{D},|y-x|\le 2r} u(x,y) \le 2(1 \vee C_4). \tag{42}$$

Proof Let $B = B(x,5r)$, then

$$\begin{aligned} u(x,y) &= E_y^x\{\tau_B = \zeta;\ e_q(\tau_B)\}\\ &\quad + E_y^x\{\tau_B < \zeta;\ e_q(\tau_B) E_y^{X(\tau_B)}[e_q(\zeta)]\}. \end{aligned}\tag{43}$$

Since $x \in \overline{D}_0$, by the conditions on r in (25), we have $B(x,5r) \subset\subset D_1$, $X(\tau_B) \in D_1$ and $|X(\tau_B) - y| \ge 5r - |y-x| \ge 5r - 2r > 2r$. Hence we can apply (41) to obtain

$$E_y^{X(\tau_B)}[e_q(\zeta)] \le C_4.$$

It follows from (43) that

$$u(x,y) \le (1 \vee C_4) E_y^x\{e_q(\tau_B)\} \le 2(1 \vee C_4).$$

The factor 2 comes from (24) with $U = B$, recalling the condition $v_d(5r) < \delta(1/2)$. □

Combining (41) and (42), we obtain

$$\sup_{(x,y)\in\overline{D}_0\times\overline{D}} u(x,y) \le 2(1 \vee C_4).$$

Finally we must expand the $\overline{D}_0$ in the above inequality to D. This step is exactly the same as the last step in the proof of Theorem 7.5, using $m(U) < \delta(1/2)$. This completes the proof of Theorem 7.12. □

The following result is the extension of Theorem 7.6.

Theorem 7.19 *Let D and q be as in Theorem 7.5. Then*

$$\inf_{\substack{(x,y)\in D\times\overline{D}\\ x\neq y}} u(x,y) > 0. \tag{44}$$

Proof In view of (16), it is sufficient to prove (44) for $y \in D$ only. Note that the following argument is valid for any bounded D. For each $(x,y) \in D \times D$, $x \neq y$, we have

$$\begin{aligned} E_y^x\left\{\int_0^\zeta |q(X_t)|\mathrm{d}t\right\} &= \int_D \frac{G(x,w)|q(w)|G(w,y)}{G(x,y)}\mathrm{d}w \\ &\le C\left[\int_D g(x-w)|q(w)|\mathrm{d}w + \int_D g(y-w)|q(w)|\mathrm{d}w\right] \\ &\le 2C\sup_{x\in D}\int_D g(x-w)|q(w)|\mathrm{d}w < \infty \end{aligned}$$

by Theorems 6.5 and 6.15, and the Corollary to Proposition 3.1. Hence (44) follows by Jensen's inequality as in the proof of Theorem 7.6. □

7.4 Representation of the Conditional Gauge

In this section, we shall establish some connections between the q-Green function and the extended conditional gauge $u(x,y)$, and extend the continuity of the latter to $\overline{D} \times \overline{D}$.

We assume that D is a bounded domain, $q \in J_{\text{loc}}$ and (D,q) is gaugeable. Then the q-Green function V of D is as defined in Section 6.1 (1) and (2). When $q \equiv 0$, V reduces to G. The following relation is basic.

Theorem 7.20 *For all $(x,y) \in D \times D$, $x \neq y$,*

$$u(x,y) = 1 + \frac{1}{G(x,y)}\int_D V(x,w)q(w)G(w,y)\mathrm{d}w. \tag{45}$$

Proof We have

$$u(x,y) = E_y^x \left[\exp\left(\int_0^\zeta q(X_s) \mathrm{d}s \right) \right] = 1 + E_y^x \left[\int_0^\zeta e_q(t) q(X_t) \mathrm{d}t \right] \tag{46}$$

by a formal integration as in (4.64). This yields (45) when the conditional expectation on the right-hand side of (46) is written out, but we need the following justification.

It follows from the Fubini–Tonelli theorem that

$$\begin{aligned} E_y^x \left[\int_0^\zeta e_q(t) |q(X_t)| \mathrm{d}t \right] &= \int_0^\infty E_y^x[t < \zeta;\ e_q(t)|q(X_t)|] \mathrm{d}t \\ &= \frac{1}{G(x,y)} \int_0^\infty E^x[t < \tau_D;\ e_q(t)|q(X_t)|G(X_t,y)] \mathrm{d}t \\ &= \frac{1}{G(x,y)} \int_0^\infty \int_D u(t;x,w)|q(w)|G(w,y) \mathrm{d}w\, \mathrm{d}t \\ &= \frac{1}{G(x,y)} \int_D V(x,y)|q(w)|G(w,y) \mathrm{d}w. \end{aligned} \tag{47}$$

The right member of (47) is finite by (6.12) and the Corollary to Proposition 3.1. This is sufficient for the required justification. □

Theorem 7.21 *For all* $(x, y) \in D \times D$, $x \neq y$,

$$u(x,y) = \frac{V(x,y)}{G(x,y)}. \tag{48}$$

Consequently $u(x, y)$ *is symmetric in* $(x, y) \in D \times D$.

Proof This is a consequence of Theorems 6.3 and 7.20. □

From now on, we assume further that D is a bounded Lipschitz domain. The next comparison theorem follows from Theorems 7.12, 7.19, and 7.21.

Theorem 7.22 *If* (D, q) *is gaugeable, then there exists a constant* $C = C(D, q) \geq 1$ *such that*

$$\frac{1}{C} G(x,y) \leq V(x,y) \leq C\, G(x,y), \ (x,y) \in D \times D,\ x \neq y. \tag{49}$$

In order to extend the continuity of the conditional gauge $u(\cdot,\cdot)$ from $D \times \overline{D}$ to $\overline{D} \times \overline{D}$, we need the following lemmas:

Lemma 7.23 *Let* $(x, y, w) \in D \times D \times D$, *and*

$$f(w;x,y) = \frac{1}{G(x,y)} V(x,w) q(w) G(w,y), \quad x \neq y. \tag{50}$$

Then the set of functions $f(\cdot;x,y)$, *for all* $(x,y) \in D \times D$, $x \neq y$, *is uniformly integrable on* D.

Proof By Theorem 7.22 and the 3G Theorem (Theorems 6.5 and 6.15), we have

$$|f(w;x,y)| \le C[|q(w)|g(x-w) + |q(w)|g(y-w)].$$

Thus the lemma follows from the Corollary to Proposition 3.1. □

Now we can extend Theorem 7.20 to $D \times \overline{D}$ as follows.

Theorem 7.24 *For all* $(x,z) \in D \times \partial D$, *we have*

$$u(x,z) = 1 + \frac{1}{K(x,z)} \int_D V(x,w)q(w)K(w,z)\mathrm{d}w. \tag{51}$$

Proof The proof is similar to that of Theorem 7.20. The only thing we need to verify is the finiteness of the right-hand side of (47) when G is replaced by K, namely:

$$\frac{1}{K(x,z)} \int_D V(x,w)|q(w)|K(w,z)\mathrm{d}w < \infty. \tag{52}$$

This follows from Lemma 7.23, since by (5.8) we have

$$\lim_{\substack{y_n \to z \\ y_n \in D}} \frac{G(w,y_n)}{G(x,y_n)} = \frac{K(w,z)}{K(x,z)}.$$

□

The next lemma is elementary analysis and is stated here for clarity.

Lemma 7.25 *Let f be a function defined in a bounded open set $S \subset \mathbb{R}^d$ $(d \ge 1)$. Suppose that for any sequence of points $\{p_n\}$ in S which converges to any point in $\overline{S}$, $\{f(p_n)\}$ is a convergent sequence. Then there exists a continuous function $\overline{f}$ in $\overline{S}$ such that $\overline{f}(p) = f(p)$ for all $p \in S$.*

For any $\xi \in \partial D$, $r > 0$, $a > 0$, we define a class of function pairs as follows:

$$\begin{aligned} H_0^+(\xi,r,a) = \{(u,v):\ & u \in H_0^+(\xi,r),\ v \in H_0^+(\xi,r), \text{ and} \\ a^{-1} \le \frac{u(y)}{v(y)} & \le a \text{ for all } y \in D \cap B(\xi,r)\}, \end{aligned} \tag{53}$$

where $H_0^+(\xi,r)$ is defined in Section 6.2 as the set of strictly positive harmonic functions in $D \cap B(\xi,2r)$ which vanish continuously on $\partial D \cap B(\xi,2r)$.

Lemma 7.26 *For any given $\xi \in \partial D$, $r > 0$, $a > 0$ and $\{x_n\} \subset D$ with $x_n \to \xi$, the sequence*

$$\frac{u(x_n)}{v(x_n)}$$

converges uniformly for all $(u,v) \in H_0^+(\xi,r,a)$.

This result, due to Jerison and Kenig (1982, Theorem 7.9), is a refinement of the boundary Harnack principle (Lemma 6.8). The asserted uniformity is easily seen from its proof.

Theorem 7.27 *If (D,q) is gaugeable, then there exists a strictly positive, symmetric and continuous function $u(x,y)$ in $\overline{D}\times\overline{D}$ such that*

$$u(x,y)=\begin{cases} E_y^x[e_q(\zeta)], & (x,y)\in D\times\overline{D},\ x\neq y,\\ 1, & x=y\in\overline{D}.\end{cases} \tag{54}$$

Proof Since $\overline{D}\times\overline{D}$ is the closure of the open set

$$S=\{(x,y)\in D\times D:\ x\neq y\},$$

we may begin with the function $u(x,y)$ as defined above in S and apply Lemma 7.25. Thus, we suppose that $\{(x_n,y_n)\}$ is a sequence in S with $(x_n,y_n)\to(\xi,\eta)\in\overline{S}=\overline{D}\times\overline{D}$. We need to prove that $\{u(x_n,y_n)\}$ is a convergent real sequence. Let us write (x_n,y_n) as (x,y) and consider the various cases of (ξ,η) in $\overline{D}\times\overline{D}$ below.

(i) $(\xi,\eta)\in S$, i.e., $(\xi,\eta)\in D\times D$ with $\xi\neq\eta$. By (48) and the continuity of G and V (Theorem 2.6(iii) and Theorem 6.2(a)), we have

$$\lim_{(x,y)\to(\xi,\eta)} u(x,y)=u(\xi,\eta).$$

(ii) $\xi=\eta\in\overline{D}$. In this case we shall prove

$$u(x,y)\to 1. \tag{55}$$

In view of (45) and Lemma 7.23, it is sufficient to prove that for each $w\neq\xi$:

$$\lim_{(x,y)\to(\xi,\xi)} f(w;x,y)=0. \tag{56}$$

By (49) and (50),

$$|f(w;x,y)|\le C|q(w)|\frac{G(x,w)G(w,y)}{G(x,y)}. \tag{57}$$

In the case $d\ge 3$, we have by Theorem 6.5, (6.13):

$$\frac{G(x,w)G(w,y)}{G(x,y)}\le C\frac{|x-y|^{d-2}}{|x-w|^{d-2}|y-w|^{d-2}}\to 0$$

as $(x,y)\to(\xi,\xi)$ with $w\neq\xi$, so (56) holds.

In the case $d=2$, (56) follows from (57) and Proposition 6.26.

(iii) $\xi\in D$ and $\eta\in\partial D$. In this case we have by Lemma 7.13 and the continuity of V that for each $w\neq\xi$ in D:

$$f(w;x,y)\to\frac{V(\xi,w)K(w,\eta)}{K(\xi,\eta)}.$$

Hence by (45), (51), and Lemma 7.23, we have

$$u(x,y)\to u(\xi,\eta).$$

(iv) $\xi \in \partial D$ and $\eta \in D$. Since u is symmetric in S (Theorem 7.21), we have by case (iii) above:

$$u(x,y) = u(y,x) \to u(\eta,\xi).$$

(v) $\xi \in \partial D$, $\eta \in \partial D$ and $\xi \neq \eta$. As before, by Lemma 7.23, we need only prove that for each fixed $w \in D$, $f(w;x,y)$ converges as $(x,y) \to (\xi,\eta)$. Since $\frac{V(x,w)}{G(x,w)} = u(x,w)$ converges by case (iv), it is sufficient to prove that $Q(x,y) = \frac{G(x,w)G(w,y)}{G(x,y)}$ converges. We take a sufficiently small $r > 0$ so that

$$B(\xi,2r) \cap B(\eta,2r) = \emptyset$$

and

$$w \notin B(\xi,2r) \cup B(\eta,2r).$$

We may also assume that $x \in B(\xi,r)$ and $y \in B(\eta,r)$. Now for each $v \in D$, $G(v,\cdot)$ is harmonic in $D \backslash \{v\}$ and vanishes on ∂D; thus, using the boundary Harnack principle (Lemma 6.8) twice, we see that for all $(x,y) \in [D \cap B(\xi,r)] \times [D \cap B(\eta,r)]$:

$$Q(x,y) \approx Q(x_1,y) \approx Q(x_1,y_1),$$

where the symbol $\approx$ is used as in the proof of Theorem 7.5 and (x_1,y_1) is a fixed point in $[D \cap B(\xi,r)] \times [D \cap B(\eta,r)]$. Hence there exists $a > 0$ such that for all $x \in D \cap B(\xi,r)$:

$$(G(x,w)G(w,\cdot), G(x,\cdot)) \in H_0^+(\eta,r,a).$$

Therefore $Q(x,y)$ converges uniformly for all $x \in D \cap B(\xi,r)$ as $y \to \eta$ by Lemma 7.26. In a similar manner we can prove that $Q(x,y)$ converges uniformly for all $y \in D \cap B(\eta,r)$ as $x \to \xi$. Consequently $Q(x,y)$ converges as $(x,y) \to (\xi,\eta)$ jointly.

Collecting the various cases above, we conclude that if $\{(x_n,y_n)\}$ in S converges to any point $(\xi,\eta) \in \overline{S} = \overline{D} \times \overline{D}$, then $\{u(x_n,y_n)\}$ is a convergent real sequence. Hence by Lemma 7.25, we can extend $u(\cdot,\cdot)$ in S to a continuous function $u(\cdot,\cdot)$ in $\overline{D} \times \overline{D}$. It follows from (55) that for all $\xi \in \overline{D}$,

$$u(\xi,\xi) = 1.$$

This completes the proof of the theorem. □

We note that Theorem 7.11 is contained in Theorem 7.27, but it was proved by a different method.

As applications of the conditional gauge, we prove two results which sharpen previous ones on gaugeability for a bounded Lipschitz domain. We recall the harmonic measure defined in (1.37), for any domain D. If $A \in \mathcal{B}(\partial D)$ and there exists $x \in D$ such that $H(x,A) > 0$, then this inequality holds for all $x \in D$ by Proposition 1.10. If z is a regular boundary point, and B is any ball with center

z, then $A = (\partial D) \cap B$ is such a set. This is a consequence of Theorem 1.23 since $\lim_{D \ni x \to z} H(x, A) = 1$. We recall the notation in (4.37) so that u_1 denotes the gauge, and note that $h_{1_A}(x) = H(x, A)$.

Proposition 7.28 *Let D be a bounded Lipschitz domain and $q \in J_{\text{loc}}$. Suppose there exists a set $A \in \mathcal{B}(\partial D)$ such that $h_{1_A} \not\equiv 0$ and $u_{1_A} \not\equiv \infty$ in D, then u_1 is bounded in $\overline{D}$.*

Proof Let $u_{1_A}(x_0) < \infty$, then by (5.18):

$$E^{x_0}\{e_q(\tau_D) 1_A(X(\tau_D))\} = \int_A u(x_0, z) H(x_0, \mathrm{d}z) < \infty.$$

Since $H(x_0, A) > 0$ by the discussion above, it follows that there exists $z_0 \in A$ such that $u(x_0, z_0) < \infty$. Hence (D, q) is gaugeable by Theorem 7.5. □

The second result is a sharpening of Theorem 4.17.

Proposition 7.29 *Let D and q be as in Proposition 7.28, B any ball with center at any boundary point and $A = (\partial D) \cap \overline{B}$. Suppose there exists a function u which is finite and continuous in $D \cup A$ such that u is q-harmonic in D and $u > 0$ in $D \cup A$, then (D, q) is gaugeable.*

Proof Let $\delta = \inf_A u > 0$. By the Appendix to Chapter 1, there exist domains $D_n \uparrow\uparrow D$ such that $D_1 \cap B \neq \emptyset$. Let

$$E_n = D_n \cup [D \cap B].$$

Since the infimum of u in both D_n and $\overline{D \cap B}$ is strictly positive, we have $\inf_{E_n} u > 0$. Similarly $\sup_{E_n} u < \infty$. Therefore by Theorem 4.17, (E_n, q) is gaugeable. Since $u \in C(\overline{E}_n)$ we have by Theorem 4.15, for any $x \in E_1$:

$$\begin{aligned} u(x) &= E^x\{e_q(\tau_{E_n}) u(X(\tau_{E_n}))\} \\ &\geq E^x\{e_q(\tau_{E_n}) 1_A u(X(\tau_{E_n}))\} \\ &\geq \delta E^x\{e_q(\tau_{E_n}) 1_A(X(\tau_{E_n}))\}. \end{aligned}$$

Since A is relatively open in ∂D, and $X(\tau_{E_n})$ converges to $X(\tau_D)$, we have

$$\underline{\lim}_n 1_A(X(\tau_{E_n})) \geq 1_A(X(\tau_D)).$$

Hence it follows by Fatou's lemma that

$$\begin{aligned} u(x) &\geq \delta E^x\{\underline{\lim}_n e_q(\tau_{E_n}) 1_A(X(\tau_{E_n}))\} \\ &\geq \delta E^x\{e_q(\tau_D) 1_A(X(\tau_D))\} = \delta u_{1_A}(x). \end{aligned}$$

Thus $u_{1_A}(x) < \infty$; since $h_{1_A} > 0$ by a remark above, we conclude by Proposition 7.28 that u_1 is bounded in $\overline{D}$. □

It is interesting to compare Proposition 7.29 with the following result.

Proposition 7.30 *Let D be an arbitrary domain and $q \in J$. Suppose there exists a bounded q-harmonic function u in D and a constant $\delta > 0$ such that for all $z \in (\partial D)_r$:*

$$\underline{\lim}_{D \ni x \to z} u(x) \geq \delta.$$

Then (D, q) is gaugeable. If $m(D) < \infty$, the boundedness of u may be omitted from the hypothesis.

For a proof of this see Chung (1987b).

Notes on Chapter 7

Theorem 7.5 was first proved by Falkner (1983) for $q \in \mathcal{B}_b$ and a class of domains including C^2 domains. A simpler proof was given in Chung (1984a). Partial extensions of the result to $q \in J$, and C^2 or $C^{1,1}$ domains were given by Zhao (1984, 1986). Theorem 7.5 is in Chung (1983/84, 1984/85). It reduced the problem for a general pair (D, q) to a single *cordon-sanitaire* estimate which is stated as Lemma 7.1 here. For a bounded Lipschitz domain, this estimate is quickly verified by the 3G Theorems in Sections 6.2 and 6.3. But there are other conditions under which the estimate holds, and then the conditional gauge theorem will also hold. Falkner (1987) gave such a condition which, though technique-driven and therefore unappealing, is shown to include the case where D is bounded and satisfies a 'uniform interior cone condition' (see Proposition 1.22 for an exterior cone), and $q \in L^p(D)$ for some $p > d/2$. Other suitable conditions remain to be discovered. As Falkner pointed out, the example in Cranston and McConnell (1983) mentioned after Theorem 5.7 serves also as one for which the conditional gauge theorem is false, when q is taken to be a sufficiently small constant. For an extension of these results when the Schrödinger equation is enlarged with a gradient term to $\frac{1}{2}\Delta + b \cdot \nabla$, see Cranston and Zhao (1987).

Theorem 7.12 was given in Cranston, Fabes and Zhao (1988) but its proof has been radically revised and consolidated. The passage from a boundary to an interior point in the conditioning entailed a more complicated connecting argument, which had been blithely disposed of by intonation of the two little words 'Harnack chain'. Anyone who takes a light view of such details should begin by tackling an exercise such as the following: if D is an open connected set and C a closed subset of D, when is $C \backslash D$ connected? Fortunately we do not need such general results.

Earlier versions of results such as Theorems 7.21 and 7.27 under a stronger assumption on D appeared in Zhao (1986). In its present form the latter result was treated in Cranston, Fabes and Zhao (1988). We have rewritten a good part of this, but have omitted some remarks which are relevant to a more general framework (Martin boundary etc.) and certain analytical counterparts, which cannot be discussed adequately in this book. Theorems 7.28 and 7.29 are improvements on earlier results by R.J. Williams (1985) and N. Falkner (1983).

8. Various Related Developments

8.1 Variation of Gauge

In this section, D will denote a bounded domain in $\mathbb{R}^d$, $d \geq 2$, $q \in J_{\text{loc}}$.

The gauge for (D, q), as defined at the beginning of Section 4.3, depends of course on both D and q. In this section we study the question of its finiteness when q is fixed and D varies. Accordingly, we set

$$u_D(x) = E^x\{e_q(\tau_D)\},$$

and simply write $u_D < \infty$ to signify its finiteness at some point in D, equivalent to its being bounded in $\overline{D}$ by the gauge theorem (Theorem 5.19). By Proposition 4.5 and its corollary, for a Green-bounded domain D, if $u_D < \infty$ then $u_G < \infty$ for any domain $G \subset D$. But does there exist a domain $G \supsetneqq D$, such that $u_G < \infty$? Does there exist a domain $G \supset\supset D$ such that $u_G < \infty$? An affirmative answer to the first question is given in Theorem 8.2 below. A negative answer to the second question will then be given, for a regular D. However, in Section 8.2, we show that under a general condition only slightly stronger than regularity the answer is affirmative (Theorem 8.7).

We recall the notation $u(D, q, f; x)$ from Section 4.1. The function f is defined on ∂D but we may extend the notation to any f to mean its restriction to ∂D. In particular, let $z \in \partial D$, $\varepsilon > 0$, and set

$$\phi_\varepsilon(x) = u(D, q, 1_{B(z,\varepsilon)}; x). \tag{1}$$

If $u_D < \infty$, then $\lim_{\varepsilon \to 0} \phi_\varepsilon(x) = 0$ for all $x \in D$, since the singleton $\{z\}$ is a polar set in $\mathbb{R}^d$ for $d \geq 2$. By Theorem 4.7(ii), for each $\varepsilon > 0$, ϕ_ε is continuous in D. Hence the convergence of ϕ_ε as $\varepsilon \downarrow 0$ is uniform in each compact subset of D by Dini's theorem. However, this is not sufficient for our later use, and we need the following strengthening.

Lemma 8.1 *Suppose $u_D < \infty$. Let A be a compact subset of $\overline{D}$ and $z \notin A$. Then $\lim_{\varepsilon \to 0} \phi_\varepsilon(x) = 0$ uniformly for $x \in A$.*

Proof There exists $\varepsilon > 0$ such that A is disjoint from $\overline{B(z, \varepsilon)}$. Let $a \in A$. By Proposition 4.11, we can find a number $0 < r < d(A, \overline{B(z, \varepsilon)})$ such that

$$\sup_{x \in B(a,r)} E^x[e_q(\tau_{B(a,r)})] < \infty. \tag{2}$$

Writing τ_r for $\tau_{B(a,r)}$ and $\tilde{\phi}_\varepsilon = 1_D\phi_\varepsilon$, we have by the strong Markov property, for each $x \in B(a,r)$:

$$\begin{aligned}\phi_\varepsilon(x) &= E^x[\tau_r < \tau_D; e_q(\tau_r)\phi_\varepsilon(X(\tau_r))] \\ &\leq E^x[e_q(\tau_r)\tilde{\phi}_\varepsilon(X(\tau_r))] = u(B(a,r), q, \tilde{\phi}_\varepsilon; x). \end{aligned} \tag{3}$$

Since $\phi_\varepsilon \leq u_D$, ϕ_ε is bounded in D by Theorem 5.19, and $\tilde{\phi}_\varepsilon$ is bounded in $\mathbb{R}^d$. Hence, by applying Theorem 4.7(ii) to $B(a,r)$, it follows from (2) that the function of x in the last member of (3) is continuous in $B(a,r)$. Therefore, by Dini's theorem, it converges uniformly in $B(a, \frac{r}{2})$ to zero as $\varepsilon \downarrow 0$. A fortiori $\lim_{\varepsilon\to 0}\phi_\varepsilon = 0$ uniformly in $B(a, \frac{r}{2})$. This being true for every a in the compact set A, the lemma follows. □

In what follows, $\| \;\|$ denotes $\| \;\|_\infty$ unless otherwise specified.

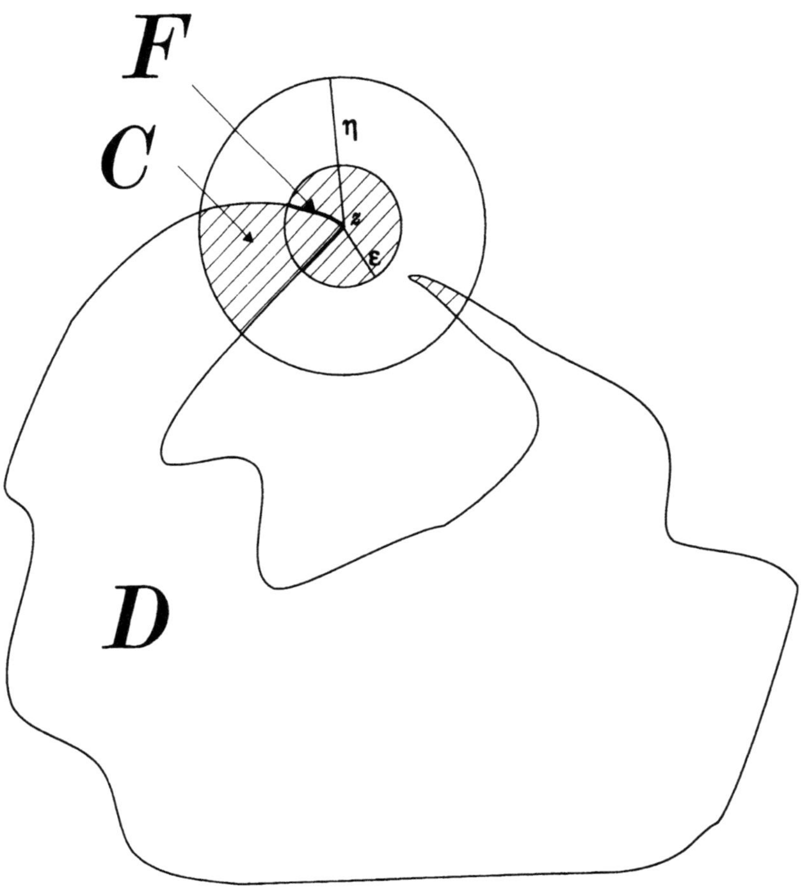

Fig. 8.1.

Theorem 8.2 *Suppose $u_D < \infty$. For any $z \in \partial D$ there exists $\varepsilon > 0$ such that if $G = D \cup B(z, \varepsilon)$, then $u_G < \infty$. Furthermore, for any $\delta > 0$ there exists $\varepsilon(z, \delta)$ such that $\|u_G\| < \|u_D\| + \delta$ if $\varepsilon < \varepsilon(z, \delta)$.*

Proof For any given $0 < \beta < \frac{1}{3}$, by Proposition 4.11 and its proof, there exists $\eta > 0$ such that

$$\sup_{x \in B(z,\eta)} E^x[e_{|q|}(\tau_{B(z,\eta)})] < 1 + \beta. \tag{4}$$

Let

$$A = \overline{D} \cap B(z, \eta)^c.$$

We apply Lemma 8.1 to find ε so that $0 < \varepsilon < \eta$ and

$$\sup_{x \in A} \phi_\varepsilon(x) < \beta. \tag{5}$$

For this ε, let G be the extension of D given in the statement of this theorem, and

$$\begin{aligned} C &= G \cap B(z, \eta), \\ F &= (\partial D) \cap B(z, \varepsilon). \end{aligned}$$

Define $T_0 = 0$, and for $n \geq 1$,

$$\begin{aligned} T_{2n-1} &= T_{2n-2} + \tau_D \circ \theta_{T_{2n-2}}, \\ T_{2n} &= T_{2n-1} + \tau_C \circ \theta_{T_{2n-1}}. \end{aligned}$$

On $\{T_{2n-1} < \tau_G\}$, we have $X(T_{2n-1}) \in F$, while on $\{T_{2n} < \tau_G\}$, we have $X(T_{2n}) \in A$. Let $T_\infty = \lim_n \uparrow T_n$. On $\{T_\infty < \infty\}$, the path of the Brownian motion undergoes infinitely many oscillations of distance exceeding $\frac{\eta-\varepsilon}{2}$ before the time T_∞, since $d(F, A) = \eta - \varepsilon$. The continuity of paths implies that $T_\infty = \infty$ a.s. Since $\tau_G < \infty$ a.s., we have for all $x \in G$:

$$P^x[\cup_{n=1}^\infty (\tau_G = T_n)] = 1. \tag{6}$$

If follows from (4) that

$$\sup_{x \in F} E^x[e_q(\tau_C)] \leq 1 + \beta. \tag{7}$$

Let $x \in D \backslash C \subset A$. Applying the strong Markov property repeatedly to T_n, $n \geq 1$, and using (5) and (7), we obtain

$$\begin{aligned} E^x\{\tau_G = T_{2n-1}; e_q(\tau_G)\} &\leq [\beta(1+\beta)]^{n-1}\|u_D\|; \\ E^x\{\tau_G = T_{2n}; e_q(\tau_G)\} &\leq [\beta(1+\beta)]^n. \end{aligned} \tag{8}$$

Summing (8) over $n \geq 1$ and using (6), we obtain

$$u_G(x) \leq \frac{\|u_D\| + \beta(1+\beta)}{1 - \beta(1+\beta)}. \tag{9}$$

For $x \in C$, we have by (4) and (9),

$$\begin{aligned} u_G(x) &= E^x\{\tau_C < \tau_G; e_q(\tau_C)u(X(\tau_C))\} + E^x\{\tau_C = \tau_G; e_q(\tau_C)\} \\ &\le E^x\{e_q(\tau_C)\} \sup_{y \in D \cap \partial B(z,\eta)} u_G(y) \\ &\le (1+\beta)\frac{\|u_D\| + \beta(1+\beta)}{1-\beta(1+\beta)}. \end{aligned} \tag{10}$$

Thus it follows from (9) and (10) that $u_G < \infty$ and for any given $\delta > 0$ we can take a sufficiently small $\beta > 0$ so that

$$\|u_G\| \le \|u_D\| + \delta.$$

□

Next, we construct a counterexample to the existence of a domain $G \supset\supset D$ with $u_G < \infty$. We begin with a simple proposition which can be proved by applying our previous results.

Proposition 8.3 *There exists a number $r_1 > 0$ such that*

$$E^0\{\exp[\tau_{B(0,r)}]\} < \infty \tag{11}$$

if and only if $r < r_1$.

Proof As in Theorem 4.19, let $\lambda(D, q)$ denote the principal eigenvalue of $\frac{\Delta}{2} + q$ in D. It is easy to see by scaling that for $r > 0$,

$$\lambda(B(0,r),0) = \frac{1}{r^2}\lambda(B(0,1),0).$$

By Theorem 2.7, $\lambda(B(0,1),0) < 0$. Let $r_1 = \sqrt{-\lambda(B(0,1),0)}$; then

$$\lambda(B(0,r),1) = 1 + \lambda(B(0,r),0) = 1 - \left(\frac{r_1}{r}\right)^2.$$

Hence $\lambda(B(0,r),1) < 0$ if and only if $r < r_1$. This is equivalent to the assertion of the proposition by the equivalence theorem (Theorem 4.19).

Remark The value of r_1 is the first root of a Bessel function.

Example Let $q \equiv 1$ in $\mathbb{R}^2$ and let r_1 be the number given in Proposition 8.3. We take $r_3 < r_2 < r_1$ and let $B_i = B(0, r_i)$, $i = 1, 2, 3$, and $C = B_1 \backslash \overline{B}_3$. We may take $r_1 - r_3$ to be so small that

$$\sup_{x \in C} u(C, 1, 1; x) < 2. \tag{12}$$

We take a sequence of arcs A_n on the circle ∂B_2 shrinking to the point $z \in \partial B_2$. The continuous functions $u(B_2, 1, 1_{A_n}; x)$ decrease to zero as $n \to \infty$. By Dini's theorem, we may choose A so that

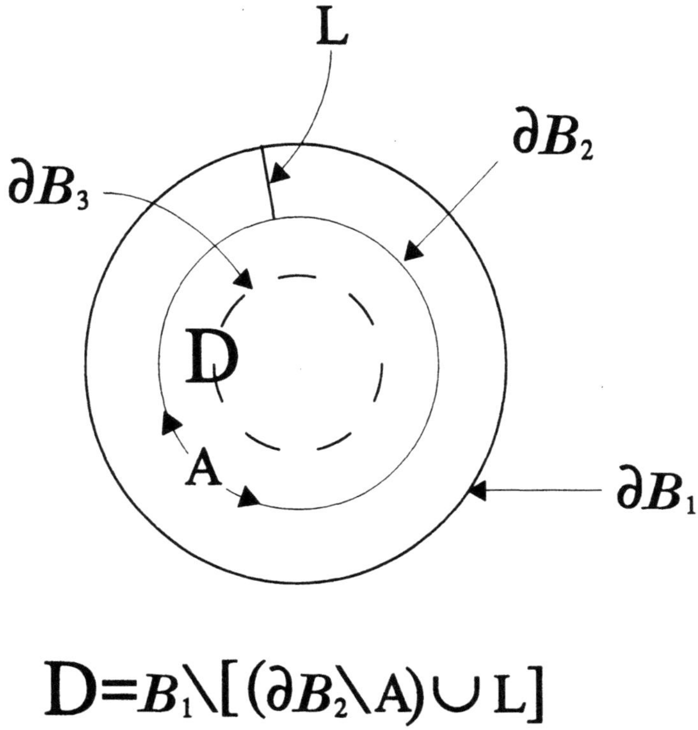

Fig. 8.2.

$$\sup_{x\in\overline{B}_3} u(B_2, 1, 1_A; x) < \frac{1}{3}. \tag{13}$$

Let L be a closed line segment connecting ∂B_1 and $\partial B_2\backslash A$. We now define

$$D = B_1\backslash[(\partial B_2\backslash A) \cup L].$$

Then D is a simply connected domain. L is only used to ensure this. It is easy to see that D is a regular domain. If we denote $(\partial B_2\backslash A) \cup L$ by E, then $E \subset \partial D$, $D \cup E = B_1 \subset \overline{D}$, and $u_{B_1} = \infty$ by the definition of r_1. Since it is clear that any domain G containing $\overline{D}$ must contain B_1, it follows that $u_G = \infty$. We now show that $u_D < \infty$ by an argument similar to the proof of Theorem 8.2.

Let $T_0 = 0$, and for $n \geq 1$:

$$\begin{aligned} T_{2n-1} &= T_{2n-2} + \tau_{B_2} \circ \theta_{T_{2n-2}}, \\ T_{2n} &= T_{2n-1} + \tau_C \circ \theta_{T_{2n-1}}. \end{aligned}$$

Since $T_\infty \equiv \lim_n \uparrow T_n = \infty$ and $\tau_D < \infty$ a.s., we have

$$E^0[e_1(\tau_D)] = \sum_{n=1}^{\infty} E^0[\tau_D = T_n; e_1(\tau_D)]. \tag{14}$$

Using (12) and (13), we have

$$E^0[\tau_D = T_{2n-1}; e_1(\tau_D)] \le \left(\frac{2}{3}\right)^{n-1} \|u_{B_2}\|,$$

and

$$E^0[\tau_D = T_{2n}; e_1(\tau_D)] \le \left(\frac{2}{3}\right)^{n}.$$

It follows from these inequalities and (14) that $E^0[e_1(\tau_D)] < \infty$; in other words, $u_D < \infty$.

The fact that B_1 has infinite gauge while D has finite gauge reflects the fact that B_1 and D are quite different geometrically though not in terms of measure. □

8.2 Variation of the Principal Eigenvalue

For a fixed $q \in J$, let

$$\lambda_D = \lambda(D, q) \tag{15}$$

denote the λ_1 in (3.79) and (3.81). If $m(D) < \infty$, it is the principal eigenvalue of the Schrödinger operator $\frac{\Delta}{2} + q$ on D. Let u_D denote the gauge for (D, q). By Theorem 4.19, the two assertions '$u_D < \infty$' and '$\lambda_D < 0$' are equivalent if $m(D) < \infty$. Hence questions concerning the variation of u_D with D may be formulated as questions relating to the variation of λ_D with D. The latter problem is known in classical analysis, particularly in the case $q = 0$. The standard reference is Courant and Hilbert (1953), in which complicated analytic conditions are imposed. In this section, in Theorem 8.6, we shall establish a sort of semi-continuity of λ_D as a function of the domain D, under a general probabilistic condition.

We begin with an easy consequence of Proposition 3.29.

Proposition 8.4 *If U and D are bounded domains such that $U \subset D$, then $\lambda_U \le \lambda_D$. Furthermore, for any given D and $\varepsilon > 0$, there exists a domain $U \subset\subset D$ such that*

$$\lambda_U > \lambda_D - \varepsilon. \tag{16}$$

Proof The first assertion is trivial by (3.81). To prove the second assertion, we consider a ϕ in the class indicated in (3.81) such that

$$-\frac{1}{2}\int_D |\nabla\phi|^2 dx + \int_D q|\phi|^2 dx > \lambda_D - \varepsilon. \tag{17}$$

Since supp(ϕ) is a compact set in D, there exists a subdomain U such that $U \subset\subset D$ and supp$(\phi) \subset U$. But then by (3.81) applied to U, the left member of (17) does not exceed λ_U, and so (16) is true. □

Corollary 8.5 *If there exists a strictly positive q-harmonic function in D, then* $\lambda_D \leq 0$.

Proof For any subdomain $U \subset\subset D$, a strictly positive q-harmonic function is bounded by two positive constants in U. Hence $\lambda_U < 0$ by Theorems 4.17 and 4.19. If $\lambda_D > 0$, then by the second assertion of Proposition 8.4, there exists $U \subset\subset D$ such that $\lambda_U > 0$, which is a contradiction. □

Proposition 8.4 may be regarded as asserting the continuity from below of λ_D as a function of D, when the domains are ordered by strict inclusion. The example in Section 8.1 shows that λ_D is in general not continuous from above, since there exists a domain with $\lambda_D < 0$ such that for any domain $U \supset\supset D$ we have $\lambda_U \geq 0$. We now proceed to show that if D_0 satisfies a somewhat stronger condition than the usual regularity, then the function λ_D of D is indeed continuous from above at $D = D_0$.

Definition *A domain D is said to be* strongly regular *iff for each* $z \in \partial D$:

$$P^z\{\tau_{\overline{D}} = 0\} = 1. \tag{18}$$

For such a domain D and any $x \in \mathbb{R}^d$, we have

$$P^x\{\tau_D < \tau_{\overline{D}}\} = E^x\left\{P^{X(\tau_D)}(0 < \tau_{\overline{D}})\right\} = 0,$$

hence

$$P^x\{\tau_D = \tau_{\overline{D}}\} = 1. \tag{19}$$

A domain which satisfies an exterior cone condition at each of its boundary points is strongly regular. This result is contained in the proof of Proposition 1.22.

Theorem 8.6 *Let D be a bounded and strongly regular domain. For any* $\varepsilon > 0$ *there exists* $U \supset\supset D$ *such that*

$$\lambda_U < \lambda_D + \varepsilon. \tag{20}$$

Proof By Theorem 3.17, for any $t > 0$ and any bounded and regular D, the function $T_t 1$ defined in (3.34) belongs to $C_0(D)$; since it vanishes outside D, it is continuous in $\mathbb{R}^d$. Its sup-norm $\|T_t 1\|$ is also the norm $\|T_t\|$ of the operator T_t considered on $C_0(D)$. By the spectral radius theorem (see e.g. Yosida (1980, page 212)), the principal eigenvalue is given by

$$\lambda_D = \inf_{t>0} \frac{\ln \|T_t\|}{t}. \tag{21}$$

Hence for any $\varepsilon > 0$ there exists $t > 0$ such that

$$\frac{\ln \|T_t\|}{t} < \lambda_D + \varepsilon. \tag{22}$$

Fix this t from now on. Since D is bounded, there exist bounded and regular domains D_n such that $D_{n+1} \subset\subset D_n$ and $\cap_n D_n = \overline{D}$; see the Appendix to Chapter 1. Now by (19) and Proposition 1.20, we have for each $x \in \mathbb{R}^d$, P^x-a.s.

$$\begin{aligned}(t < \tau_D) &\subset \cap_n (t < \tau_{D_n}) \\ &\subset (t \le \tau_{\overline{D}}) = (t \le \tau_D) = (t < \tau_D).\end{aligned}$$

It follows that P^x-a.s.

$$1_{(t<\tau_{D_n})} \downarrow 1_{(t<\tau_D)}. \tag{23}$$

We let $T_t^{(n)}$ denote the above T_t when D is replaced by D_n. Then by monotone convergence we have in $\mathbb{R}^d$:

$$\lim_{n\to\infty} \downarrow T_t^{(n)} 1 = T_t 1. \tag{24}$$

Since the functions in (24) are continuous in $\mathbb{R}^d$, the convergence is uniform on the compact set $\overline{D_1}$, by Dini's theorem. It then follows that

$$\lim_{n\to\infty} \|T_t^{(n)}\| = \|T_t\|. \tag{25}$$

Consequently (22) holds when T_t is replaced by $T_t^{(n)}$ for a sufficiently large n, so that we obtain by (21),

$$\lambda_{D_n} \le \frac{\ln \|T_t^n\|}{t} < \lambda_D + \varepsilon.$$

Thus (20) holds with $U = D_n$. □

One immediate consequence of Theorem 8.6 is the following extension theorem, which gives a positive answer to the second question posed in Section 8.1.

Theorem 8.7 *Let D be a bounded, strongly regular domain. If $u_D < \infty$, then there exists a domain $U \supset\supset D$ such that $u_U < \infty$.*

Let us note here that any domain $\subsetneq \mathbb{R}^1$ is strongly regular, and the preceding result is true even for a half line; see Theorem 9.5.

Combining Theorems 8.4 and 8.7, we may state the following result on the continuous variation of λ_D with D.

Let D be a bounded, strongly regular domain. For each $\varepsilon > 0$, there exists a 'neighborhood' (D_1, D_2) of D such that $D_1 \subset\subset D \subset\subset D_2$ and for any domain U such that $D_1 \subset U \subset D_2$ we have

$$|\lambda(U) - \lambda(D)| < \varepsilon.$$

We now turn to the variation of $\lambda(D, q)$ with q when D is fixed. Let $1_D q$ and $1_D q_n$ belong to J; we shall denote these by q and q_n, respectively, in what follows. Consider the following condition. There exists $t > 0$ such that

$$\lim_{n\to\infty}\sup_x \int_0^t P_s|q_n - q|(x)\mathrm{d}s = 0. \tag{26}$$

Note that the integral in (26) may be written as $E^x\left\{\int_0^t |q_n - q|(X_s)\mathrm{d}s\right\}$. This is an analogue of the equivalent definition of J given in (3.13) or (3.15). It is easy to see from the semigroup property of $\{P_t\}$ that if (26) holds for some $t > 0$, then it holds for all $t < \infty$. Furthermore, since by hypothesis $q_n - q$ vanishes outside D, we have for all $x \in \mathbb{R}^d$:

$$\begin{aligned} E^x\left\{\int_0^t |q_n - q|(X_s)\mathrm{d}s\right\} &\leq E^x\left\{T_D \leq t; E^{X(T_D)}\left[\int_0^t |q_n - q|(X_s)\mathrm{d}s\right]\right\} \\ &\leq \sup_{x\in\overline{D}} E^x\left\{\int_0^t |q_n - q|(X_s)\mathrm{d}s\right\}. \end{aligned}$$

Hence the supremum in (26) taken over $\mathbb{R}^d$ is the same as that taken over $\overline{D}$.

Theorem 8.8 *Let $m(D) < \infty$ and assume that (26) holds. If (D, q) is gaugeable, then there exists an integer $N \geq 1$ such that (D, q_n) is gaugeable for $n \geq N$.*

Proof By Theorem 4.19, (4.33) holds, hence there exists $t > 0$ such that

$$\|T_t\|_\infty < 1. \tag{27}$$

This is the condition (vi) given after Theorem 4.19. We let $T_t^{(n)}$ denote the T_t when q is replaced by q_n. Then we have by the Cauchy–Schwarz inequality:

$$\begin{aligned} |T_t^{(n)}1(x) - T_t1(x)| &= |E^x\{t < \tau_D; e_q(t)[e_{q_n-q}(t) - 1]\}| \\ &\leq E^x\{t < \tau_D; e_{2q}(t)\}^{1/2} E^x\{t < \tau_D; [e_{q_n-q}(t) - 1]^2\}^{1/2}. \end{aligned}$$

The first factor in the last member of the above is bounded by (3.22). Since $(\mathrm{e}^r - 1)^2 \leq \mathrm{e}^{2|r|} - 1$ for any real number r, the second factor does not exceed

$$\sup_x E^x\{e_{2|q_n-q|}(t)\} - 1.$$

Using the argument in Lemma 3.7 we have

$$1 \leq E^x\left\{e_{2|q_n-q|}(t)\right\} \leq \frac{1}{1-\alpha},$$

where

$$\alpha = 2\sup_x \int_0^t P_s|q_n - q|\mathrm{d}s$$

which converges to zero as $n \to \infty$ by (26). Hence $T_t^{(n)}1$ converges uniformly to T_t1 in $\mathbb{R}^d$ as $n \to \infty$, and so there exists N such that $\|T_t^{(n)}\|_\infty = \|T_t^{(n)}1\|_\infty < 1$ for $n \geq N$. Thus the condition (vi) cited above holds for (D, q_n), which implies that the latter is gaugeable. □

Here is an important application of Theorem 8.8, which may be called the 'supergauge theorem'.

Theorem 8.9 *Let D be a domain in $\mathbb{R}^d$ ($d \geq 1$) with $m(D) < \infty$, and $q \in J$. If (D, q) is gaugeable, then there exists a number $\delta > 0$ such that $(D, (1+\delta)q)$ is gaugeable.*

Proof Let $q_n = (1 + \frac{1}{n})q$, $n \geq 1$. Then for $t > 0$ we have

$$\sup_x \int_0^t P_s|q_n - q|(x)\mathrm{d}s = \frac{1}{n} \sup_x \int_0^t P_s|q|(x)\mathrm{d}s.$$

It follows from (3.23) that (26) is satisfied. Hence there exists $N > 0$ such that $(D, (1 + \frac{1}{N})q)$ is gaugeable by Theorem 8.8. □

8.3 Principal Eigenfunction and Sharp Variation

We begin with a simple but useful result for q-harmonic functions. When $q = 0$, i.e. for a harmonic function, it is a consequence of the maximum principle (Proposition 1.10).

Proposition 8.10 *Let D be an arbitrary domain, and let u be q-harmonic and ≥ 0 in D. Then either $u \equiv 0$ or $u > 0$ in D.*

Proof Suppose that $x \in D$ and $u(x) = 0$. By Proposition 4.11, there exists $r_0 > 0$ such that for all $0 < r \leq r_0$, $(B(x, r), |q|)$ is gaugeable. Hence by Proposition 4.13 (the representation theorem in a special case), we have for $0 < r \leq r_0$:

$$0 = u(x) = E^x\{e_q(\tau_{B(x,r)})u(X(\tau_{B(x,r)})\}.$$

Since $e_q(\tau_{B(x,r)}) > 0$ it follows that $u(X(\tau_{B(x,r)}) = 0$ P^x-a.s. Since $X(\tau_{B(x,r)})$ is uniformly distributed on $\partial B(x, r)$ and u is continuous there by definition, we have $u(y) = 0$ for all $y \in \partial B(x, r)$. Thus $u \equiv 0$ in $B(x, r_0)$. This means that the set of x in D for which $u(x) = 0$ is an open set; however, it is also closed in D by the continuity of u. Hence it is either empty or equal to D, because D is connected. □

In what follows in this section, D is an arbitrary domain in $\mathbb{R}^d$ ($d \geq 1$), and $q \in J$.

We consider the semigroup $\{T_t\}$ (see (3.34)) in $L^2(D)$. Suppose that $\lambda_1 = \lambda(D, q)$ is defined as in (3.79), and that there exists a function ϕ_1 satisfying the following conditions:

$$\phi_1 \in W_0^{1,2}(D); \quad \left(\frac{\Delta}{2} + q\right)\phi_1 = \lambda_1\phi_1; \quad \|\phi_1\|_2 = 1. \tag{28}$$

Then ϕ_1 is called the principal eigenfunction associated with the principal eigenvalue λ_1; the uniqueness of ϕ_1 will be proved shortly. This case prevails when $m(D) < \infty$, as reviewed before Theorem 4.19, but it can occur under other circumstances and the discussion below applies in general. We have for all $t > 0$:

$$T_t\phi_1 = e^{\lambda_1 t}\phi_1. \tag{29}$$

This is a general result from the spectral theory for the self-adjoint operator T_t; see the review at the end of Section 2.4. Here, as elsewhere later, the equations in (28) and (29) hold in D, in the sense of $L^2(D)$. Another result from the spectral theory asserts that

$$\|T_t\|_2^2 = e^{2\lambda_1 t}. \tag{30}$$

See e.g. Yosida (1980).

Our first result below is well known in analysis, but the proof is new and uses an essential probabilistic argument.

Theorem 8.11 *Either $\phi_1 > 0$ m-a.e. in D or $\phi_1 < 0$ m-a.e. in D.*

Proof We shall write λ for λ_1 and ϕ for ϕ_1. By Theorem 5.21, a weak solution of $[\frac{1}{2} + (q - \lambda)]\phi = 0$ may be taken to be continuous in D. Now let

$$G = \{x \in D : \phi(x) > 0\}.$$

Suppose that G is neither empty nor equal to D, then by Proposition 8.10, the set $\{x \in D : \phi(x) < 0\}$ is nonempty. Since this is an open set we have

$$m(D) > m(G). \tag{31}$$

We may assume that G is connected, since otherwise we may replace G by one of its components. Let T_t^G denote the old T_t when D is replaced by G; for clarity let T_t^D denote the old T_t. The following result is crucial to the proof.

Lemma *We have*

$$\forall x \in G: \; T_t^D\phi(x) = T_t^G\phi(x), \tag{32}$$

where on the right-hand side we use the same ϕ to denote the restriction of ϕ in G.

Proof of Lemma Since $\tau_G \leq \tau_D$, we have for all $x \in G$:

$$\begin{aligned} T_t^D\phi(x) &= E^x\{t < \tau_G; e_q(t)\phi(X_t)\} + E^x\{\tau_G \leq t < \tau_D; e_q(t)\phi(X_t)\} \\ &= T_t^G\phi(x) + \psi(x), \end{aligned}$$

where

$$\psi(x) = E^x\{\tau_G < t < \tau_D; e_q(t)\phi(X_t)\}$$

because $P^x\{\tau_G = t\} = 0$ by Proposition 1.20. It remains to prove that $\psi(x) = 0$ for all $x \in G$. The argument below is a ramified form of the strong Markov property which will be described in detail. Using the standard approximation of τ_G as in the proof of Theorem 1.2, we set for $n \geq 1$:

$$\tau_n = \frac{[2^n \tau_G + 1]}{2^n},$$

where $[x]$ denotes the greatest integer $\leq x$; then $\tau_n \downarrow\downarrow \tau_G$. Let

$$\psi_n(x) = E^x\{\tau_n < t < \tau_D; e_q(t)\phi(X_t)\};$$

then we have for all $x \in G$:

$$\lim_{n\to\infty} \psi_n(x) = \psi(x). \tag{33}$$

For each constant $r < t$, we have by the Markov property and (29):

$$\begin{aligned} E^x\{t < \tau_D; e_q(t)\phi(X_t)|\mathcal{F}_r\} &= 1_{\{r<\tau_D\}}e_q(r) \\ &\quad \times E^{X(r)}\{t-r < \tau_D; e_q(t-r)\phi(X_{t-r})\} \\ &= 1_{\{r<\tau_D\}}e_q(r)T^D_{t-r}\phi(X(r)) \\ &= 1_{\{r<\tau_D\}}e_q(r)\mathrm{e}^{\lambda(t-r)}\phi(X(r)). \end{aligned} \tag{34}$$

Since τ_n is countably valued, we can apply (34) on each $\{\tau_n = r\}$ to obtain that on $\{\tau_n < t\}$:

$$E^x\{t < \tau_D; e_q(t)\phi(X_t)|\mathcal{F}_{\tau_n}\} = 1_{\{\tau_n<\tau_D\}}e_q(\tau_n)\mathrm{e}^{\lambda(t-\tau_n)}\phi(X(\tau_n)). \tag{35}$$

Since $E^x\{t < \tau_D; e_q(t)|\phi(X_t)|\} = T^D_t|\phi|(x) < \infty$ because $\phi \in L^2(D)$, it follows from a well-known property of conditioning (used in the proof of Proposition 4.14) that the sequence of random variables on the right-hand side of (35) is uniformly integrable as n varies. Multiplying both sides of (35) by $1_{\{\tau_n<t\}}$ and taking the expectation, we obtain

$$\psi_n(x) = E^x\{\tau_n < t \wedge \tau_D; e_q(\tau_n)\mathrm{e}^{\lambda(t-\tau_n)}\phi(X(\tau_n))\}.$$

It follows from the uniform integrability that

$$\lim_{n\to\infty} \psi_n(x) = E^x\{\tau_G < t \wedge \tau_D; e_q(\tau_G)\mathrm{e}^{\lambda(t-\tau_G)}\phi(X(\tau_G))\}. \tag{36}$$

But $\phi(X(\tau_G)) = 0$ by the definition of G and the continuity of $t \to \phi(X(t))$. Therefore $\psi(x) = 0$ from (33) and (36). The Lemma is proved. □

Now we let $\phi^* = 1_G\phi$ in D. Then since $\phi^* \geq 0$ and $G \subset D$, it is trivial that

$$\forall x \in G : T^D_t\phi^*(x) \geq T^G_t\phi(x). \tag{37}$$

Furthermore, since for each $x \in D$:

$$E^x\{t < \tau_D; X_t \in G\} = \int_G p^D(t; x, y)\mathrm{d}y > 0$$

by Theorem 2.4, and $e_q(t)\phi^*(X_t) > 0$ on $\{X_t \in G\}$, we have

$$\forall x \in D : T^D_t\phi^*(x) > 0. \tag{38}$$

Combining (37) and (38) and using (31), we obtain

$$\int_D (T_t^D \phi^*(x))^2 \mathrm{d}x > \int_G (T_t^G \phi(x))^2 \mathrm{d}x. \tag{39}$$

It follows from the Lemma and (29) that the right member of (39) is equal to

$$\int_G \mathrm{e}^{2\lambda t} \phi(x)^2 \mathrm{d}x.$$

On the other hand, since $\|\phi^*\|_2 \leq \|\phi\|_2 = 1$, the left member of (39) is not greater than

$$\|T_t^D\|_2^2 \|\phi^*\|_2^2 = \mathrm{e}^{2\lambda t} \int_D \phi^*(x)^2 \mathrm{d}x = \int_G \mathrm{e}^{2\lambda t} \phi(x)^2 \mathrm{d}x$$

by virtue of (30). Therefore (39) is false, and we have obtained the desired contradiction. Thus either $G = \emptyset$ or $G = D$. In the former case $\phi < 0$, while in the latter case $\phi > 0$. □

Corollary *There exists a unique continuous positive solution of (28).*

Proof Let ϕ_1 and ϕ_2 be two such solutions. If $\phi_1 \not\equiv \phi_2$, then $\frac{\phi_1 - \phi_2}{\|\phi_1 - \phi_2\|_2}$ is also a solution, and we have by the theorem either $\phi_1 - \phi_2 > 0$ or $\phi_1 - \phi_2 < 0$ in D, which contradicts the last condition in (28) applied to both ϕ_2 and ϕ_1. □

Theorem 8.11 plays an important role in classical analysis. Here we shall apply it in relation to the 'sharp' variation of $\lambda_D = \lambda(D, q)$ with D. In the next two results we consider two domains G and D, $G \subset D$, and assume that the existence of ϕ_1 as specified in (28) is guaranteed when D is replaced by G (this need not be true for D itself). In other words, we assume that λ_G is an eigenvalue.

Proposition 8.12 *Let G be a subdomain of D and $\lambda_G = \lambda(G, q)$ be the principal eigenvalue. If there exists $x \in D$ such that*

$$P^x\{\tau_G < \tau_D\} > 0, \tag{40}$$

then

$$\lambda_G < \lambda_D. \tag{41}$$

Proof Let ϕ be the principal eigenfunction for G as specified in (28). Then by Theorem 8.11 applied to G, we may suppose that $\phi > 0$ in G. Let $\phi^* = \phi$ in G, and $= 0$ in $D \backslash G$. We shall prove that (39) holds with the new interpretation of ϕ. If so, then using (30) with T_t and λ_1 replaced by T_t^D and λ_D, and (29) with T_t and λ_1 replaced by T_t^G and λ_G, we obtain from (39):

$$\mathrm{e}^{2\lambda_D t} = \|T_t^D\|_2^2 \geq \|T_t^D \phi^*\|_2^2 > \int_G \mathrm{e}^{2\lambda_G t} \phi(x)^2 \mathrm{d}x = \mathrm{e}^{2\lambda_G t}$$

from which (41) follows.

To prove (39), we first consider the case where (31) is assumed. Since both (37) and (38) are seen to be true with the new ϕ^*, (39) follows at once.

Now we consider the case in which $m(D) = m(G)$. It follows from (40) that for some $x \in D$ there exists $r > 0$ such that

$$P_r^D 1(x) - P_r^G 1(x) = P^x \{\tau_G \le r < \tau_D\} > 0, \tag{42}$$

where we have used the notation of (2.1). Since $P_r^D 1$ is continuous in D and $P_r^G 1 = P^\bullet(r < \tau_G)$ is upper semi-continuous in $\mathbb{R}^d$ (Proposition 1.19), there exists a ball $B \subset D$ such that (42) holds for all $x \in B$. Using this, $m(G \cap B) = m(B) > 0$ and the fact that $p_r^D(\cdot,\cdot) \ge p_r^G(\cdot,\cdot) > 0$ in $G \times G$, we obtain for $t = 2r$ and all $x \in G$:

$$P_t^D 1(x) = \int_D p_r^D(x,y) P_r^D 1(y) \mathrm{d}y > \int_G p_r^G(x,y) P_r^G 1(y) \mathrm{d}y = P_t^G 1(x);$$

in other words,

$$P^x\{t < \tau_D\} > P^x\{t < \tau_G\}. \tag{43}$$

Since we have assumed that $m(D \backslash G) = 0$, we have $X_t \in G$, and so $e_q(t)\phi^*(X_t) > 0$ almost surely on $\{t < \tau_D\}$. Therefore (43) implies that

$$\forall x \in G : T_t^D \phi^*(x) > T_t^G \phi(x)$$

which implies (39). Proposition 8.12 is proved. □

The condition (40) might appear to be an *ad hoc* technical assumption, but this is by no means the case. However, to elucidate it fully, we need to resort to a fundamental result in probabilistic potential theory. Let T_A denote the hitting time of the set A as defined in (1.28), but with the Brownian motion. We have the following astonishing dichotomy.

Either for all $x \in \mathbb{R}^d$ we have

$$P^x\{T_A < \infty\} = 0, \tag{44}$$

or there exists $x \in A$ such that

$$P^x\{T_A = 0\} = 1. \tag{45}$$

In the first case A is called a polar set (the term has been used before, see e.g. Section 1.6). This result is true for a class of subsets of $\mathbb{R}^d$ including the Borel sets $\mathcal{B}^d$. We recall from Proposition 1.15 that if A is either open or closed, then $T_A \in \mathcal{F}$ and so the probabilities in (44) and (45) are defined. For a general A in $\mathcal{B}^d$, this may not be true, as forewarned in Section 1.5! However, there exists a tribe $\tilde{\mathcal{F}}$ containing $\mathcal{F}$ such that each measure P^x, $x \in \mathbb{R}^d$, can be extended to $\tilde{\mathcal{F}}$, and such that $T_A \in \tilde{\mathcal{F}}$, so that the above probabilities are still defined. For a detailed discussion of these matters of measurability, see Chung (1982a, pages 59–65). The above dichotomy is contained in the Corollary on page 223 of Chung (1982a), but its proof requires several substantial results in that book.

We are now ready to enhance Proposition 8.12 to a full theorem.

Theorem 8.13 *We have*

$$\lambda_G = \lambda_D \ \ or \ \ \lambda_G < \lambda_D$$

according to whether $D \backslash G$ is a polar set or not.

Proof Let $A = D\backslash G$, then $\{\tau_G < \tau_D\} = \{T_A < \tau_D\}$. If A is polar, then it follows from (44) that for all $x \in \mathbb{R}^d$:

$$P^x\{\tau_G < \tau_D\} = P^x\{T_A < \tau_D\} \leq P^x\{T_A < \infty\} = 0.$$

Therefore $\tau_G = \tau_D$ almost surely, and so the two operators T_t^D and T_t^G are identical by their definitions. Hence $\lambda_D = \lambda_G$ by (30).

If A is not polar, then by (45) there exists $x \in A \subset D$ such that

$$P^x\{\tau_G < \tau_D\} = P^x\{T_A < \tau_D\} = P^x\{0 < \tau_D\} = 1.$$

Thus (40) is satisfied, and so $\lambda_G < \lambda_D$ by Proposition 8.12. □

Example Let D be a domain in $\mathbb{R}^d$ with $m(D) < \infty$. If C is a closed variety of dimension $d-1$ contained in D such that $D\backslash C$ is connected, then $\lambda_{D\backslash C} < \lambda_C$. If in the above $d-1$ is changed to $d-2$, then $\lambda_{D\backslash C} = \lambda_C$.

This follows from Theorem 8.13 because it is well known that a variety of dimension $d-2$ in $\mathbb{R}^d$ is polar, while a variety of dimension $d-1$ in $\mathbb{R}^d$ is not polar; see e.g. Chung (1982a, page 152, Exercise 2). We shall now give a proof for a specific case without resorting to the above general dichotomy. Suppose that D is a bounded domain in $\mathbb{R}^2$ and C is a closed line segment contained in D. Consider the domain $G = D\backslash C$. At each $z \in C$, there is a subsegment in C containing z, which is a 'truncated flat cone' in the sense of Theorem 4 on page 165 of Chung (1982a); hence, by that theorem z is a regular point of ∂G. Now let

$$f(z) = \begin{cases} 1, & z \in C, \\ 0, & z \in \partial D; \end{cases}$$

and for $x \in G$,

$$\begin{aligned} h(x) &= E^x\{f(X(\tau_G))\} \\ &= P^x\{\tau_G < \tau_D\}. \end{aligned}$$

Then by Theorem 1.23, h is harmonic in G, and as $x \in G$, $x \to z \in C$, $h(x) \to 1$. This implies the condition (40). The same argument is valid for any $d \geq 2$,

In particular, consider the disk $B(0, r_1)$ in $\mathbb{R}^2$ where r_1 is given in Proposition 8.3, with C being part of a radius of the disk; then $\lambda_{B\backslash C} < \lambda_B = 0$ and so $u_{B\backslash C} < \infty$ while $u_B = \infty$. This is a much simpler example than the one constructed in the example of Section 8.1. But did we waste our time constructing the latter?

8.4 Boundary Harnack Principle and Application

In this section, we shall extend the boundary Harnack principle (Lemma 6.8) for harmonic functions to the case of q-harmonic functions.

For a domain D and a relatively open set $E \subset \partial D$, let $S(D, E, q)$ denote the class of strictly positive q-harmonic functions in D which are continuous in $D \cup E$ and vanish on E.

Theorem 8.14 *Let D be a bounded Lipschitz domain in $\mathbb{R}^d$, $d \geq 2$, and $q_i \in J_{\text{loc}}$, $i = 1, 2$. Then for any relatively open nonempty set $E \subset \partial D$ and subdomain $D_0 \subset D$ such that $(\partial D_0) \cap (\partial D) \subset E$, there exists a strictly positive constant $C = C(D, E, D_0, q_1, q_2)$ with the following property. For any $u_i \in S(D, E, q_i)$, $i = 1, 2$, we have for all x and y in D_0:*

$$\frac{u_1(x)}{u_2(x)} \leq C \frac{u_1(y)}{u_2(y)}. \tag{46}$$

Proof By the discussion at the beginning of Section 5.2, there exist constants $C > 0$ and $s_0 > 0$ with the following property. For each $z \in \partial D$, there exists a coordinate system $(\xi, \eta) \in \mathbb{R}^{d-1} \times \mathbb{R}^1$ and a function ϕ satisfying

$$D \cap B(z, s_0) = \{(\xi, \eta) : \phi(\xi) < \eta\} \cap B(z, s_0), \tag{47}$$

and

$$|\phi(\xi_1) - \phi(\xi_2)| \leq C|\xi_1 - \xi_2| \tag{48}$$

in $B(z, s_0)$. By Proposition 4.11, we may choose s_0 small enough that $(B(z, s_0), |q_i|)$ is gaugeable, $i = 1, 2$. We denote $(\partial D_0) \cap (\partial D)$ by E_0. Let ρ_0 denote the distance from E_0 to $(\partial D) \backslash E$ ($\rho_0 = \infty$ if $\partial D = E$), and let

$$r = \frac{s_0 \wedge \rho_0 \wedge r_0}{8(C+1)^2}, \tag{49}$$

where r_0 is given in Lemma 6.8.

For each $z = (\xi_0, \phi(\xi_0)) \in E_0$, let the cylinder set $L(z, r, (C+1)r)$ defined in Section 5.2 be denoted more simply as

$$L(z, r) = \{(\xi, \eta) : |\xi - \xi_0| < r,\ \phi(\xi) < \eta < \phi(\xi) + (C+1)r\}. \tag{50}$$

Then using (48) we can verify the inequalities:

$$D \cap B(z, r) \subset L(z, r) \subset D \cap B(z, 2(C+1)r). \tag{51}$$

Since E_0 is a compact set, there exists a finite set of points $\{z_k,\ 1 \leq k \leq N\}$ in E_0 such that

$$E_0 \subset \bigcup_{k=1}^{N} B\left(z_k, \frac{r}{2}\right). \tag{52}$$

The balls may be ordered so that each intersects the next. Let

$$D_r = \left\{ x \in D_0 : \rho(x, E_0) < \frac{r}{2} \right\}. \tag{53}$$

Then by (51)–(53) we have

$$D_r \subset \bigcup_{k=1}^{N} [D \cap B(z_k, r)] \subset \bigcup_{k=1}^{N} L(z_k, r). \tag{54}$$

Let

$$K_r = \left\{ x \in \overline{D}_0 : \ \rho(x, E_0) \geq \frac{r}{4} \right\};$$

then D_r intersects K_r and $D_r \cup K_r \supset D_0$.

We first prove (46) for all x and y in $L(z, r)$, where $z \in E_0$. Let

$$L = L(z, r), \quad G = L(z, 4(C+1)r).$$

Then by (51)

$$D \cap B(z, 4(C+1)r) \subset G \subset D \cap B(z, 8(C+1)^2 r), \tag{55}$$

and

$$(\partial G) \cap (\partial D) \subset E \tag{56}$$

because $8(C+1)^2 r \leq \rho_0$.

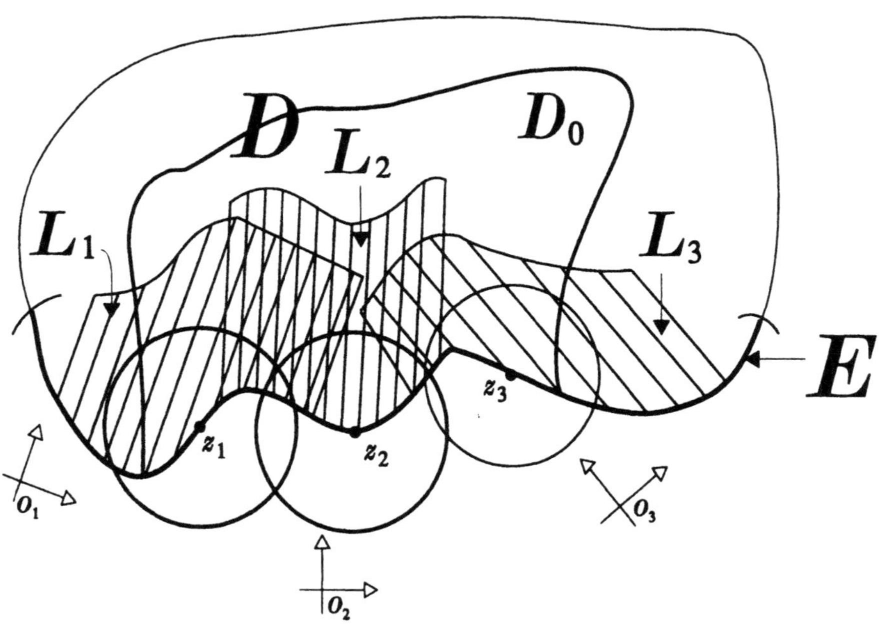

Fig. 8.3.

By the choice of s_0, (49) and (55), (G, q_i) is gaugeable. By (56) and the definition of $S(D, E, q_i)$, we have $u_i \in C(\overline{G})$. Therefore we can apply Theorem 4.15 to represent u_i as follows:

$$\begin{aligned} u_i(x) &= E^x\{e_{q_i}(\tau_G)u_i(X(\tau_G))\} \\ &= \int_{\partial G} u_i(z)H_q(x, \mathrm{d}z), \end{aligned} \tag{57}$$

where H_q is the q-harmonic measure for G (see (5.19)). Now let

$$h_i(x) = \int_{\partial G} u_i(z)H_0(x, \mathrm{d}z). \tag{58}$$

Since u_i is bounded and continuous on ∂G by (56), h_i is bounded and harmonic in G. Since u_i vanishes on $(\partial G) \cap (\partial D)$, so does h_i by Theorem 1.23, because G being Lipschitzian is regular. Thus by (55) and (56), $h_i \in H_0^+(z, 2(C+1)r)$, where H_0^+ is as in Lemma 6.8. Since $2(C+1)r \leq r_0$, we can apply Lemma 6.8 to obtain for all x and y in $D \cap B(z, 2(C+1)r)$:

$$\frac{h_1(x)}{h_2(x)} \leq C_0 \frac{h_1(y)}{h_2(y)}, \tag{59}$$

where C_0 is the constant C in Lemma 6.8 which depends only on D. Since G is a Lipschitz domain and (G, q_i) is gaugeable, it follows from (57), (58), and Theorem 7.7 that there exist constants $C_i > 0$ such that for all $x \in G$:

$$C_i^{-1}h_i(x) \leq u_i(x) \leq C_i h_i(x). \tag{60}$$

Both (59) and (60) hold in L. Combining these we obtain for all x and y in L:

$$\frac{u_1(x)}{u_2(x)} \leq C_0 C_1^2 C_2^2 \frac{u_1(y)}{u_2(y)}.$$

This establishes (46) for all x and y in $L(z, r)$. Now it is trivial that if such a relation holds in two intersecting sets, then it holds in their union. The constant C may vary but its dependence remains the same. Hence by (51) and (54), (46) holds in D_r. On the other hand, it holds in K_r by Theorem 5.18 applied to D and K_r, followed by the taking of quotients. Therefore it holds for $D_r \cup K_r$ which contains D_0. □

We close this section with an application of Theorem 8.14 to the gauge function u for a general domain. We recall from the Corollary to Theorem 5.18 that if $u \not\equiv \infty$ in D then $u < \infty$ in D, indeed it is a continuous q-harmonic function in D. If furthermore (D, q) is gaugeable, then we have by Theorem 4.7(iii), for any $z \in (\partial D)_r$:

$$\lim_{\overline{D} \ni x \to z} u(x) = 1. \tag{61}$$

This result is also true for a half line in $\mathbb{R}^1$, by Theorem 9.15. The question as to whether it is true for a general domain D in $\mathbb{R}^d$ $(d \geq 2)$ under the sole assumption that u is finite in D was raised by Chung. A partial answer has been given by Zhao as follows.

Theorem 8.15 *The result (61) is true provided D is locally Lipschitzian at z.*

Proof Take z to be the origin of the local coordinate system. As in the proof of Theorem 8.14, let L denote the cylinder $L(z,r)$ given in (50). Then $L \subset D$ by (51). By Proposition 4.11, we may take r to be so small that $(L,|q|)$ is gaugeable. We have by the strong Markov property:

$$u(x) = E^x\{\tau_L = \tau_D; e_q(\tau_L)\} + E^x\{\tau_L < \tau_D; e_q(\tau_L)u(X(\tau_L))\}. \tag{62}$$

The first term on the right-hand side of (62) may be expressed as

$$u_1(x) = E^x\{e_q(\tau_L)f_1(X(\tau_L))\},$$

where $f_1 = 1_{(\partial L)\cap(\partial D)}$. Since (L,q) is gaugeable, f_1 is continuous at z, and z is regular, we have by Theorem 4.7(iii):

$$\lim_{\overline{L}\ni x\to z} u_1(x) = 1. \tag{63}$$

The second term on the right-hand side of (62) may be expressed as

$$u_2(x) = E^x\{e_q(\tau_L)f_2(X(\tau_L))\}, \tag{64}$$

where $f_2 = 1_{(\partial L)\setminus(\partial D)}u$. Since (L,q) is gaugeable and $u_2 \not\equiv \infty$ in L because $u \not\equiv \infty$, u_2 is continuous and q-harmonic in L by the Corollary to Theorem 5.18. However, it may not be bounded and that is the crux of the matter, as it is for the initial u. But let us now replace f_2 by $f_2 \wedge n$ in (64) and denote the resulting function by v_n. Since (L,q) is gaugeable, we can apply Theorem 4.7(iii) to v_n to conclude that it vanishes continuously on $(\partial L)\cap(\partial D)$ since L is a regular domain, being Lipschitzian by a previous remark. Thus $v_n \in S(L,(\partial L)\cap(\partial D),q)$ for all n. Therefore we can apply Theorem 8.14 to v_n and v_1 to obtain

$$\frac{v_n(x)}{v_n(x_0)} \le C\frac{v_1(x)}{v_1(x_0)}, \tag{65}$$

where x_0 is any fixed point in L, and the constant C is that asserted in Theorem 8.14. Letting $n \to \infty$ we obtain by monotone convergence:

$$\frac{u_2(x)}{u_2(x_0)} \le C\frac{v_1(x)}{v_1(x_0)}. \tag{66}$$

It follows that u_2 vanishes continuously on $(\partial L)\cap(\partial D)$, as does v_1; in particular,

$$\lim_{\overline{L}\ni x\to z} u_2(x) = 0. \tag{67}$$

Since $u = u_1 + u_2$, (61) follows from (63) and (67). Indeed, we have proved (67) for a neighborhood of z on ∂D. □

At the seminar on stochastic processes in March 1992, we proposed the problem of finding an example for which (61) is false when $q \equiv 1$, when the Lipschitzian assumption is dropped. K. Burdzy gave such an example (oral communication). But mathematics marches on: what is a necessary and sufficient condition for the validity of (61)? After all, Wiener (1924) gave such a condition for the regularity of ∂D at a given z.

8.5 Schrödinger Equation in the Classical Setting

In this section we give a summary of some of our main results in a special form, when they are recognizable as extensions from the Laplacian to the Schrödinger case. This requires a smoother domain than we have assumed before.

For a bounded C^2 domain D in $\mathbb{R}^d$, $d \geq 2$, by Proposition 5.13, we have

$$\begin{aligned} H(x, \mathrm{d}z) &= \overline{K}(x, z)\sigma(\mathrm{d}z) \\ &= \frac{1}{2}\frac{\partial}{\partial n_z}G(x, z)\sigma(\mathrm{d}z), \end{aligned} \tag{68}$$

where H is the harmonic measure, $\overline{K}$ is the Poisson kernel and G is the Green function of D and n_z is the unit inner normal vector at $z \in \partial D$.

We now assume that $q \in J_{\mathrm{loc}}$ and (D, q) is gaugeable.

We recall the conditional gauge defined in Section 5.2:

$$u(x, z) = E_z^x[e_q(\tau_D)], \quad (x, z) \in D \times \partial D;$$

and the q-harmonic measure given in (5.19):

$$H_q(x, \mathrm{d}z) = u(x, z)H(x, \mathrm{d}z), \quad (x, z) \in D \times \partial D. \tag{69}$$

In the present case, we have by (68) and (69),

$$H_q(x, \mathrm{d}z) = u(x, z)\overline{K}(x, z)\sigma(\mathrm{d}z). \tag{70}$$

Hence, if we set

$$\begin{aligned} \overline{K}_q(x, z) &= u(x, z)\overline{K}(x, z) \\ &= \frac{1}{2}u(x, z)\frac{\partial}{\partial n_z}G(x, z), \end{aligned} \tag{71}$$

we may call $\overline{K}_q(\cdot, \cdot)$ the q-Poisson kernel, and we have

$$H_q(x, \mathrm{d}z) = \overline{K}_q(x, z)\sigma(\mathrm{d}z)$$

as an extension of (68).

Now, by Theorem 7.21, the q-Green function V and the 0-Green function G are related by

$$V(x,y) = u(x,y)G(x,y), \quad (x,y) \in D \times D, \quad x \neq y, \tag{72}$$

where $u(x,y)$ is defined in Section 7.3. It follows from (72) and the continuity of $u(\cdot,\cdot)$ on $D \times \overline{D}$ (Theorem 7.27) that

$$\begin{aligned} \frac{\partial V}{\partial n_z}(x,z) &= \lim_{y \to z} \frac{V(x,y)}{\rho(y)} \\ &= \lim_{y \to z} u(x,y) \lim_{y \to z} \frac{G(x,y)}{\rho(y)} \\ &= u(x,z) \frac{\partial G}{\partial n_z}(x,z), \end{aligned} \tag{73}$$

where $\rho(y)$ denotes the distance from y to ∂D.

Summarizing the above formulae, we have

Theorem 8.16 *The normal derivative of the q-Green function $V(x,\cdot)$ exists:*

$$\frac{\partial}{\partial n_z} V(x,z) = \lim_{y \to z} \frac{V(x,y)}{\rho(y)}, \quad (x,z) \in D \times \partial D.$$

Given any $f \in C(\partial D)$, the solution u_f of the Dirichlet problem

$$\begin{cases} (\frac{\Delta}{2} + q)u = 0 \text{ in } D \\ u|_{\partial D} = f \end{cases} \tag{74}$$

may be represented by

$$u_f(x) = \frac{1}{2} \int_{\partial D} \left[\frac{\partial}{\partial n_z} V(x,z) \right] f(z) \sigma(dz). \tag{75}$$

The following so-called 'strong Harnack inequality' for q-harmonic functions in a C^2 domain is an easy consequence of (70) and the conditional gauge theorem.

Theorem 8.17 *For any subdomain $D_0 \subset\subset D$, there exists a constant $C = C(D, D_0, q) > 0$ such that for any $f \in C(\partial D)$, the solution u_f of (74) satisfies the following inequality:*

$$\sup_{x \in D_0} |u_f(x)| \leq C \int_{\partial D} |f(z)| \sigma(dz). \tag{76}$$

If, in addition, $f \geq 0$, then

$$\inf_{x \in D_0} u_f(x) \geq C^{-1} \int_{\partial D} f(z) \sigma(dz). \tag{77}$$

Proof For any $f \in C(\partial D)$, we have by (70):

$$u_f(x) = \int_{\partial D} u(x,z) \overline{K}(x,z) f(z) \sigma(dz). \tag{78}$$

By Theorem 7.7 there exists a constant $C_1 > 1$ such that

$$C_1^{-1} \leq u(x,z) \leq C_1, \quad (x,z) \in D \times \partial D. \tag{79}$$

Since $\overline{K}(\cdot,\cdot)$ is continuous and strictly positive in $D \times \partial D$, there exists a constant $C_2 > 1$ such that

$$C_2^{-1} \leq \overline{K}(x,z) \leq C_2, \quad (x,z) \in \overline{D}_0 \times \partial D. \tag{80}$$

Thus (76) and (77) follow from (78)–(80). □

8.6 Dirichlet Problem and Truncated Gauge

In this section, let D be a bounded regular domain in $\mathbb{R}^d$ $(d \geq 1)$, $q \in J_{\text{loc}}$ and $f \in C(\partial D)$. We consider the Dirichlet problem for the Schrödinger equation, namely that of finding $u \in C(\overline{D})$ satisfying

$$\begin{aligned} \left(\frac{\Delta}{2} + q\right) u &= 0 \text{ in } D \\ u &= f \text{ on } \partial D. \end{aligned} \tag{81}$$

This problem is treated in Section 4.4. Our main result (Theorem 4.7) is that if (D,q) is gaugeable, then the unique solution is given by

$$u_f(x) = E^x\{e_q(\tau_D) f(X(\tau_D))\}. \tag{82}$$

In the theory of elliptic partial differential equations, this problem is solved under the more general assumption that zero is not an eigenvalue of the operator $\frac{\Delta}{2} + q$. In this section we shall present the probabilistic solution to this extended problem by means of a truncated gauge.

We recall from (3.39), the potential operator

$$V = \int_0^\infty T_t dt$$

for the killed Feynman–Kac semigroup. We begin by deriving another expression for the u in (81), which is hidden in the proof of Theorem 4.19 in the case $f \equiv 1$.

Theorem 8.18 *Suppose that (D,q) is gaugeable, and*

$$h(x) = E^x\{f(X(\tau_D))\}. \tag{83}$$

Then we have in D:

$$u = V(qh) + h. \tag{84}$$

Proof Since h is bounded in D and $V(|q|) < \infty$ by Theorem 4.19, the following calculation is valid by the Markov property and Fubini's theorem

$$\begin{aligned} E^x\{[e_q(\tau_D)-1)]f(X(\tau_D))\} &= E^x\left\{\int_0^{\tau_D} q(X_t)e_q(t)E^{X_t}[f(X(\tau_D))]\mathrm{d}t\right\} \\ &= E^x\left\{\int_0^{\tau_D} e_q(t)q(X_t)h(X_t)\mathrm{d}t\right\} = V(qh); \end{aligned}$$

whence (84) follows. □

Now, formally speaking, we have

$$(I-T_t)V = \left(\int_0^\infty - \int_t^\infty\right) T_s \mathrm{d}s = \int_0^t T_s \mathrm{d}s.$$

We know from Proposition 3.16 that $\int_0^t T_s|q|\mathrm{d}s$ is bounded for each $t < \infty$. Therefore the following function is bounded without the assumption of gaugeability:

$$u_t = \int_0^t T_s(qh)\mathrm{d}s + (I-T_t)h. \tag{85}$$

It turns out that u_t has a simple probabilistic representation.

Proposition 8.19 *We have*

$$u_t(x) = E^x\{\tau_D \le t; e_q(\tau_D)f(X(\tau_D))\}. \tag{86}$$

Proof We have

$$\begin{aligned} \int_0^t T_s(qh)\mathrm{d}s &= \int_0^t E^x\{s<\tau_D; e_q(s)(qh)(X_s)\}\mathrm{d}s \\ &= E^x\left\{\int_0^{\tau_D\wedge t} e_q(s)q(X_s)E^{X_s}[f(X(\tau_D))]\mathrm{d}s\right\} \\ &= E^x\left\{\int_0^{\tau_D\wedge t} e_q(s)q(X_s)\mathrm{d}s\cdot f(X(\tau_D))\right\} \\ &= E^x\{[e_q(\tau_D\wedge t)-1]f(X(\tau_D))\} \\ &= E^x\{t<\tau_D; [e_q(t)-1]f(X(\tau_D))\} \\ &\quad + E^x\{\tau_D\le t; [e_q(\tau_D)-1]f(X(\tau_D))\} \\ &= T_th(x) - E^x\{t<\tau_D; f(X(\tau_D))\} \\ &\quad + E^x\{\tau_D\le t; [e_q(\tau_D)-1]f(X(\tau_D))\} \\ &= T_th(x) - h(x) + E^x\{\tau_D\le t; e_q(\tau_D)f(X(\tau_D))\}. \end{aligned}$$

This is equivalent to (86). □

Instead of the usual assumption about eigenvalues mentioned above, we shall assume that for a given $t > 0$, the equation

$$T_t\phi = \phi \tag{87}$$

does not have a solution ϕ in $L^1(D)$ or $C_0(D)$. This is the same as saying that 1 is not an eigenvalue of T_t, which is equivalent to the usual assumption mentioned above because $1 = e^{0t}$. By the Fredholm–Riesz alternative theorem (see e.g. Gilbarg and Trudinger (1977, page 79)), the operator $I - T_t$ considered in $L^1(D)$ or $C_0(D)$ has a bounded inverse $(I - T_t)^{-1}$. We can therefore set

$$v_t = (I - T_t)^{-1} u_t. \tag{88}$$

Theorem 8.20 *For each $t > 0$, under the stated assumption regarding (87), v_t is the unique solution of (81).*

Proof The uniqueness follows from the remark above, since the difference between two solutions of (81) is an eigenfunction corresponding to the eigenvalue zero.

Now for fixed t let

$$w = (I - T_t)^{-1} \int_0^t T_s(qh)\mathrm{d}s, \tag{89}$$

so that $v_t = w + h$. Since $qh \in L^1(D)$, we have $(I - T_t)^{-1}(qh) \in L^1(D)$. Since each T_s commutes with $(I - T_t)^{-1}$, it follows from (89) that w belongs to $D(A_1)$, where A_1 is the infinitesimal generator of $\{T_t\}$ in $L^1(D)$. Hence we have by standard semigroup theory:

$$\begin{aligned} A_1 w &= \int_0^t \frac{\mathrm{d}}{\mathrm{d}s} T_s[(I - T_t)^{-1}(qh)]\mathrm{d}s \\ &= (T_t - I)(I - T_t)^{-1}(qh) = -qh. \end{aligned}$$

Since $\int_0^t T_s(qh)\mathrm{d}s \in C_0(D)$, we have $w \in C_0(D)$. Hence, by Proposition 3.24,

$$\left(\frac{\Delta}{2} + q\right)w = A_1 w = -qh.$$

Now $\Delta h = 0$ and $h|_{\partial D} = f$ by Theorem 1.23, hence

$$\left(\frac{\Delta}{2} + q\right) v_t = \left(\frac{\Delta}{2} + q\right)(w + h) = 0,$$

and v_t is a solution of (81). □

We now rewrite (86) and (88) together as follows:

$$(I - T_t)v_t = E^x\{\tau_D \le t; e_q(\tau_D) f(X(\tau_D))\}. \tag{90}$$

Suppose that (D, q) is gaugeable. Then by Theorem 4.19, zero is not an eigenvalue of $\frac{\Delta}{2} + q$, hence the assumption of Theorem 8.20 holds for all $t > 0$. The conclusion then implies that v_t in fact does not depend on t and may be denoted by v, and $v \in C(\overline{D})$. Hence, it follows from (vii) after Theorem 4.19 that $\lim_{t\to\infty} T_t v = 0$. Letting $t \to \infty$ in (90), we obtain $v = u$ where u is the u_f in (82). Thus we have retrieved our earlier result in Theorem 4.7. Indeed, we can go a step further. If u is bounded, we have

$$\begin{aligned} T_t u &= E^{\bullet}\{t < \tau_D; e_q(t)u(X_t)\} \\ &= E^{\bullet}\left\{t < \tau_D; e_q(t)\exp\left(\int_t^{\tau_D} q(X_s)\mathrm{d}s\right) f(X(\tau_D))\right\} \\ &= E^{\bullet}\{t < \tau_D; e_q(\tau_D)f(X(\tau_D))\} \end{aligned}$$

so that

$$(I - T_t)u = u_t. \tag{91}$$

Now it follows from (4.66) that $\|T_t\| < 1$ for all large values of t, and so

$$(I - T_t)^{-1} = \sum_{n=0}^{\infty} (T_t)^n$$

without the Fredholm–Riesz theorem. On the other hand, we have by the Markov property:

$$(T_t)^n u_t = E^{\cdot}\{nt < \tau_D \leq (n+1)t; e_q(\tau_D)f(X(\tau_D))\}.$$

Therefore, in this case the relation unravels as the banality

$$u(x) = \sum_{n=0}^{\infty} E^x\{nt < \tau_D \leq (n+1)t; e_q(\tau_D)f(X(\tau_D))\}.$$

Notes on Chapter 8

The contents of Section 8.1 are improvements of the results in Chung, Durrett and Zhao (1983). The extension from q in $\mathcal{B}_b$ to J is easy.

The variation of the eigenvalues with the domain is a classical problem. One source frequently cited is the venerable text Courant and Hilbert (1953). Unfortunately the treatment there is not rigorous; see for instance page 419 there. This gap was discovered and led to a number of papers mentioned in Chung, Li and Williams (1986). Here the famous Theorem 8.11 has a brand new probabilistic proof found by Zhao during a final revision of the manuscript. For a recent treatment of this result from the point of view of operator theory, see Goelden (1977). The definitive result of Theorem 8.13 is probably also new, as are some of the precise formulations of the results in Section 8.2. The condition of strong regularity in Theorem 8.6 is a curious one. According to Martine Labrèche, it was used in an implicit way in Keldysh's work on potential theory.

Theorem 8.8 is new. A version of Theorem 8.9 using the eigen connection (see Theorem 4.19(v)) is given in Aizenman and Simon (1982). Our proof is totally different and is a special case of a result in Chung and Rao (1988).

Proposition 8.10 arose as a problem about the 'nodes' of quantum wave functions considered by Hoffman-Ostenhofs and Simon. Empirically there is an observable flash on the screen when the function changes sign. Several 'complicated' proofs of the result were published prior to Chung and Rao (1981).

Yet in a later paper (Aizenman and Simon, 1982) the authors still 'emphasize that [they] know of no direct application of the solution of (the Dirichlet boundary value) problem to quantum mechanics'. See the comments in Chung (1985b).

The probabilistic proof of the old Theorem 8.11 is due to Zhao. So far as we are aware, no such attempts have been made before. Theorem 8.13 may also be new in its generality, though it is futile to compare such a result with any non-probabilistic precursor.

Theorem 8.14 was stated in Cranston, Fabes and Zhao (1988) with a cursory proof. The small amount of Lipschitzian geometry required does not seem well known and so we have taken pains to make it explicit.

Theorem 8.15 is given in Zhao (1990a) but the present bowdlerized presentation is due to Chung.

Theorem 8.17 was one of two major results in Aizenman and Simon (1982), proved when D is a ball by time-reversing as mentioned in Notes on Chapter 5. The proof given here, which reduces it to an application of the conditional gauge theorem for a ball, was given in Zhao (1983).

The solution of the Dirichlet problem beyond gaugeability given in Section 8.6 was posed by Chung in his lectures in Beijing 1985. Ma and later Ma and Zhao solved the problem after a fashion in Ma and Zhao (1987), but there was a dispensable eigen expansion which polluted the atmosphere. This has been cleaned up in the present version with a further elucidation of the role of gaugeability.

9. The Case of One Dimension

In the case of $\mathbb{R}^1$, the special geometry leads to new questions and concepts. At the same time, simpler analysis is often available and yields fuller results. We shall emphasize those aspects of the general theory which do not have ready extensions to higher dimension. In particular, we shall treat the gaugeability of an infinite interval.

9.1 Fundamental Expectations

We begin by reviewing a number of essential properties of the standard Brownian motion in $\mathbb{R}^1$, which we shall use later with only an occasional cross reference. For details, see Chung (1981 and 1982b).

The hitting time T_A is defined as in Section 1.5, but clearly in $\mathbb{R}^1$ we can always reduce A to a point and write T_x for $T_{\{x\}}$. Thus, if $I = (y, z)$ is a finite interval, and τ_I the usual exit time from I, then for $x \in I$, under P^x we have $\tau_I = T_y \wedge T_z$ by continuity of paths. In what follows, unspecified x, y, z, a, b, c, etc. denote elements of $\mathbb{R}^1$.

(i) For any x and y, we have

$$P^x\{T_y < \infty\} = 1.$$

If $x \neq y$, then $E^x\{T_y\} = \infty$. The first property is usually referred to as 'recurrence'.

(ii) If $y < x < z$, we have

$$\begin{aligned} P^x\{T_y < T_z\} &= \frac{z-x}{z-y}; \\ P^x\{T_z < T_y\} &= \frac{x-y}{z-y}; \\ E^x\{T_y \wedge T_z\} &= (x-y)(z-x). \end{aligned}$$

These are solutions of the Dirichlet boundary value problem and the Poisson equation for the Laplace operator, which reduces to $\mathrm{d}^2/\mathrm{d}x^2$.

(iii) For any x, we have

$$P^x\{T_{(-\infty,x)} = 0\} = P^x\{T_{(x,\infty)} = 0\} = P^x\{T_x = 0\} = 1.$$

In the terminology of regularity introduced in Section 1.5, this means that the point x is regular for both (x, ∞) and $(-\infty, x)$, and (consequently) for itself. This is a fundamental property not shared by Brownian motion in $\mathbb{R}^d$, $d \geq 2$.

(iv) The Green function for the finite interval (a, b) is given by

$$G(x,y) = \begin{cases} 0 & \text{for } x \notin (a,b) \text{ or } y \notin (a,b); \\ \frac{2}{b-a}(x-a)(b-y) & \text{for } a \leq x \leq y \leq b; \\ \frac{2}{b-a}(y-a)(b-x) & \text{for } a \leq y \leq x \leq b. \end{cases}$$

(v) The Green function for the infinite interval (a, ∞) is given by

$$G(x,y) = 2[(x-a) \wedge (y-a)], \quad \text{for } x, y \in (a, \infty).$$

Throughout this chapter, we assume $q \in J$ unless otherwise specified. By Proposition 3.1, this implies that $q \in L^1_{\text{loc}}(\mathbb{R}^1)$. Now for each $D \in \mathcal{B}^1$, it is easy to see by the definition of J that $q \in L^1(D)$ implies that $1_D q \in J$; hence $q \in L^1_{\text{loc}}(\mathbb{R}^1)$ is equivalent to $q \in J_{\text{loc}}$. Moreover, for a fixed finite interval D, and any result which does not depend on the values of q outside D, the assumption $q \in J$ may be replaced by $q \in L^1(D)$. Thus all the results in this section except Proposition 9.3 require only $q \in L^1_{\text{loc}}(\mathbb{R}^1)$, although the quantity $u(x, y)$ involves the half line $(-\infty, y)$ or (y, ∞). This will be scrutinized in the proof of Proposition 9.1.

We now introduce the fundamental expectation below which has no analogue in higher dimensions, for arbitrary x and y in $\mathbb{R}^1$:

$$u(x,y) = E^x\{e_q(T_y)\} = E^x\left\{\exp\left[\int_0^{T_y} q(X_t)\mathrm{d}t\right]\right\}. \tag{1}$$

One preliminary but vital result is that the function u is strictly positive although it may be infinite. In fact, we need a stronger result.

Proposition 9.1 *Let $x \in (y, z)$. Then*

$$E^x\{T_y < T_z;\, e_q(T_y)\} > 0. \tag{2}$$

Proof In what follows, we shall omit the phrase 'almost surely' when the context clearly requires it. Since $T_y < \infty$, we have

$$\left|\int_0^{T_y} q(X_t)\mathrm{d}t\right| < \infty$$

under any P^x by the remarks at the beginning of Section 4.1. This implies that $e_q(T_y) > 0$. On the other hand, $P^x\{T_y < T_z\} > 0$. Proposition 9.1 follows from these two properties. $\square$.

Corollary to Proposition 9.1 *For all x and y, $0 < u(x,y) \leq \infty$.*

Proposition 9.2 *If $x < y < z$ or $x > y > z$, then*

$$u(x,y)u(y,z) = u(x,z). \tag{3}$$

Proof Both sides of (3) may be ∞, but by virtue of the preceding corollary, the ambiguity $0 \cdot \infty$ cannot occur. The relation is a consequence of the strong Markov property, but since the latter is usually only stated for finite quantities, we shall spell out the detail. For positive numbers M and N, we have

$$E^x\left\{(e_q(T_y) \wedge M)\left(\exp\left[\int_{T_y}^{T_z} q(X_t)\mathrm{d}t\right] \wedge N\right)\middle| \mathcal{F}_{T_y}\right\}$$
$$= (e_q(T_y) \wedge M)E^y\{e_q(T_z) \wedge N\}.$$

Letting $M \uparrow \infty$, $N \uparrow \infty$, we obtain by monotone convergence

$$E^x\{e_q(T_z)|\mathcal{F}_{T_y}\} = e_q(T_y)u(y,z).$$

Taking expectations, we obtain (3), regardless of finiteness. □

Proposition 9.3 *Given $\varepsilon > 0$, there exists $\delta = \delta(q, \varepsilon)$ such that if I is an interval with $m(I) < \delta$, then*

$$\sup_{x \in I} E^x\{e_{|q|}(\tau_I)\} < 1 + \varepsilon. \tag{4}$$

Proof Let $I = (a,b)$, then by the explicit formula for the Green function, we see that for all $x \in I$:

$$E^x\left\{\int_0^{\tau_I} |q(X_t)|\mathrm{d}t\right\} \leq 2m(I)\int_I |q(y)|\mathrm{d}y. \tag{5}$$

By Proposition 3.1, we have

$$M = \sup_{m(I)\leq 1} \int_I |q(y)|\mathrm{d}y < \infty.$$

Hence, if $m(I) \leq 1 \wedge (4M)^{-1}\varepsilon$, the quantity in (5) does not exceed $\varepsilon/2$. Therefore, by Lemma 3.7, for all $x \in I$:

$$E^x\{e_{|q|}(\tau_I)\} \leq \frac{1}{1-\frac{\varepsilon}{2}} < 1 + \varepsilon,$$

provided $\varepsilon < 1$. □

Remark We need $q \in J$ through Proposition 3.1, but if D is a fixed finite interval and I is restricted to subintervals of D, then $q \in L^1(D)$ is sufficient.

We can now proceed to the crucial question of the finiteness of u. It is trivial that if $q \leq 0$ (no need for J), then $u(x,y) \leq 1$ for all x and y. At the opposite end of the scale, if q is a constant > 0, then $e_q(T_y) > qT_y$, and consequently,

$u(x, y) = \infty$ for all $x \neq y$ by the second assertion in (i) above. The fact that $u(x, x) = 1$ is a fluke due to the regularity reviewed in (iii).

Proposition 9.4 *For any fixed $y \in \mathbb{R}^1$, if $u(x, y) < \infty$ for some $x < y$, then $u(x, y) < \infty$ for all $x < y$. If $u(x, y) < \infty$ for some $x > y$, then $u(x, y) < \infty$ for all $x > y$.*

Proof The two cases are entirely similar, so we need only prove the former. If $x < z < y$, then by Proposition 9.2,

$$u(x, z)u(z, y) = u(x, y) < \infty;$$

since $u(x, z) > 0$ by the Corollary to Proposition 9.1, it follows that $u(z, y) < \infty$. Next, suppose that $z < x < y$. Then, by the strong Markov property we have

$$\begin{aligned} \infty > u(x, y) &\geq E^x\{T_z < T_y; e_q(T_y)\} \\ &= E^x\left\{T_z < T_y; e_q(T_z)E^{X(T_z)}[e_q(T_y)]\right\} \\ &= E^x\{T_z < T_y; e_q(T_z)\}u(z, y), \end{aligned}$$

since $X(T_z) = z$ by continuity of the path. Hence $u(z, y) < \infty$ follows by Proposition 9.1 with y and z interchanged. □

Proposition 9.4 asserts that given any y, the set

$$\{x < y : u(x, y) < \infty\}$$

is either empty or equal to $(-\infty, y)$. The set of y for which the latter case holds will now be determined together with its dual.

Theorem 9.5 *The set*

$$\{y \in \mathbb{R}^1 : \forall x \leq y,\ u(x, y) < \infty\} \tag{6}$$

is either empty, or equal to the whole line $\mathbb{R}^1$, or to a half line $(-\infty, \beta)$ with $\beta \in \mathbb{R}^1$. The set

$$\{y \in \mathbb{R}^1 : \forall x \geq y,\ u(x, y) < \infty\}$$

is either empty, or equal to the whole line $\mathbb{R}^1$, or to a half line (α, ∞) with $\alpha \in \mathbb{R}^1$.

Proof Denote the set in (6) by B. If $B \neq \emptyset$, let $y_0 \in B$. Then if $x < y < y_0$, we have $u(x, y_0) < \infty$ by Proposition 9.4, and

$$u(x, y)u(y, y_0) = u(x, y_0)$$

by Proposition 9.2. Hence $u(x, y) < \infty$ by the Corollary to Proposition 9.1. This means that $y \in B$, in other words, that $y_0 \in B$ implies that $(-\infty, y_0) \subset B$. It follows that either $B = \mathbb{R}^1$, or $B = (-\infty, \beta)$ for some $\beta \in \mathbb{R}^1$, or $B = (-\infty, \beta]$ for some $\beta \in \mathbb{R}^1$. The theorem will be proved by excluding the last possibility. This amounts to proving that given $y \in B$, there exists $z > y$ such that $z \in B$. Let $x < y < w$ be such that $w - x$ is so small that

$$E^y\{e_{|q|}(\tau_{(x,w)})\} < \infty. \tag{7}$$

This is possible by Proposition 9.3. Next, since $\lim_{z\downarrow y} P^y\{T_x < T_z\} = 0$, it follows from (7) that there exists $z \in (y, w)$ such that

$$\begin{aligned} E^y\{T_x < T_z; e_{|q|}(\tau_{(x,z)})\} &\leq E^y\{T_x < T_z; e_{|q|}(\tau_{(x,w)})\} \\ &< \frac{1}{u(x,y)}. \end{aligned} \tag{8}$$

We shall prove that $u(x, z) < \infty$ as desired.

Writing $\tau = \tau_{(x,z)} = T_x \wedge T_z$ under P^y, we define a sequence of hitting times by induction as follows:

$$\begin{aligned} S_0 &= 0; \\ S_{2n+1} &= \begin{cases} S_{2n} + \tau \circ \theta_{S_{2n}} & \text{on } \{X(S_{2n}) = y\} \\ S_{2n} & \text{on } \{X(S_{2n}) = z\}; \end{cases} \\ S_{2n+2} &= \begin{cases} S_{2n+1} + T_y \circ \theta_{S_{2n+1}} & \text{on } \{X(S_{2n+1}) = x\} \\ S_{2n+1} & \text{on } \{X(S_{2n-1}) = z\} \end{cases} \qquad (n \geq 0). \end{aligned}$$

Let $N = \min\{n \geq 0 : X(S_n) = z\}$; then $N < \infty$ by the recurrence property (i), and N is an odd integer by definition. We have by repeated use of the strong Markov property:

$$\begin{aligned} u(y,z) &= \sum_{n=0}^{\infty} E^y\{N = 2n+1; e_q(S_{2n+1})\} \\ &= \sum_{n=0}^{\infty} E^y\left\{N = 2n+1; \exp\left[\sum_{k=0}^{2n} \int_{S_k}^{S_{k+1}} q(X_t)\mathrm{d}t\right]\right\} \\ &= \sum_{n=0}^{\infty} (E^y\{T_x < T_z; e_q(T_x)\} E^x\{e_q(T_y)\})^n E^y\{T_z < T_x; e_q(T_z)\}. \end{aligned}$$

The factor above raised to the nth power is strictly less than 1 by (8), while the factor following it is not greater than the expectation in (7). Therefore, the infinite series converges and $u(y, z) < \infty$. Hence, $z \in B$ and the proof of the first assertion of the theorem is complete. The proof of the second assertion is similar. □

Remark The main point of the above proof implies that if $(-\infty, b)$, $b \in \mathbb{R}^1$ is gaugeable, then there exists $c > b$ such that $(-\infty, c)$ is gaugeable. An analogous argument shows that this assertion is also true when $-\infty$ is replaced by any $a \in \mathbb{R}^1$. Thus the result includes Theorem 8.2 in $\mathbb{R}^1$, but the proof is quite different.

In order to cover all cases in the Theorem 9.5, we set

$$\begin{aligned} \alpha &= \inf\{y \in \mathbb{R}^1 : u(x,y) < \infty \text{ for all } x \geq y\}; \\ \beta &= \sup\{y \in \mathbb{R}^1 : u(x,y) < \infty \text{ for all } x \leq y\}. \end{aligned}$$

The theorem asserts that we may change the qualifier 'all' to 'some' without changing the value of α or β. It also asserts that neither the infimum nor the supremum above is ever attained in $\mathbb{R}^1$. This is by no means obvious *a priori*, and is in contrast with, e.g. the behavior of the abscissa of convergence of a power series or generating function. The result is an analogue of Theorem 8.2 which may perhaps be extended to an unbounded domain in a similar manner.

Returning to the previous examples, in the new notation we note that: if $q \leq 0$, then $\alpha = -\infty$, $\beta = +\infty$; if q is a constant > 0, then $\alpha = +\infty$, $\beta = -\infty$.

Examples in which $-\infty < \beta < \alpha < \infty$ and $-\infty < \alpha \leq \beta < \infty$ will be given in Sections 9.2 and 9.3.

After finiteness, the next important question is the continuity of u. This is related to the continuity of the hitting time T_y as a function of y. We shall prove the following basic lemma to illustrate the method.

Lemma 9.6 *If $y_n \to y \in \mathbb{R}^1$, then $T_{y_n} \to T_y$ almost surely.*

Proof We have to consider monotone sequences $\{y_n\}$ and the relative position of the starting point x with respect to y. There are two essentially different cases according as y_n moves away from x or towards x. If y_n moves away from x, we may suppose that $x < y_n < y$ for all n, and $y_n \uparrow y$. Then, under P^x, T_{y_n}' increases to a limit S, and $X(T_{y_n}) \to X(S)$ by continuity of paths, so that $X(S) = y$. Since $X(t) < y_n$ for $t \in [0, T_{y_n})$ for all n, we have $X(t) < y$ for $t \in [0, S)$. Thus, $S = T_y$ as desired.

If y_n moves towards x, we may suppose that $x < y < y_n$ for all n and $y_n \downarrow y$. As before, let $T_{y_n} \downarrow S$, then $X(S) = y$. Clearly $S \geq T_y$, but we need the regularity property of (iii) above to show that $S \leq T_y$, as follows. We have

$$\begin{aligned} P^x\{T_y < S\} &= E^x\{T_y < S;\, P^{X(T_y)}[T_{(y,\infty)} = 0]\} \\ &\leq P^x\{T_{(y,\infty)} < S\}. \end{aligned} \tag{9}$$

Since $X(t) < y_n$ for $t \in [0, T_{y_n})$ for all n, we have $X(t) \leq y$ for $t \in [0, S]$. This implies that the last probability in (9) is equal to zero, and so the proof is complete. □

Proposition 9.7 *If $x < \beta$ and $x_n \uparrow x$, or if $x > \alpha$ and $x_n \downarrow x$, then*

$$\lim_{n\to\infty} u(x_n, x) = 1. \tag{10}$$

If $x < \beta$ and $x_n \downarrow x$, or if $x > \alpha$ and $x_n \uparrow x$, then we have

$$\lim_{n\to\infty} u(x, x_n) = 1. \tag{11}$$

Proof The two cases in each of the above sentences are proved in the same way. Consider the first case in the first sentence, where $u(x_n, x) < \infty$ for all n by the definition of β. Using the strong Markov property, we have (omitting q in e_q):

$$\begin{aligned} E^{x_1}\{e(T_x)|\mathcal{F}(T_{x_n})\} &= e(T_{x_n})E^{X(T_{x_n})}\{e(T_x)\} \\ &= e(T_{x_n})u(x_n, x). \end{aligned}$$

Letting $n \to \infty$, we obtain by a martingale convergence theorem (see Chung (1974)), since $e(T_{x_n}) \to e(T_x)$ by Lemma 9.6,

$$E^{x_1}\{e(T_x)| \vee_{n=1}^{\infty} \mathcal{F}(T_{x_n})\} = e(T_x) \lim_{n\to\infty} u(x_n, x).$$

Taking E^{x_1} in the above, we have

$$u(x_1, x) = u(x_1, x) \lim_{n\to\infty} u(x_n, x).$$

This implies (10) because $u(x_1, x) > 0$.

Next we prove the second case in the second sentence of the proposition. We may suppose that $x_1 > \alpha$ so that $u(x_n, x_1) < \infty$ for all n. We then have

$$E^x\{e(T_{x_1})|\mathcal{F}(T_{x_n})\} = e(T_{x_n})u(x_n, x_1).$$

By another martingale convergence theorem (see Chung (1974)), we obtain

$$E^x\{e(T_{x_1})| \wedge_{n=1}^{\infty} \mathcal{F}(T_{x_n})\} = e(T_x) \lim_{n\to\infty} u(x_n, x_1),$$

and consequently

$$u(x, x_1) = u(x, x) \lim_{n\to\infty} u(x_n, x_1) = \lim_{n\to\infty} u(x_n, x_1).$$

By Propositions 9.1 and 9.2, we then have

$$\lim_{n\to\infty} u(x, x_n) = \lim_{n\to\infty} \frac{u(x, x_1)}{u(x_n, x_1)} = 1.$$

□

It should be observed that in the proof of Proposition 9.7, as well as in that of Lemma 9.6, it is the temporal order of T_{x_n}, not the spatial order of x_n, that makes the difference.

The local continuity of u in Proposition 9.7, together with the functional equation (3) imply that u is continuous.

Theorem 9.8 *If $x_0 \neq y_0$ and $u(x_0, y_0) < \infty$, then the function $u(\cdot, \cdot)$ is continuous at (x_0, y_0).*

Proof We may assume that $x_0 < y_0$. The result follows from Propositions 9.1, 9.2 and 9.7, and the following list of cases with $x_n \to x$, $y_n \to y$.

If $x_n < x < y_n < y$, then

$$u(x_n, y_n) = \frac{u(x_n, x)}{u(y_n, y)} u(x, y).$$

If $x_n < x < y < y_n$, then

$$u(x_n, y_n) = u(x_n, x)u(x, y)u(y, y_n).$$

If $x < x_n < y_n < y$, then

$$u(x_n, y_n) = \frac{u(x,y)}{u(x,x_n)u(y_n,y)}.$$

If $x < x_n < y < y_n$, then

$$u(x_n, y_n) = u(x,y)\frac{u(y,y_n)}{u(x,x_n)}.$$

□

Readers who (like the authors) consider this argument tedious are invited to discover a rigorous shorter proof.

Supplement to Theorem 9.8 *If $\beta < \infty$, then for each $x < \beta$*

$$\lim_{y\uparrow\beta} u(x,y) = +\infty.$$

If $\alpha > -\infty$, then for each $x > \alpha$

$$\lim_{y\downarrow\alpha} u(x,y) = +\infty.$$

Proof By Theorem 9.5, $u(x,\beta) = +\infty$ for each $x < \beta$. By Lemma 9.6, if $y \uparrow \beta$, then $T_y \to T_\beta$ and so $e_q(T_y) \to e_q(T_\beta)$ a.s. Therefore it follows from Fatou's lemma that

$$\underline{\lim}_{y\uparrow\beta} E^x\{e_q(T_y)\} \geq E^x\{e_q(T_\beta)\} = u(x,\beta) = +\infty.$$

This establishes the first assertion; the second is similar. □

9.2 Gauge for a Finite or Infinite Interval

We first consider a finite open interval $D = (a,b)$ with $\partial D = \{a\} \cup \{b\}$. Both boundary points are regular for D by property (iii) of Section 9.1. If $x \in D$, then under P^x we have $\tau_D = T_a \wedge T_b$ and $X(\tau_D) = a$ or b. Let $q \in L^1(D)$. We set

$$\begin{aligned} w_a(x) &= E^x\{X(\tau_D) = a; e_q(\tau_D)\} = E^x\{T_a < T_b; e_q(T_a)\} \\ w_b(x) &= E^x\{X(\tau_D) = b; e_q(\tau_D)\} = E^x\{T_b < T_a; e_q(T_b)\}. \end{aligned} \tag{12}$$

Then for any (finite-valued) function f defined on ∂D, we have, recalling the notation in Theorem 4.7:

$$u_f(x) = E^x\{e_q(\tau_D)f(X(\tau_D))\} = f(a)w_a(x) + f(b)w_b(x). \tag{13}$$

In particular, the gauge for (D,q) is $u_1 = w_a + w_b$. These formulae remain valid for $x = a$ or b by regularity, so that

$$w_a(a) = 1, \quad w_a(b) = 0; \quad w_b(a) = 0, \quad w_b(b) = 1. \tag{14}$$

We can express the conditional gauge by means of w_a and w_b, and the hitting probabilities given in (ii) of Section 9.1. In fact, the conditional gauge theorem

(Theorem 7.5) can be proved for any finite interval by the general method, with suitable modifications (see the Appendix at the end of this section). However, it takes on a simpler form with a more direct proof, as follows.

Theorem 9.9 *Let $q \in L^1(D)$. If $w_a(x) < \infty$ for some $x \in D$, then we also have $w_b(x) < \infty$; and vice versa. In either case, u_1 is bounded in $\overline{D}$.*

Proof We may choose c and d so that $a < x < c < d < b$, and

$$E^d\{e_{|q|}(\tau_{(c,b)})\} < \infty. \tag{15}$$

This is possible by Proposition 9.3 and the Remark following it. Now we use a successive hitting argument similar to that used in the proof of Theorem 9.5. This results in

$$\begin{aligned}\infty &> w_a(x)\\ &= E^x\{T_a < T_d; e_q(T_a)\}\\ &\quad + \sum_{n=0}^{\infty} E^x\{T_d < T_a; e_q(T_d)\}\big(E^d\{T_c < T_b; e_q(T_c)\}\big)^{n+1}\\ &\quad \times \big(E^c\{T_d < T_a; e_q(T_d)\}\big)^n E^c\{T_a < T_d; e_q(T_a)\}.\end{aligned}$$

By Proposition 9.1, all the expectations in the above series are strictly positive, hence finite, and the convergence of the series implies that

$$E^d\{T_c < T_b; e_q(T_c))\}E^c\{T_d < T_a; e_q(T_d)\} < 1.$$

By the same token, we have

$$\begin{aligned}w_b(x) &= \sum_{n=0}^{\infty} E^x\{T_d < T_a; e_q(T_d)\}(E^d\{T_c < T_b; e_q(T_c)\}\\ &\quad \times E^c\{T_d < T_a; e_q(T_d)\})^n \cdot E^d\{T_b < T_c; e_q(T_b)\}.\end{aligned}$$

The last expectation in the above is finite by (15) (it is essential that we have $|q|$ rather than q there). Comparing the two infinite series, we conclude that $w_b(x) < \infty$. Thus $u_1(x) < \infty$ and the boundedness of u_1 follows from the general gauge theorem (Theorem 5.20). But we shall give a simpler direct proof.

Assume that for some $x \in D$ we have $u_1(x) < \infty$. Then by the strong Markov property, for any $y \in D$:

$$\infty > u_1(x) \geq E^x\{T_y < \tau_D; e_q(T_y)\}u_1(y).$$

Hence it is sufficient to show that the last expectation in the above is bounded away from zero for all y. We proceed as in Lemma 7.3, using Jensen's inequality for the conditional expectation:

$$\begin{aligned}&P^x\{T_y < \tau_D\}E^x\{e_q(T_y)|T_y < \tau_D\}\\ &\geq P^x\{T_y < \tau_D\}\exp\left[\frac{-E^x\{\int_0^{\tau_D}|q(X_t)|\mathrm{d}t\}}{P^x\{T_y < \tau_D\}}\right].\end{aligned}$$

By (ii) of Section 9.1,

$$C_1(x) = \inf_{y \in D} P^x\{T_y < \tau_D\} = \frac{(x-a) \wedge (b-x)}{b-a} > 0.$$

Next, since $G_D|q| < \infty$ by Theorem 4.3 or direct verification using (iv) of Section 9.1,

$$C_2(x) = -E^x\left\{\int_0^{\tau_D} |q(X_t)|\mathrm{d}t\right\} > -\infty.$$

Hence for all $y \in D$:

$$E^x\{T_y < \tau_D; e_q(T_y)\} \geq C_1(x)\exp\left(\frac{C_2(x)}{C_1(x)}\right) > 0.$$

□

We now set $w = w_a$ or w_b and apply Theorem 4.7. The simpler analysis in $\mathbb{R}^1$ will produce further information. Note firstly that if $w \in L^\infty(D)$, then $G_D(|qw|) < \infty$ provided that $q \in L^1(D)$. This is the condition (4.41) which justifies the calculations following it there.

Theorem 9.10 *Let $q \in L^1(D)$ and suppose that (D, q) is gaugeable. Then $w \in C^{(1)}(\overline{D})$, w' is absolutely continuous in D and w is a weak solution of the equation*

$$\phi'' + 2q\phi = 0 \tag{16}$$

in D. If, in addition, $q \in C(D)$, then $w \in C^{(2)}(D)$, and w is a strict solution of (16). If furthermore $q \in C(\overline{D})$, then $w \in C^{(2)}(\overline{D})$ and (16) holds even at ∂D.

Proof By (4.38), we have with $G = G_D$:

$$w(x) = h(x) + G(qw)(x)$$

where

$$h(x) = \frac{b-x}{b-a} \quad \text{or} \quad \frac{x-a}{b-a}$$

according as $w = w_a$ or w_b. By the explicit formula for G given in (iv) of Section 9.1, we have

$$G(qw)(x) = \frac{2(b-x)}{b-a}\int_a^x (y-a)q(y)w(y)\mathrm{d}y + \frac{2(x-a)}{b-a}\int_x^b (b-y)q(y)w(y)\mathrm{d}y.$$

Since w is bounded by Theorem 9.9, it is clear that we can differentiate the above. After simplification, we obtain for $x \in \overline{D}$:

$$w'(x) = C - 2\int_a^x q(y)w(y)\mathrm{d}y, \tag{17}$$

where

$$C = \frac{2b}{b-a}\int_a^b q(y)w(y)\mathrm{d}y - \frac{2}{b-a}\int_a^b yq(y)w(y)\mathrm{d}y \mp \frac{1}{b-a}.$$

Hence $w \in C^{(1)}(\overline{D})$. Furthermore, w' is absolutely continuous in D, and we have a.e.

$$w''(x) = -2q(x)w(x).$$

Together with the absolute continuity of w', this implies by standard analysis (partial integration and Lebesgue's theorem on differentiation) that w is a weak solution of (16). The case when q is also continuous is easily verified by (17). □

The Schrödinger equation (16) in $\mathbb{R}^1$ is a case of the Sturm–Liouville equations, as described in Section 3.1. In conformity with our general terminology in previous chapters, we shall call a continuous weak solution of (16) in any domain D a q-harmonic function in D. The following characterization of such a function is more explicit; in particular the extension to $\overline{D}$ is most convenient.

Proposition 9.11 *Let $q \in J_{\text{loc}}$. For any domain D in $\mathbb{R}^1$, a function ϕ is q-harmonic in D if and only if it can be extended to $\overline{D}$ so that $\phi \in C^1(\overline{D})$, ϕ' is absolutely continuous in any compact subinterval of D, and for m-a.e. x in D we have*

$$\phi''(x) + 2q(x)\phi(x) = 0.$$

If in addition $q \in C(\overline{D})$, then $\phi \in C^2(\overline{D})$ and the above equation holds in $\overline{D}$.

Proof The extensibility is the bright new feature and will be proved by a transparent probabilistic method. Since all the properties in question are local ones, we may suppose that $D = (a, b)$ and $(D, |q|)$ is gaugeable by Proposition 9.3. Let $[c, d] \subset D$, then $\phi \in C([c, d])$. Hence by the general representation theorem (Theorem 4.15), for any $x \in E = (c, d)$:

$$\begin{aligned}\phi(x) &= E^x\{e_q(\tau_E)\phi(X(\tau_E))\}\\ &= E^x\{T_c < T_d; e_q(\tau_E)\}\phi(c) + E^x\{T_d < T_c; e_q(\tau_E)\}\phi(d). \qquad (18)\end{aligned}$$

We now fix x and d in the above, set $U = (a, d)$ and let $c \downarrow a$. Then a.s. $T_c \to T_a$ and $\tau_E \to \tau_U$ on $\{T_c < T_d\}$ by Lemma 9.6, and $e_q(\tau_E) \to e_q(\tau_U)$. Since $|e_q(\tau_E)| \le e_{|q|}(\tau_D)$, we have by dominated convergence:

$$\begin{aligned}\lim_{c\downarrow a} E^x\{T_c < T_d; e_q(\tau_E)\} &= E^x\{T_a < T_d; e_q(\tau_U)\},\\ \lim_{c\downarrow a} E^x\{T_d < T_c; e_q(\tau_E)\} &= E^x\{T_d < T_a; e_q(\tau_U)\}.\end{aligned}$$

Thus all the quantities in (18) apart from $\phi(c)$ converge to finite and strictly positive limits (Proposition 9.1) and therefore $\lim_{c\downarrow a}\phi(c)$ exists! Of course the limit will be denoted by $\phi(a)$. Since ϕ is continuous in (a, b) by definition, the extension is continuous in $[a, b]$. Now we can apply the given representation theorem, which essentially requires that $\phi \in C(\overline{D})$, to obtain

$$\phi(x) = E^x\{e_q(\tau_D)\phi(X(\tau_D))\}, \quad x \in \overline{D}.$$

We recall that D is regular by (iii) of Section 9.1. Thus $\phi = u_\phi$ in the notation of (13) and so the rest of the theorem is contained in Theorem 9.10.

The last assertion of the proposition follows from the corresponding properties of w in Theorem 9.10. □

In view of the Proposition 9.11, we shall say that ϕ is q-harmonic in $\overline{D}$ (as well as in D). This will simplify the language when an endpoint of the interval is needed.

According to the theory of linear differential equations of the second order, at any $x_0 \in \mathbb{R}^1$ there exists a unique strict solution ϕ of (16) with given initial values $\phi(x_0)$ and $\phi'(x_0)$, provided that the function q satisfies some analytic assumption such as being Lipschitzian. For weak solutions with a more general q as treated here, such results must be scrutinized for their validity. But our discussion above aided a little by classical analysis will yield the fundamental theorems without recourse to the literature, as we now proceed to show.

Theorem 9.12 *Let $q \in J$. Given any three constants a, c_0 and c_1, there is a unique q-harmonic function ϕ in $\mathbb{R}^1$ satisfying*

$$\phi(a) = c_0, \quad \phi'(a) = c_1. \tag{19}$$

Proof By Proposition 9.3, there exists $\delta_0 > 0$ such that if $D = (a, b)$ with $b - a \leq \delta_0$, then

$$\sup_{x \in \mathbb{R}^1} E^x\{e_{|q|}(\tau_D)\} < 2.$$

Hence (D, q) is gaugeable, and furthermore we have

$$0 < w_a \leq 2, \quad 0 < w_b \leq 2,$$

where w_a and w_b are given in (12). Next we have by (17),

$$w_b'(a+) = \frac{2}{b-a} \int_a^b (b-y)q(y)w_b(y)dy + \frac{1}{b-a},$$

where the one-sided derivative at a is indicated. Since $q \in J$, we have

$$\left| \int_a^b (b-y)q(y)w_b(y)dy \right| \leq 2(b-a)M$$

where

$$M = \sup_{x \in \mathbb{R}^1} \int_x^{x+\delta_0} |q(y)|dy < \infty$$

by Proposition 3.1. It follows that there exists $0 < \delta \leq \delta_0$ such that if $b - a \leq \delta$, then $w_b'(a+) > 0$. Note that it is essential that these estimates are uniform with respect to a in $\mathbb{R}^1$, and that is where $q \in J$ is needed.

Now we define ϕ in $[a, b]$ with $b - a = \delta$, as follows:

$$\phi = c_0 w_a + \frac{c_1 - c_0 w_a'(a+)}{w_b'(a+)} w_b.$$

Then ϕ is q-harmonic in (a, b) and (19) is verified. In particular, if $c_0 = c_1 = 0$, then $\phi \equiv 0$ in $[a, b]$. Thus we have proved that there is a unique q-harmonic function ϕ in (a, b) satisfying

$$\phi(a+) = c_0, \quad \phi'(a+) = c_1.$$

To proceed, we set $a_1 = (a + b)/2$ and $b_1 = a_1 + \delta$. The above argument applied to (a_1, b_1) shows that there exists a unique q-harmonic function ϕ_1 in (a_1, b_1) satisfying

$$\phi_1(a_1+) = \phi(a_1), \quad \phi_1'(a_1+) = \phi'(a_1).$$

It follows from the uniqueness property which we have just proved that $\phi \equiv \phi_1$ in (a_1, b), and therefore ϕ_1 gives an extension of ϕ to (a, b_1). Since δ is a fixed number, this extension can be continued to yield an extension of ϕ to (a, ∞). In a similar way, ϕ can be extended to $(-\infty, a)$ and so to all of $\mathbb{R}^1$. The uniqueness of this full extension is implied in the step-by-step construction. □

The proof above is unconventional in that it uses the two 'fundamental' solutions w_a and w_b without any explicit mention of their linear independence. We shall now give the conventional treatment for another general result based on the latter notion and that of Wronskian. For two differentiable functions ϕ_1 and ϕ_2 on an interval, their Wronskian is the function

$$W(\phi_1, \phi_2) = \phi_1 \phi_2' - \phi_1' \phi_2. \tag{20}$$

Although any q-harmonic function can be defined in $\mathbb{R}^1$ by virtue of Theorem 9.12, the following discussion will not use this fact and is valid in an arbitrary interval.

Lemma *Let ϕ_1 and ϕ_2 be q-harmonic. Then their Wronskian vanishes identically if it vanishes anywhere. This is the case if and only if they are linearly dependent.*

Proof Let W denote the Wronskian. By Proposition 9.11, W is continuous, and

$$W' = \phi_1 \phi_2'' - \phi_1'' \phi_2 = 0$$

m-a.e.. Hence W is a constant and the first assertion is proved. To prove the second assertion, we need the uniqueness part of Theorem 9.12, in other words, if ϕ is q-harmonic and

$$\phi(x_0) = \phi'(x_0) = 0 \tag{21}$$

for some x_0, then $\phi \equiv 0$. This result was proved in a modified form with the one-sided derivative, but we shall give its standard proof in the spirit of this discussion. It is sufficient to prove that $\phi \equiv 0$ in a neighborhood (a, b) of x_0 which is gaugeable. Now $W(w, \phi)$ vanishes at x_0 by (20), where $w = w_a$ or w_b; hence it vanishes in (a, b). Since $w > 0$ by Proposition 9.1, $(\phi w^{-1})' = W(w, \phi)w^{-2} = 0$, and consequently ϕ is a constant multiple of w. This being true for both w_a and w_b, we must have $\phi \equiv 0$ by (14).

Now let $W(\phi_1, \phi_2) = 0$. If $\phi_1 \equiv 0$, then of course ϕ_1 and ϕ_2 are linearly dependent. Otherwise, let $\phi_1(x_0) \neq 0$ and set

$$\phi = -\frac{\phi_2(x_0)}{\phi_1(x_0)}\phi_1 + \phi_2.$$

It is immediately clear that (21) is satisfied. Hence, $\phi \equiv 0$ and the linear dependence of ϕ_1 and ϕ_2 is exhibited. The Lemma is proved. □

It is now a simple matter to establish the following fundamental result from the general theory of linear differential equations, which is valid for any interval in which the functions are defined.

Theorem 9.13 *Let ϕ_1 and ϕ_2 be two linearly independent q-harmonic functions. Then any q-harmonic function is a linear combination of these two.*

Proof Take any x_0 in the domain of definition. Then $W(\phi_1, \phi_2)(x_0) \neq 0$ by the above lemma. Hence, for any q-harmonic function ϕ, the two equations below can be solved for a_1 and a_2:

$$\begin{aligned} a_1\phi_1(x_0) + a_2\phi_2(x_0) &= \phi(x_0) \\ a_1\phi_1'(x_0) + a_2\phi_2'(x_0) &= \phi'(x_0). \end{aligned}$$

Then the function $\phi - (a_1\phi_1 + a_2\phi_2)$ is q-harmonic and satisfies (21). Hence it is identically zero as proved above. □

For an interval (a, b) which is small enough to be gaugeable, the representability of any q-harmonic function in (a, b) as a linear combination of w_a and w_b is a particular case of the Theorem 9.13.

We now proceed to the case of an infinite interval. Without loss of generality, we may suppose this interval is $\mathbb{R}_0 = (0, \infty)$. The process appropriate for $\mathbb{R}_0$ is the Brownian motion stopped (or 'absorbed') at 0, i.e.

$$\{X(t \wedge T_0), t \geq 0\}.$$

We recall from (i) of Section 9.1 that $T_0 < \infty$ a.s.. Properly speaking, the corresponding state space is $\overline{\mathbb{R}}_0 = [0, \infty)$. There is no point at infinity and the sole boundary point is 0, which plays the same role as ∂ used before. We shall keep the notation X_t for the stopped process. This process has been intensively studied. Its transition probability density is given by

$$p(t; x, y) - p(t; x, -y), \tag{22}$$

where $p(t; x, y)$ is given in (1.11) with $d = 1$. The Green function given in Section 9.1 can be computed using the general formula (2.14). Furthermore, under each P^x, $x \in \mathbb{R}_0$, T_0 has the distribution

$$\begin{aligned} P^x(T_0 < t) &= \sqrt{\frac{2}{\pi t}} \int_x^\infty \mathrm{e}^{-\frac{y^2}{2t}} \mathrm{d}y, \\ P^x\{T_0 \in \mathrm{d}t\} &= \frac{x}{\sqrt{2\pi t^3}} \mathrm{e}^{-\frac{x^2}{2t}} \mathrm{d}t. \end{aligned} \tag{23}$$

See e.g. Chung (1982a, page 153, Exercise 12).

If $x \geq y \geq 0$, our notation $u(x, y)$ in Section 9.1 needs no change; but if $0 < x < y$, we will signify the new setting as follows:

$$_0u(x, y) = E^x\{T_y < T_0; e_q(T_y)\}. \tag{24}$$

Observe that the gauge for $(\mathbb{R}_0, q)$ is

$$u(x, 0) = E^x\{e_q(\tau_{\mathbb{R}_0})\} = E^x\{e_q(T_0)\}, \quad x \in \mathbb{R}_0.$$

Theorem 9.14 *We have $u(x, 0) < \infty$ for all $x \in \mathbb{R}_0$ if and only if $\alpha < 0$. If this is the case, then for any $D = (a, b)$ with $0 \leq a < b < \infty$, (D, q) is gaugeable; and for $0 < x < y < \infty$, we have*

$$_0u(x, y)u(y, x) < 1. \tag{25}$$

Conversely, if (25) holds for a pair (x, y) with $0 < x < y$, then $u(x, 0) < \infty$ for all $x \in \mathbb{R}_0$.

Proof The first sentence follows from the definition of α and the remark following it. To prove the second assertion, we observe that if $x \in D$, then

$$\begin{aligned} w_a(x) &= E^x\{T_a < T_b; e_q(\tau_D)\} \\ &= E^x\{T_a < T_b; e_q(T_a)\} \leq u(x, a) < \infty. \end{aligned}$$

Hence (D, q) is gaugeable by Theorem 9.9.

To prove (25), we define a sequence of successive hitting times as follows. Let $S_0 = 0$; for $n \geq 1$:

$$\begin{aligned} S_{2n-1} &= S_{2n-2} + (T_0 \wedge T_y) \circ \theta_{S_{2n-2}} \\ S_{2n} &= \begin{cases} S_{2n-1} + T_x \circ \theta_{S_{2n-1}} & \text{on } \{X(S_{2n-1}) = y\}; \\ S_{2n-1} & \text{on } \{X(S_{2n-1}) = 0\}. \end{cases} \end{aligned}$$

We also set

$$N = \min\{n \geq 0 : X(S_{2n+1}) = 0\}.$$

Since $P^x\{T_0 < \infty\} = 1$, we have $P^x\{N < \infty\} = 1$. Now we have as in the proof of Theorem 9.9,

$$\begin{aligned} &E^x\{N = n; e_q(T_0)\} \\ &= (E^x\{T_y < T_0; e_q(T_y)\}E^y\{e_q(T_x)\})^n E^x\{T_0 < T_y; e_q(T_0)\} \\ &= ({}_0u(x, y)u(y, x))^n E^x\{T_0 < T_y; e_q(T_0)\}. \end{aligned}$$

The sum of the above terms over $n \geq 0$ is $u(x, 0) < \infty$, and the last expectation in the above is strictly positive by Proposition 9.1. Hence (25) must be true.

Conversely, suppose that (25) holds for some x and y with $0 < x < y$. Observing that ${}_0u(x, y)$ is just the w_b in (12) with $a = 0$ and $b = y$, we see that ${}_0u(x, y) < \infty$ implies that

$$E^x\{T_0 < T_y; e_q(T_0)\} < \infty.$$

Turning round the above argument we now see that $u(x, 0) < \infty$. Hence $u(x, 0) < \infty$ for all $x \in \mathbb{R}_0$ by the first sentence of the theorem. □

The state of affairs described in the preceding theorem suggests the following extension of gaugeability. For any domain D and $q \in \mathcal{B}(D)$ such that $e_q(\tau_D)$ is well defined, we say that (D, q) is pseudo-gaugeable iff the gauge

$$u(x) = E^x\{\tau_D < \infty; e_q(\tau_D)\}, \quad x \in D,$$

is finite in D. When $D = \mathbb{R}_0$, $u(x) = u(x, 0)$ in the previous notation.

Theorem 9.15 *If $(\mathbb{R}_0, q)$ is pseudo-gaugeable, then $u \in C^{(1)}(\overline{\mathbb{R}}_0)$ and is q-harmonic and > 0 in $\overline{\mathbb{R}}_0$ with $u(0) = 1$. Moreover, u' is absolutely continuous in $[0, b]$ for any $b \in \mathbb{R}_0$. If v is any positive q-harmonic function in $\overline{\mathbb{R}}_0$ with $v(0) = 1$, then $v \geq u$ in $\overline{\mathbb{R}}_0$. If in addition v/u is bounded, then $v = u$.*

Proof That $u > 0$ is proved in Proposition 9.1. Next, for any $b \in \mathbb{R}_0$, let $w_0^{(0,b)}$ and $w_b^{(0,b)}$ denote the fundamental expectations in (12) when $D = (0, b)$. We have by the strong Markov property, for $x \in (0, b)$:

$$\begin{aligned} u(x) &= E^x\{T_0 < T_b; e_q(T_0)\} + E^x\{T_b < T_0; e_q(T_b)\}u(b) \\ &= w_0^{(0,b)}(x) + w_b^{(0,b)}(x)u(b). \end{aligned} \tag{26}$$

This representation, which is valid for all b, establishes the first sentence of the theorem by virtue of Theorem 9.10.

Now let v be q-harmonic and positive in $\overline{\mathbb{R}}_0$ with $v(0) = 1$. By the representation theorem (Theorem 4.15), we have for all x in $[0, b]$:

$$\begin{aligned} v(x) &= E^x\{e_q(\tau_{(0,b)})v(X(\tau_{(0,b)}))\} \\ &= w_0^{(0,b)}(x) + w_b^{(0,b)}(x)v(b). \end{aligned} \tag{27}$$

Since $E^x\{e_q(T_0)\} < \infty$ and $P^x\{T_0 < T_b\} \to 1$ as $b \to \infty$, we have

$$u(x) = \lim_{b \to \infty} E^x\{T_0 < T_b; e_q(T_0)\} = \lim_{b \to \infty} w_0^{(0,b)}(x). \tag{28}$$

Comparing this with (27), we see that $u(x) \leq v(x)$, proving the second sentence of the theorem. Finally, if we compare (26) and (28), we see that

$$\lim_{b \to \infty} w_b^{(0,b)}(x)u(b) = 0. \tag{29}$$

If v/u is bounded, then we can replace u by v in (29); the resulting relation, together with (27) and (28), shows that $v = u$. □

The next result is an analogue of Theorems 4.17 and 4.18, but both the hypothesis and conclusion there are stronger than in the following result.

Theorem 9.16 $(\mathbb{R}_0, q)$ *is pseudo-gaugeable if and only if there exists a q-harmonic function in* $\overline{\mathbb{R}}_0$ *which is strictly positive there.*

Proof If $(\mathbb{R}_0, q)$ is pseudo-gaugeable, then the gauge u is such a function by Theorem 9.15. Conversely, let v be q-harmonic and > 0 in $\overline{\mathbb{R}}_0$. Then by definition it is continuous; hence, it satisfies the condition (4.61) in $D = (0, b)$ for any $b > 0$. Therefore, by Theorem 4.17, (D, q) is gaugeable and consequently by Theorem 4.15, we have for all $x \in D$:

$$\begin{aligned} v(x) &= E^x\{e_q(\tau_D)v(X(\tau_D))\} \\ &\geq v(0)E^x\{T_0 < T_b; e_q(T_0)\}. \end{aligned}$$

Fix x and let $b \to \infty$. We obtain by (28),

$$v(x) \geq v(0)u(x).$$

Hence, $u(x) < \infty$ for all $x \in \mathbb{R}_0$, and $(\mathbb{R}_0, q)$ is pseudo-gaugeable. □

We borrow the following result from the theory of Sturm–Liouville equations (see Hartman (1982)). Note, however, that our q is more general than in the usual treatment which assumes q to be continuous and considers strict solutions of (16).

Theorem 9.17 *Let* ϕ *be positive q-harmonic in* $\overline{\mathbb{R}}_0 = [0, \infty)$ *with* $\phi(0) > 0$. *Define*

$$\psi(x) = \phi(x) \int_0^x \frac{dy}{\phi(y)^2} \quad \text{or} \quad \psi(x) = \phi(x) \int_x^\infty \frac{dy}{\phi(y)^2} \tag{30}$$

according as

$$\int_0^\infty \frac{dy}{\phi(y)^2} = \infty \quad \text{or} \quad \int_0^\infty \frac{dy}{\phi(y)^2} < \infty. \tag{31}$$

Then ψ *is positive q-harmonic in* $\overline{\mathbb{R}}_0$. *Any q-harmonic function in* $\overline{\mathbb{R}}_0$ *is a linear combination of* ϕ *and* ψ.

We shall call ψ the *complement* to ϕ in either case.

Proof By Proposition 8.10, $\phi > 0$ in $\overline{\mathbb{R}}_0 = [0, \infty)$. Since $\phi \in C^{(1)}(\overline{\mathbb{R}}_0)$, so is ψ; indeed in the first case

$$\psi'(x) = \phi'(x) \int_0^x \frac{dy}{\phi(y)^2} + \frac{1}{\phi(x)}, \tag{32}$$

while in the second case

$$\psi'(x) = \phi'(x) \int_x^\infty \frac{dy}{\phi(y)^2} - \frac{1}{\phi(x)}. \tag{33}$$

Since ϕ is q-harmonic, ϕ' is locally absolutely continuous by Proposition 9.11; hence, so is ψ' by (33). By (31)–(33), we have

$$\begin{aligned} \phi\psi' - \psi\phi' &= \pm 1, \\ \phi\psi'' - \psi\phi'' &= 0, \end{aligned} \tag{34}$$

a.e. since both ϕ'' and ψ'' exist a.e. It follows that

$$\psi'' = \frac{\phi''\psi}{\phi} = -2q\psi$$

a.e. Therefore by Proposition 9.11, ψ is q-harmonic. The last assertion of the theorem follows from (34) and Theorem 9.13. □

Using Theorem 9.17, we can determine the gauge u of (D, q) when it is pseudo-gaugeable.

Theorem 9.18 *Let ϕ and ψ be as in Theorem 9.17. Under the first alternative in (31), $u = \phi/\phi(0)$; under the second alternative, $u = \psi/\psi(0)$.*

Proof We have $u = A\phi + B\psi$ where A and B are constants. In the first case, $\psi(x)/\phi(x) \to \infty$ as $x \to \infty$. By Theorem 9.15, $u \leq \phi/\phi(0)$. Hence $B = 0$. In the second case, $\psi(x)/\phi(x) \to 0$ as $x \to \infty$. By Theorem 9.15, $u \leq \psi/\psi(0)$. Hence $A = 0$. □

When (D, q) is pseudo-gaugeable, let v denote the complement to the gauge u.

Proposition 9.19 *We have*

$$\int_0^\infty \frac{dy}{u(y)^2} = \infty, \quad i.e. \quad \lim_{x\to\infty} \frac{v(x)}{u(x)} = \infty. \tag{35}$$

The function v is unbounded in $\mathbb{R}_0$. If u is bounded, then any bounded positive q-harmonic function is a constant multiple of u.

Proof Let u be the ϕ in Theorem 9.17. If the second case in (31) were true, then the complement v to u would have the property that $v(x)/u(x) \to 0$ as $x \to \infty$. This contradicts the minimality of u, and so the first case must hold for u, i.e., (35) is true, and

$$v(x) = u(x) \int_0^x \frac{dy}{u(y)^2}.$$

Suppose v is bounded by B so that

$$F(x) \equiv \int_0^x \frac{dy}{u(y)^2} \leq \frac{B}{u(x)}.$$

Then

$$F'(x) = \frac{1}{u(x)^2} \geq \frac{F(x)^2}{B^2} \quad \text{or} \quad -\left(\frac{1}{F(x)}\right)' \geq \frac{1}{B^2}.$$

This implies that

$$\frac{1}{F(1)} \geq \frac{1}{F(1)} - \frac{1}{F(x)} \geq \frac{x-1}{B^2},$$

which is absurd for large x. Hence, v cannot be bounded in $\mathbb{R}_0$.

Now suppose that u is bounded and that there is another bounded positive q-harmonic function ϕ which is not a constant multiple of u, then ϕ is linearly independent of u. Hence, by Theorem 9.13, any q-harmonic function would be bounded, This contradicts the unboundedness of v. □

We add the following curious result communicated by K.B. Erickson.

Proposition 9.20 *If (D, q) is pseudo-gaugeable, then*

$$E^x\{e_q(T_0)\} = \underline{\lim}_{b\to\infty} E^x\{e_q(T_0 \wedge T_b)\}. \tag{36}$$

Proof If we replace '=' by '$\leq$' in (36), then the result is true by Fatou's lemma. By Proposition 9.19, we may choose $b_n \to \infty$ such that $v(b_n) \to \infty$. Then for any $x \in (0, b_n)$, since $v(x) \geq w_{b_n}^{(0,b_n)}(x)v(b_n)$, we have

$$\lim_n w_{b_n}^{(0,b_n)}(x) = 0.$$

Since

$$E^x\{e_q(T_0 \wedge T_b)\} = w_0^{(0,b)}(x) + w_b^{(0,b)}(x),$$

it follows by (28) that

$$\lim_{n\to\infty} E^x\{e_q(T_0 \wedge T_{b_n})\} = \lim_{n\to\infty} w_0^{(0,b_n)}(x) = u(x).$$

This shows that if we replace '=' by '$\geq$' in (36), the result is also true. □

What is the gauge for $(\mathbb{R}^1, q)$? Since there is no boundary, it cannot be defined by the old formula. We must content ourselves with the expectations $u(x, y)$ studied in Section 9.1. Pseudo-gaugeability (or gaugeability) should mean that these expectations are finite (or bounded) for all x and y. The former is equivalent to the proposition that '$\alpha = -\infty$ and $\beta = \infty$'. It is a pleasant surprise that the connective 'and' in the last sentence may be replaced by 'or'. This is part of the following theorem which was historically the first result obtained in this area as a whole. Its analogue for $\mathbb{R}_0$ given in Theorem 9.14 was an afterthought a decade later, occasioned by the drafting of this book.

Theorem 9.21 *The following assertions are equivalent:*

(i) $\alpha = -\infty$;

(ii) $\beta = \infty$;

(iii) *for all x and y in $\mathbb{R}^1$,*

$$u(x, y)u(y, x) \leq 1; \tag{37}$$

(iv) *(37) holds for some x and some y, $x \neq y$.*

Proof Since $u(x, y) > 0$, $u(y, x) > 0$ by the Corollary to Proposition 9.1, it is clear that (iii) implies both (i) and (ii). We shall first prove that (ii) implies (iii); that (i) implies (iii) is completely similar and so the equivalence of (i), (ii) and (iii) ensues.

Let $x < y < z$. Under (ii) we have

$$\infty > u(y, z) = \sum_{n=0}^{\infty} (E^y\{T_x < T_z; e_q(T_x)\}u(x, y))^n E^y\{T_z < T_x; e_q(T_z)\}.$$

Since the series is convergent and $u(x, y) > 0$,

$$E^y\{T_x < T_z; e_q(T_x)\} < \frac{1}{u(x, y)}.$$

It follows that

$$u(y, x) = \lim_{z \to \infty} E^y\{T_x < T_z; e_q(T_x)\} \leq \frac{1}{u(x, y)}$$

which is (37).

It only remains to prove that (iv) implies (i). Suppose that (37) holds for a pair (x, y) with $x < y$; we shall prove that $u(y, z) < \infty$ for any $z > y$. We have

$$u(y, z) = \sum_{n=0}^{\infty} (E^y\{T_x < T_z; e_q(T_x)\}u(x, y))^n E^y\{T_z < T_x; e_q(T_z)\}. \tag{38}$$

Since $u(y, x) < \infty$ by assumption, and $P^y\{T_x < T_z\} < 1$, it follows that

$$E^y\{T_x < T_z; e_q(T_x)\} < u(y, x), \tag{39}$$

so that the series in (38) converges apart from the last factor there. But, by Theorem 9.9 applied with $a = x$ and $b = z$, we see that the finiteness of the left member of (39) implies that of the last factor in (38). Hence $u(y, z) < \infty$. This being true for any $z > y$, we conclude that $\beta = \infty$. □

It is curious that the equivalence of (i) and (ii) in the theorem signifies a certain reversibility which is by no means intuitively suggested, especially if one realizes that the function q is quite general and carries no sort of directional symmetry. Indeed, the problem had its origin in mathematical physics, and was a moot case at the time it was proved.

Appendix to Section 9.2

The general proof of the conditional gauge theorem (Theorem 7.5) is valid with some changes in detail. Let $D = (a, b)$ be a finite interval. Then by Section 9.1(ii), we have

$$K(x, a) = \frac{b - x}{b - a}, \quad K(x, b) = \frac{x - a}{b - a},$$

and $G(x, y)$ is given explicitly in Section 9.1(iv). Hence by a direct computation we obtain the one-dimensional case of the corollary to the 3G Theorem (Corollary 6.13) as follows:

$$\frac{G(x, y)K(y, z)}{K(x, z)} \le 4(b - a).$$

Thus Lemma 7.1 is true. Now let $a < a_1 < a_2 < b_2 < b_1 < b$, $D_0 = (a_1, b_1)$ and $U = (a, a_2) \cup (b_2, b)$ with $m(U)$ sufficiently small. Lemma 7.3 is obviously false as stated, but it can be changed as follows. For $y = a_1$ and $z = a$, or $y = b_1$ and $z = b$, we have

$$0 < E_z^y\{\tau_D = \tau_U; e_q(\tau_D)\} < \infty.$$

The rest of the proof of Theorem 7.5 goes through as before.

Since the one-dimensional Poisson kernel $K(\cdot, \cdot)$ has the special form shown above, we have by Proposition 5.11 or 5.12,

$$E_a^x\{e_q(\tau_D)\} = \frac{b - a}{b - x} w_a(x),$$

and

$$E_b^x\{e_q(\tau_D)\} = \frac{b - a}{x - a} w_b(x);$$

but we can give a direct proof of these equalities as follows. For any $M > 0$, $a_n \downarrow\downarrow a$ we have by the bounded convergence theorem,

$$\begin{aligned} E_a^x\{e_q(\tau_D) \wedge M\} &= \lim_n E_a^x\{e_q(T_{a_n}) \wedge M\} \\ &= \frac{b - a}{b - x} \lim_n E^x\{T_{a_n} < T_b; \frac{b - a_n}{b - a}[e_q(T_{a_n}) \wedge M]\} \\ &= \frac{b - a}{b - x} E^x\{T_a < T_b; [e_q(T_a) \wedge M]\}. \end{aligned}$$

Letting $M \to \infty$, we obtain the first equality. The proof for the second equality is similar.

In view of the two equations above, Theorem 9.9 is equivalent to the one-dimensional conditional gauge theorem. Indeed, it follows from Theorem 9.10 that

$$E_b^a\{e_q(\tau_D)\} = (b-a)w_b'(a),$$

and

$$E_a^b\{e_q(\tau_D)\} = (a-b)w_a'(b).$$

This is the one-dimensional version of Theorem 7.27.

9.3 Special Cases and Examples

In this section we first establish a condition under which the gauge of an infinite interval is bounded. Then we treat the special cases where q is nonpositive, or eventually positive, or periodic. Finally, by way of illustrating the power of the probabilistic method, we shall derive a number of old results using the new theory developed in preceding sections.

Let $\mathbb{R}_a = (a, \infty)$, where $a \in \mathbb{R}^1$. The gauge for $(\mathbb{R}_a, q)$ is then

$$u(x,a) = E^x\{e_q(T_a)\}, \quad x \in \mathbb{R}_a.$$

The basic results proved for $\mathbb{R}_0$ are valid for $\mathbb{R}_a$ when we replace $q(x)$ by $q(a+x)$. As in Sections 9.1 and 9.2, the assumption that $q \in L^1_{\text{loc}}(\mathbb{R}_a)$, which is equivalent to $1_{\mathbb{R}_a} q \in J_{\text{loc}}$, is sufficient, and will not be repeated. But other assumptions will be made on q for specific results below. In particular, the condition (40) is well used in analysis.

Theorem 9.22 *Let $q \not\equiv 0$, and*

$$\int_{-\infty}^{\infty} |yq(y)| \mathrm{d}y < \infty. \tag{40}$$

Then

$$\begin{aligned} \alpha &\le \inf\{a \in \mathbb{R}^1 : 2\int_a^\infty (y-a)|q(y)|\mathrm{d}y < 1\} < \infty; \\ \beta &\ge \sup\{b \in \mathbb{R}^1 : 2\int_{-\infty}^b (b-y)|q(y)|\mathrm{d}y < 1\} > -\infty. \end{aligned} \tag{41}$$

Moreover, for any $a > \alpha$ and $b < \beta$, we have

$$\begin{aligned} 0 < \lim_{x\to\infty} u(x,a) < \infty; \\ 0 < \lim_{x\to-\infty} u(x,b) < \infty. \end{aligned} \tag{42}$$

Proof It is sufficient to prove the assertion involving α; the dual assertion may be proved in the same way, or simply by considering the process $-X$ instead of X and $q(-y)$ instead of $q(y)$.

Retracing the calculations in Lemma 3.7, we set for $a \in \mathbb{R}^1$ and $x \ge a$:

$$
\begin{aligned}
M_n(x,a) &= \frac{1}{n!}E^x\left\{\left(\int_0^{T_a}|q(X_t)|\mathrm{d}t\right)^n\right\}; \\
M_n(a) &= \sup_{x\geq a} M_n(x,a).
\end{aligned}
$$

Then

$$M_n(a) \leq (M_1(a))^n.$$

Next by the explicit formula for $G_{\mathbb{R}_a}$ given in (v) of Section 9.1, we have

$$M_1(x,a) = G_{\mathbb{R}_a}|q|(x) = 2\int_a^\infty [(x-a)\wedge(y-a)]|q(y)|\mathrm{d}y,$$

and consequently

$$M_1(a) = 2\int_a^\infty (y-a)|q(y)|\mathrm{d}y.$$

It follows from (40) that

$$\lim_{a\to-\infty} M_1(a) = \infty \quad\text{and}\quad \lim_{a\to\infty} M_1(a) = 0. \tag{43}$$

In fact, $M_1'(a) = -2\int_a^\infty |q(y)|\mathrm{d}y \leq 0$, so that $M_1(a)$ is a nonincreasing function of a. Let

$$a_0 = \inf\{a \in \mathbb{R}^1 : M_1(a) < 1\}.$$

By (43) and the continuity of M_1, we have $a_0 \in \mathbb{R}^1$, $M_1(a_0) = 1$ and $M_1(a) < 1$ for all $a > a_0$. If $a > a_0$, we have for all $x \geq a$,

$$
\begin{aligned}
u(x,a) &\leq 1 + \sum_{n=1}^\infty M_n(x,a) \leq 1 + \sum_{n=1}^\infty M_n(a) \\
&\leq 1 + \sum_{n=1}^\infty (M_1(a))^n = \frac{1}{1-M_1(a)} < \infty.
\end{aligned}
$$

Hence by definition, $\alpha \leq a_0$, proving (41).

Next, we have by (43)

$$\lim_{a\to\infty}\left[\sup_{x\geq a}|u(x,a)-1|\right] \leq \lim_{a\to\infty}\frac{M_1(a)}{1-M_1(a)} = 0. \tag{44}$$

It follows from this that there exists $a_0 > 0$ such that

$$\sup_{x\geq a_0} u(x,a_0) \geq \frac{1}{2},$$

and

$$\lim_{x\geq y\to\infty} |u(x,a_0)-u(y,a_0)| = 0.$$

Consequently

$$u(\infty,a_0) = \lim_{x\to\infty} u(x,a_0)$$

exists, is finite and > 0. The same is then true when a_0 is replaced by any $a > \alpha$, by the Corollary to Proposition 9.1 and Proposition 9.2. □

Theorem 9.22 gives a sufficient condition for the gaugeability of $(\mathbb{R}_0, q)$. The gauge $u(x) = u(x, 0)$ is not only bounded in $\mathbb{R}_0$, but strictly positive and continuous there as in Theorem 4.7. One more property is useful and will be stated below under more general conditions.

Proposition 9.23 *Suppose $q \in L^1(\mathbb{R}_0)$ and $(\mathbb{R}_0, q)$ is gaugeable, then*

$$u(x) = 1 + G(qu)(x), \quad x \in \mathbb{R}_0, \tag{45}$$

where G is the Green function for $\mathbb{R}_0$.

Proof This is, of course, the extension of (4.38) with $f \equiv 1$. We recall that for $x \in (0,b)$, where $b \in \mathbb{R}_0$, and we have written w^b for $w_0^{(0,b)}$:

$$w^b(x) = \frac{b-x}{b} + G_{(0,b)}(qw^b)(x).$$

By (28), $u(x) = \lim_{b\to\infty} w^b(x)$. Since u is bounded by hypothesis, we may let $b \to \infty$ in the above equation to obtain the conclusion, using dominated convergence and the fact that

$$\lim_{b\to\infty} G_{(0,b)}(|q|) = G_{(0,\infty)}(|q|) < \infty.$$

The last inequality follows from the explicit formula for $G_{(0,\infty)}$ given in (v) of Section 9.1. □

Remark If $(\mathbb{R}_0, q)$ is pseudo-gaugeable, the representation (45) holds provided $G(|qu|) < \infty$.

We conclude this section by giving a numerical example in which both α and β are finite.

Example For any $c \geq \frac{\pi}{4\sqrt{2}}$, let $q = 1_{(-c,c)}$. We shall show that in this case $\alpha = c - \frac{\pi}{2\sqrt{2}}$ and $\beta = -\alpha$.

To see this, let $c - \frac{\pi}{2\sqrt{2}} < a < c$, then $a \in (-c, c)$ by the choice of c. Define a function v_a in $\mathbb{R}_a$ as follows:

$$\begin{aligned} v_a(x) &= \frac{\cos(\sqrt{2}(c-x))}{\cos(\sqrt{2}(c-a))} \text{ if } a < x < c; \\ &= \frac{1}{\cos(\sqrt{2}(c-a))} \text{ if } c \leq x < \infty. \end{aligned}$$

The $v_a \in C^1(\mathbb{R}^a)$ and $v_a'' + \frac{1}{2}qv_a = 0$ in $\mathbb{R}_a$ except at c. Thus v_a is a positive q-harmonic function in $\mathbb{R}_a$, and therefore by Theorem 9.16, $(\mathbb{R}_a, q)$ is pseudo-gaugeable. The fact that $u(\cdot, a)$ is finite when a is arbitrarily near $c - \frac{\pi}{2\sqrt{2}}$ implies that $\alpha \leq c - \frac{\pi}{2\sqrt{2}}$.

Next, since v_a is bounded, it follows from Proposition 9.19 that $u(x, a) = v_a(x)$ in $\mathbb{R}_a$. Hence

$$\lim_{a \downarrow c - \frac{\pi}{2\sqrt{2}}} u(c, a) = \lim_{a \downarrow c - \frac{\pi}{2\sqrt{2}}} \frac{1}{\cos(\sqrt{2}(c - a))} = \infty.$$

This cannot be true if $\alpha < c - \frac{\pi}{2\sqrt{2}}$, by Theorem 9.8. Therefore we have proved that $\alpha = c - \frac{\pi}{2\sqrt{2}}$.

Since q is an even function, it is obvious that $\beta = -\alpha$. In particular we have the following possibilities:

$$\begin{aligned} \alpha < 0 < \beta \quad & \text{if} \quad \frac{\pi}{4\sqrt{2}} \leq c < \frac{\pi}{2\sqrt{2}}; \\ \alpha = 0 = \beta \quad & \text{if} \quad c = \frac{\pi}{2\sqrt{2}}; \\ \beta < 0 < \alpha \quad & \text{if} \quad \frac{\pi}{2\sqrt{2}} \leq c < \infty. \end{aligned}$$

Now what happens if $0 < c < \frac{\pi}{4\sqrt{2}}$? We leave this to the reader.

We now give a sufficient condition for conclusions opposite to those described in Theorem 9.22. We may confine ourselves to one half of the dual results.

Proposition 9.24 *Suppose that $q \in J_{\text{loc}}$ and that $q(x) \geq 0$ for all sufficiently large x. If*

$$\int_0^\infty q(y)\text{d}y = \infty, \tag{46}$$

then $\alpha = \infty$. If

$$\int_0^\infty yq(y)\text{d}y = \infty, \tag{47}$$

then either $\alpha = \infty$ or for any $a > \alpha$ we have

$$\lim_{x \to \infty} u(x, a) = \infty.$$

Proof Let $q(x) \geq 0$ for $x \geq a$. We have by Jensen's inequality, for $x > a$:

$$u(x, a) \geq \exp\left(E^x\left\{\int_0^{T_a} q(X_t)\text{d}t\right\}\right);$$

then by (46)

$$E^x\left\{\int_0^{T_a} q(X_t)\mathrm{d}t\right\} = 2\int_a^\infty [(x-a)\wedge(y-a)]q(y)\mathrm{d}y$$
$$\geq 2(x-a)\int_x^\infty q(y)\mathrm{d}y = \infty.$$

Hence $u(x,a) = \infty$. Since a may be arbitrarily large, this implies that $\alpha = \infty$. Next, suppose that (46) is false but (47) is true. Then

$$E^x\left\{\int_0^{T_a} q(X_t)\mathrm{d}t\right\} \geq 2\int_a^x (y-a)q(y)\mathrm{d}y.$$

This diverges to ∞ as $x \to \infty$, proving the second assertion of the proposition. □

We note that in the second case above, a general condition to ensure that $\alpha < \infty$ does not exist. In the literature of Sturm–Liouville equations, many special conditions are studied; see e.g. Hartman (1982).

Next we state and prove a classical result using our methods. When $q \in \mathcal{B}$ and $q \leq 0$, of course the gauge for $(\mathbb{R}_0, q)$ is bounded by 1, but we still need the fundamental Proposition 9.1, as we shall soon see. Note that the latter is true if $q \in L^1_{\text{loc}}(\mathbb{R})$ as remarked before.

Proposition 9.25 *Suppose that $q \leq 0$ and $q \in L^1(\mathbb{R}_0)$. Then $0 < u(x,0) \leq 1$ for $x \in \mathbb{R}_0$, and is nonincreasing. It is q-harmonic and continuously differentiable in $\mathbb{R}_0$. A necessary and sufficient condition for*

$$\lim_{x\to\infty} u(x,0) > 0 \tag{48}$$

is

$$\int_0^\infty yq(y)\mathrm{d}y > -\infty. \tag{49}$$

Proof We have $u(x,0) > 0$ for all $x \in \mathbb{R}_0$ by the Corollary to Proposition 9.1. If $x_2 > x_1 > 0$, then

$$u(x_2,0) = u(x_2,x_1)u(x_1,0) \leq u(x_1,0),$$

since it is trivial that $u(x_2,x_1) \leq 1$. The function $u(\cdot,0)$ is q-harmonic and belongs to $C^{(1)}(\mathbb{R}_0)$ by Theorem 9.15. Finally, suppose that (49) is true, then

$$\lim_{x\to\infty} E^x\left\{\int_0^{T_0} q(X_t)\mathrm{d}t\right\} = 2\lim_{x\to\infty}\int_0^\infty (x\wedge y)q(y)\mathrm{d}y = c > -\infty.$$

Hence, by Jensen's inequality, the limit in (48) is at least $\mathrm{e}^c > 0$. On the other hand, suppose that $u(x,0) \geq \delta > 0$ for all $x \in \mathbb{R}_0$. Using Proposition 9.23, we then have

$$\begin{aligned}\delta \le u(x,0) &= 1+2\int_0^\infty (x\wedge y)q(y)u(y,0)\mathrm{d}y\\ &\le 1+2\delta\int_0^x yq(y)\mathrm{d}y,\end{aligned}$$

from which (49) follows; indeed the integral there is bounded below by $\frac{\delta-1}{2\delta}$. □

We take this occasion to signal a possibility hitherto ignored, namely when the basic assumption that q is locally integrable is omitted.

Proposition 9.26 *If* $q \le 0$ *in* $\mathbb{R}_0$*, and* $\int_0^\delta xq(x)\mathrm{d}x = -\infty$ *for some* $\delta > 0$*, then* $u(x,0) = 0$ *for all* $x > 0$.

Proof According to a result due to Feller (see e.g. Freedman (1971, Theorem (132))), $\int_0^\delta x|q(x)|\mathrm{d}x = \infty$ for some $\delta > 0$ implies that $\int_0^{T_0} q(X_t)\mathrm{d}t = -\infty$, P^x-a.s. for any $x > 0$, so that $e_q(T_0) = 0$ and $u(x,0) = 0$. □

Remark The example $q(x) = -x^{-1}\mathrm{e}^{-x}$ shows that the condition in the proposition cannot be weakened to $\int_0^\delta q(x)\mathrm{d}x = -\infty$. In fact, it is known that for each $x > 0$, $P^x\left(\int_0^{T_0} q(X_t)\mathrm{d}t = -\infty\right)$ is either 1 or 0 according as $\int_0^\delta xq(x)\mathrm{d}x = -\infty$ or $> -\infty$ (see e.g. Freedman (1971, Theorem 2.66, Lemma 2.78, and Theorem 2.132)).

As another special case, suppose that $q \in \mathcal{B}$ and q is periodic with period ρ, $0 < \int_0^\rho |q(y)|\mathrm{d}y < \infty$. Let

$$c = \int_0^\rho q(y)\mathrm{d}y. \tag{50}$$

In this case, the basic structure of the Brownian motion implies that for all $x \in \mathbb{R}_0$:

$$u(x+\rho, x) = u(\rho, 0). \tag{51}$$

Writing $u(x)$ for $u(x,0)$, we have by (51) and Proposition 9.2 that

$$u(x) = \mathrm{e}^{\lambda x} r(x), \tag{52}$$

where

$$\lambda = \frac{1}{\rho}\ln u(\rho),$$

and r is periodic with period ρ. Now if $(\mathbb{R}_0, q)$ is pseudo-gaugeable, then u is a solution of (16) by Theorem 9.15. Hence, using (52) followed by a simple computation, we have:

$$-2\int_0^\rho q(y)\mathrm{d}y = \int_0^\rho \frac{u''(y)}{u(y)}\mathrm{d}y = \lambda^2 + \int_0^\rho \frac{r'(y)^2}{r(y)^2}\mathrm{d}y.$$

This is strictly positive if $\lambda \neq 0$. If $\lambda = 0$, then $r = u$, and $r(\rho) = 1$, hence the last integral in the above is strictly positive because r is not constant. We have therefore proved the following result due to K.B. Erickson.

Proposition 9.27 *If $c \geq 0$, then $(\mathbb{R}_0, q)$ is not pseudo-gaugeable, i.e. $u(x,0) \equiv \infty$ for $x \in \mathbb{R}_0$.*

When $c > 0$, the above result can also be proved by Jensen's inequality. In this case, it is easy to show by periodicity that

$$\int_1^\infty q(y)\mathrm{d}y = \infty.$$

Then

$$E^1 \left\{ \int_0^{T_0} q(X_t)\mathrm{d}t \right\} = G_{(0,\infty)}q(1) = 2\int_0^1 yq(y)\mathrm{d}y + 2\int_1^\infty q(y)\mathrm{d}y = \infty.$$

Hence by Jensen's inequality we have $u(1,0) = \infty$, which is sufficient by Proposition 9.4. The case $c = 0$ escapes this method.

According to Erickson, $c < 0$ is not sufficient for the pseudo-gaugeability of $(\mathbb{R}_0, q)$.

The rest of this section is devoted to a selection of specific examples illustrating the various possibilities; other examples can be formulated on demand, once it is realized that one can start from a solution rather than an equation! Indeed, a general positive function on $\mathbb{R}^1$ (or $\mathbb{R}^d$) is of the form

$$\phi(x) = \mathrm{e}^{f(x)},$$

where f is quite arbitrary but will be assumed to be in $C^{(2)}(\mathbb{R}^1)$. We then have

$$\phi''(x) = (f'(x)^2 + f''(x))\phi(x),$$

so that ϕ is q-harmonic with

$$q(x) = -\frac{1}{2}\left(f'(x)^2 + f''(x)\right).$$

It remains to choose f to suit the desired condition and then apply our previous theorems to determine the (pseudo-) gaugeability and to find the gauge.

Example 1 $\phi(x) = \mathrm{e}^{cx}$, $c \in \mathbb{R}^1$; $q(x) = -\frac{c^2}{2}$. Since $q \leq 0$, there is gaugeability in $\mathbb{R}_0$ with the gauge

$$u(x) = E^x\left\{\mathrm{e}^{-\frac{c^2}{2}T_0}\right\}, \quad x \in \mathbb{R}_0.$$

When $c \leq 0$, the function ϕ satisfies the first condition in (31). Hence $u = \phi$ by Theorem 9.18. We have thus obtained (without computation!) the Laplace transform of T_0:

$$E^x\left\{\mathrm{e}^{-\lambda T_0}\right\} = \mathrm{e}^{-\sqrt{2\lambda}x}. \tag{53}$$

This is a famous formula; another derivation uses the distribution of T_0 given in (23) and traditional calculus.

When $c > 0$, ϕ satisfies the second condition in (31). Hence we obtain its complement

$$\psi(x) = \mathrm{e}^{cx} \int_x^\infty \frac{\mathrm{d}y}{\mathrm{e}^{2cy}} = \frac{1}{2c}\mathrm{e}^{-cx}.$$

Thus $u(x) = \mathrm{e}^{-cx}$, as before.

Example 2 $\phi(x) = e^{\sin x}$ in $\mathbb{R}_0$; $q(x) = \frac{1}{2}\sin x - \frac{1}{2}(\cos x)^2$. Since $\mathrm{e}^{-1} \le \phi \le \mathrm{e}$, it follows from Theorem 4.17 that $(\mathbb{R}_0, q)$ is gaugeable. Since ϕ satisfies the first condition in (31), $\phi = u$. We have therefore proved that

$$E^x\left\{\exp\left(\frac{1}{2}\int_0^{T_0}\left[\sin X_t - (\cos X_t)^2\right]\mathrm{d}t\right)\right\} = \mathrm{e}^{\sin x},$$

a result first obtained by Erickson. The gauge u is bounded but does not converge as $x \to \infty$. Although q is periodic, we have $c < 0$ in (50). Proposition 9.27 does not establish gaugeability if we start with the equation rather than the solution.

Example 3 $\phi(x) = \mathrm{e}^{x \sin x}$ in $\mathbb{R}_0$. In this case, q is unbounded. Since ϕ satisfies the first condition in (31), $\phi = u$. The gauge is finite but fluctuates unboundedly as $x \to \infty$.

Example 4 Let $c \in \mathbb{R}^1$, and $q(x) = \frac{c}{x^2}$ in $\mathbb{R}^1$, so that the corresponding equation is

$$\phi''(x) + \frac{2c}{x^2}\phi(x) = 0.$$

This is an Euler equation which can be solved by making the substitution $x = \mathrm{e}^t$. But we shall proceed as before by starting with a solution:

$$\phi(x) = x^r = \mathrm{e}^{r \ln x}, \quad r \in \mathbb{R}^1, \quad x \in \mathbb{R}_a, \quad a > 0.$$

This is a solution of the equation if $r(r-1) = -2c$. The two roots are

$$r_1 = \frac{1 - \sqrt{1-8c}}{2}, \quad r_2 = \frac{1 + \sqrt{1-8c}}{2}.$$

<u>Case 1</u> $c < 0$. Then $r_1 < 0$, and

$$u(x) = E^x\{\exp[c\int_0^{T_a} X(t)^{-2}\mathrm{d}t]\} = \left(\frac{x}{a}\right)^{r_1}.$$

$(\mathbb{R}_0, q)$ is gaugeable and $\lim_{x\to\infty} u(x) = 0$.

<u>Case 2</u> $c = 0$. This case is trivial.

<u>Case 3</u> $0 < c \le \frac{1}{8}$. Then $0 < r_1 \le \frac{1}{2}$. Since $\int_a^\infty x^{-2r_1}\mathrm{d}x = \infty$, we have

$$u(x) = \left(\frac{x}{a}\right)^{r_1}, \quad x \in \mathbb{R}_a.$$

$(\mathbb{R}_0, q)$ is pseudo-gaugeable but not gaugeable.

Case 4 $c > \frac{1}{8}$. Set $\frac{1}{2}\sqrt{1-8c} = b\sqrt{-1}$. By elementary theory of differential equations, in this case the fundamental solutions are still given by x^{r_1} and x^{r_2} even though they are now complex valued. It follows that the general real-valued solution is given by

$$\phi(x) = Ax^{1/2}\cos(B + b\ln x)$$

where A and B are two real constants. Hence for an interval $I \subset \mathbb{R}_1$, if $b\ln m(I) > \pi$, then the above ϕ must vanish somewhere in I. Therefore there is no strictly positive solution in such an I, and so by Theorem 9.15, (I, q) is not pseudo-gaugeable. In particular, $\alpha = \infty$.

The situation just described is typical of the gaugeability of a finite interval. We return to this case and consider a final example where q is the negative of the q in Example 1.

Example 5 Let $q = \frac{c^2}{2}$, $c \in \mathbb{R}^1$, in $I = (a, b)$. The equation

$$\phi'' + c^2\phi = 0$$

has the general solution $A\cos(cx) + B\sin(cx)$. This is the most celebrated case of the equation, known also as the eigen equation for the Laplacian, from which the theory of Fourier series emerged.

The gauge for $((a,b), \frac{c^2}{2})$ with $0 < c < \frac{\pi}{b-a}$ is given by

$$E^x\left\{e^{\frac{c^2}{2}\tau_{(a,b)}}\right\} = \frac{\cos(c(x - \frac{a+b}{2}))}{\cos\frac{c(b-a)}{2}}, \quad x \in (a, b).$$

This formula can be derived in a number of interesting ways, one of which uses its analogue in (53); see Chung (1981). Note that the critical value c^* is determined by $c^*(b-a) = \pi$. The fact that the gauge is infinite when $c = c^*$ is a particular case of the general Theorem 8.2; it can also be proved as in Theorem 9.5 which is its extension to an infinite interval. See also Proposition 8.3 for another special case.

9.4 Local Time and Density

In one dimension, there is a renowned object associated with the Brownian motion called *local time*, invented by Paul Lévy (see Balkema and Chung (1991)). For each $y \in \mathbb{R}^1$, the local time at y is a stochastic process $\{L(t,y), 0 \le t < \infty\}$ which may be defined as follows:

$$L(t,y) = \lim_{\varepsilon \downarrow 0} \frac{1}{2\varepsilon}\int_0^t 1_{(y-\varepsilon, y+\varepsilon)}(X_s)\mathrm{d}s, \tag{54}$$

where $\{X_t, t \geq 0\}$ is the 1-dimensional Brownian motion. The existence of the limit (almost surely) is not an easy result, but we shall not need it in what follows. The fundamental property of $L(\cdot,\cdot)$ is given by the following formula which converts an integral over time into an integral over space. For any Borel set B:

$$\int_0^t 1_B(X_s)\mathrm{d}s = \int_B L(t,y)\mathrm{d}y. \tag{55}$$

Both (54) and (55) are proved in Chung and Williams (1983, Chapter 7). In addition to (55), we also need the following properties of L:

(a) for each y, $t \to L(t,y)$ is an increasing function for $t \geq 0$. The induced measure will be denoted by $L(dt,y)$. It is supported by the closed non-dense set $\{t : X(t) = y\}$.
(b) Almost surely $(t,y) \to L(t,y)$ is continuous. This implies that for any finite $\tau(\omega) \geq 0$, $y \to L(\tau(\omega),y)$ is continuous. We shall omit 'almost surely' in what follows when the context clearly requires it.
(c) For any bounded interval D and $k \geq 1$:

$$E^x\{(L(\tau_D,y))^k\} < \infty. \tag{56}$$

In fact, $L(\tau_D,y)$ has an exponential distribution; see Chung and Williams (1983).

We need a sharpening of the supergauge theorem (Theorem 8.9), which is valid in $\mathbb{R}^d$ under the same conditions, i.e. bounded and regular D and $q \in J_{\text{loc}}$.

Proposition 9.28 *For sufficiently small* $\delta > 0$, *we have*

$$\sup_{x\in D} E^x\left\{\left(\int_0^{\tau_D} e_q(t)|q(X_t)|\mathrm{d}t\right)^{1+\delta}\right\} < \infty.$$

Proof By Hölder's inequality, we have

$$\int_0^{\tau_D} e_q(t)|q(X_t)|\mathrm{d}t \leq \left[\int_0^{\tau_D} e_q(t)^{1+\delta}|q(X_t)|\mathrm{d}t\right]^{\frac{1}{1+\delta}} \left[\int_0^{\tau_D} |q(X_t)|\mathrm{d}t\right]^{\frac{\delta}{1+\delta}}.$$

Hence for $0 < \delta < 1$:

$$\begin{aligned}\left(\int_0^{\tau_D} e_q(t)|q(X_t)|\mathrm{d}t\right)^{1+\delta} &\leq \int_0^{\tau_D} e_q(t)^{1+\delta}|q(X_t)|\mathrm{d}t \cdot \left[\int_0^{\tau_D} |q(X_t)|\mathrm{d}t\right]^{\delta} \\ &\leq \int_0^{\tau_D} e_q(t)^{1+\delta}|q(X_t)|\mathrm{d}t \left[1 \vee \int_0^{\tau_D} |q(X_t)|\mathrm{d}t\right].\end{aligned}$$

Therefore, it is sufficient to prove that

$$\sup_{x\in D} E^x\left\{\int_0^{\tau_D} e_q(t)^{1+\delta}|q(X_t)|\mathrm{d}t\right\} < \infty \tag{57}$$

and

$$\sup_{x\in D} E^x\left\{\int_0^{\tau_D} e_q(t)^{1+\delta}|q(X_t)|\mathrm{d}t \int_0^{\tau_D} |q(X_t)|\mathrm{d}t\right\} < \infty. \tag{58}$$

By Theorem 8.9. for δ small enough, $(D,(1+\delta)q)$ is gaugeable, hence (57) is true by Theorem 4.19(iv), i.e. $\sup_{x\in D} V_q(|q|)(x) < \infty$ when q is replaced by $(1+\delta)q$. For the proof of (58) we write the expectation as

$$E^x\left\{\int_0^{\tau_D} e_q(t)^{1+\delta}|q(X_t)| \int_t^{\tau_D} |q(X_s)|\mathrm{d}s\mathrm{d}t\right\}$$
$$+E^x\left\{\int_0^{\tau_D} e_q(t)^{1+\delta}|q(X_t)| \int_0^{t} |q(X_s)|\mathrm{d}s\mathrm{d}t\right\}. \tag{59}$$

The first term in (59) is equal to

$$E^x\left\{\int_0^{\tau_D} e_q(t)^{1+\delta}|q(X_t)| E^{X_t}\left[\int_0^{\tau_D} |q(X_s)|\mathrm{d}s\right]\mathrm{d}t\right\}$$
$$\leq V_{(1+\delta)q}(|q|)(x)\sup_{x\in D} G|q|(x),$$

where $G = G_D$. The second term is equal to

$$E^x\left\{\int_0^{\tau_D} |q(X_s)| \int_s^{\tau_D} e_q(t)^{1+\delta}|q(X_t)|\mathrm{d}t\mathrm{d}s\right\}$$
$$= E^x\left\{\int_0^{\tau_D} |q(X_s)| E^{X_s}\left[\int_0^{\tau_D} e_q(t)^{1+\delta}|q(X_t)|\mathrm{d}t\right]\mathrm{d}s\right\}$$
$$\leq G|q|(x)\sup_{x\in D} V_{(1+\delta)q}(|q|)(x).$$

Hence the supremum of (59) over $x \in D$ does not exceed

$$2\sup_{x\in D} V_{(1+\delta)q}(|q|)(x)\sup_{x\in D} G|q|(x),$$

which is finite. □

Now let D be a finite interval in $\mathbb{R}^1$, $q \in J_{\text{loc}}$ and let

$$Z(y) = \int_0^{\tau_D} e_q(t)L(\mathrm{d}t, y).$$

Our goal is the following characterization of $Z(y)$.

Theorem 9.29 *Suppose (D,q) is gaugeable. Then for every $x \in \mathbb{R}^1$ and $y \in \mathbb{R}^1$*

$$E^x\{Z(y)\} = V(x,y), \tag{60}$$

where V is the q-Green function defined in (3.44).

Proof Integrating by parts, we have

$$Z(y) = L(\tau_D, y)e_q(\tau_D) - \int_0^{\tau_D} L(t, y)e_q(t)q(X_t)\mathrm{d}t. \tag{61}$$

It follows by Hölder's inequality that

$$E^x\left\{|L(\tau_D, y)e_q(\tau_D)|^{1+\delta}\right\} \leq E^x\left\{L(\tau_D, y)^{\frac{(1+\delta)^2}{\delta}}\right\}^{\frac{\delta}{1+\delta}} E^x\left\{e_q(\tau_D)^{(1+\delta)^2}\right\}^{\frac{1}{1+\delta}}. \tag{62}$$

Hence the expectation in the left member of (62) is finite for some $\delta > 0$ by Theorem 8.9 and (56). Next

$$\left|\int_0^{\tau_D} L(t, y)e_q(t)q(X_t)\mathrm{d}t\right| \leq L(\tau_D, y)\int_0^{\tau_D} e_q(t)|q(X_t)|\mathrm{d}t. \tag{63}$$

Hence by Hölder's inequality, Proposition 9.28 and (56), the expectation of the left member in (63) is likewise bounded. Thus we conclude by (61) that there exists $\delta > 0$ such that:

$$E^x\left\{Z(y)^{1+\delta}\right\} < \infty. \tag{64}$$

By property (b), $L(t, y)$ is continuous in y for each t. Using the bound given in (63), we see that the second term on the right-hand side of (61) is continuous in y, as is the first term there by (b). Hence $Z(y)$ is continuous in y. Since (64) implies that $Z(y)$ is uniformly integrable with respect to E^x, it follows that for each x,

$$y \to E^x[Z(y)] \quad \text{is continuous.} \tag{65}$$

Let B be a Borel subset of D. Using (55), integrating $e_q(t)$ and using Fubini's theorem:

$$\int_B \int_0^{\tau_D} e_q(t)L(\mathrm{d}t, y)\mathrm{d}y = \int_0^{\tau_D} e_q(t)1_B(X_t)\mathrm{d}t.$$

Thus

$$\int_B E^x\{Z(y)\}\mathrm{d}y = V1_B(x) = \int_B V(x, y)\mathrm{d}y.$$

Hence as a density $E^x\{Z(y)\}$ is equivalent to $V(x, y)$. For each x, they are equal for m-a.e. y. Since both are continuous in y by (65) and Theorem 6.2(a), they are identical. □

Actually, we can prove that $(x, y) \to E^x[Z(y)]$ is continuous without using the result on $V(x, y)$; see Chung (1983b).

9.5 Derivatives and Neumann's Problem

In one dimension, the solution of the Dirichlet boundary value problem given in Section 9.2 (Theorem 9.10) is differentiable. In particular, the derivatives of the two fundamental solutions may be regarded as known quantities. By Theorem 9.13, any solution can be expressed as a linear combination of two linearly independent solutions. Hence, in principle, we can solve the equation in an interval when the values of the derivatives of the solution at the two boundary points (rather than the values of the solution itself) are prescribed. The corresponding problem in $\mathbb{R}^d$ for a bounded domain is known as Neumann's problem and plays a role in classical analysis and mathematical physics. A probabilistic treatment of this problem has been developed, but this requires the construction of a new process called the reflecting Brownian motion, and a local time on the boundary which must be assumed to be rather smooth. We shall not treat this topic here; see Hsu (1985), Chung and Hsu (1986) and Papanicolaou (1990) for details. In one dimension, however, we can not only solve the problem described above, but also obtain further interesting relations between the various quantities by using the probabilistic representation based on local time given in the preceding section. We can also represent the solution of Neumann's problem by a new stopping time. This is different from the representation in higher dimensions cited above.

For simplicity of notation, we take the domain to be $I = (0,1)$. We assume that $q \in J_{\text{loc}}$, but it is sufficient to assume that $q \in L^1((0,1))$.

Assuming that (I, q) is gaugeable, we denote the two fundamental solutions of the equation

$$\phi'' + 2q\phi = 0$$

in I by w_0 and w_1, and their sum by u, which is the gauge, i.e.:

$$\begin{aligned} w_i(x) &= E^x\{X(\tau_I) = i; e_q(\tau_I)\} = E^x\{T_i < T_{1-i}; e_q(T_i)\}, \quad i = 0.1; \\ u(x) &= w_0(x) + w_1(x) = E^x\{e_q(\tau_I)\}. \end{aligned}$$

Thus, to give a probabilistic representation of the solution of the Neumann's problem, we need to express the following four derivatives:

$$w_i'(j), \quad i, j = 0, 1$$

in terms of simple probabilistic quantities.

We extend the given q from $[0,1]$ to $[-1,2]$ as follows:

$$q(x) = \begin{cases} q(-x) & \text{if } x \in [-1, 0) \\ q(2-x) & \text{if } x \in (1, 2]. \end{cases} \tag{66}$$

Let $I_0 = (-1, 1)$ and $I_1 = (0, 2)$. We now assume that both (I_0, q) and (I_1, q) are gaugeable. Quantities associated with I_i will be indicated by the subscript i, $i = 0, 1$. Thus, u_i and V_i refer to u and V when $I = I_i$.

Proposition 9.30 *We have*

$$\begin{aligned} V_0'(0,0) &= -1, \;\; V_0'(1,0) = -u_0(0); \\ V_1'(0,1) &= u_1(1), \;\; V_1'(1,1) = 1; \end{aligned} \tag{67}$$

where the derivatives are taken with respect to the first variable in $V_0(\cdot,\cdot)$ *and* $V_1(\cdot,\cdot)$.

Proof For I_0, by Theorem 6.3 and (iv) in Section 9.1, we have

$$V_0(x,y) = G_0(x,y) + \int_{-1}^{1} G_0(x,z)q(z)V_0(z,y)\mathrm{d}z,$$

where

$$G_0(x,y) = \begin{cases} (x+1)(1-y) & \text{if } -1 \le x \le y \le 1 \\ (1-x)(y+1) & \text{if } -1 \le y < x \le 1. \end{cases}$$

Differentiating, we obtain

$$V_0'(x,0) = -1 - \left(\int_{-1}^{x} - \int_{x}^{1}\right) q(z)V_0(z,0)\mathrm{d}z - \int_{-1}^{1} zq(z)V_0(z,0)\mathrm{d}z,$$

and consequently,

$$V_0'(0,0) = -1$$

by the symmetry of q and $V_0(\cdot,0)$ with respect to 0. Next, by the symmetry of V_0 in (x,y) and Fubini's theorem,

$$\begin{aligned} V_0'(1,0) &= -1 - \int_{-1}^{1} V_0(0,z)q(z)\mathrm{d}z \\ &= -1 - E^0\left\{\int_0^{\tau_{(-1,1)}} e_q(t)q(X_t)\mathrm{d}t\right\} \\ &= -E^0\left\{e_q[\tau_{(-1,1)}]\right\} = -u_0(0). \end{aligned}$$

This proves the assertions about V_0; those about V_1 are proved in the same way. □

By the representation of V_0 in Theorem 9.29, the support of $L(\cdot,0)$ and the strong Markov property, we have for $x \in I$:

$$\begin{aligned} V_0(x,0) &= E^x\left\{X(\tau_I) = 0; \int_{\tau_I}^{\tau_{I_0}} e_q(t)L(\mathrm{d}t,0)\right\} \\ &= E^x\left\{X(\tau_I) = 0; e_q(\tau_I)E^0\left[\int_0^{\tau_{I_0}} e_q(t)L(\mathrm{d}t,0)\right]\right\}, \end{aligned}$$

and similarly for $V_1(x,1)$. Thus we have

$$V_0(x,0) = w_0(x)V_0(0,0), \quad V_1(x,1) = w_1(x)V_1(1,1). \tag{68}$$

It is clear from Theorem 9.29 that $V_0(0,0) > 0$ and $V_1(1,1) > 0$. It follows from (67) and (68) that

$$\begin{aligned} w_0'(0) &= \frac{1}{V_0(0,0)}, \quad w_0'(1) = \frac{-u_0(0)}{V_0(0,0)}; \\ w_1'(0) &= \frac{u_1(1)}{V_1(1,1)}, \quad w_1'(1) = \frac{1}{V_1(1,1)}. \end{aligned} \tag{69}$$

Since the Wronskian of w_0 and w_1 is a constant, we have by (14): $w_1'(0) = w_0(0)w_1'(0) - w_0'(0)w_1(0) = w_0(1)w_1'(1) - w_0'(1)w_1(1) = -w_0'(1)$. Thus by (69),

$$u_0(0)V_1(1,1) = u_1(1)V_0(0,0). \tag{70}$$

We are now ready to solve the Neumann's problem for the Sturm–Liouville equation

$$\begin{aligned} &\phi'' + 2q\phi = 0 \\ &\phi'(0) = -f_0, \quad \phi'(1) = f_1, \end{aligned} \tag{71}$$

where the derivatives are taken in the unilateral sense as the outer normal derivatives at the boundary.

By Theorem 9.13, any q-harmonic function can be expressed as

$$\phi = \alpha w_0 + \beta w_1.$$

Hence it is sufficient to determine the constants α and β from the following equations:

$$\begin{aligned} \frac{-1}{V_0(0,0)}\alpha + \frac{u_1(1)}{V_1(1,1)}\beta &= -f_0 \\ \frac{-u_0(0)}{V_0(0,0)}\alpha + \frac{1}{V_1(1,1)}\beta &= f_1. \end{aligned} \tag{72}$$

The determinant of the linear system (72) is

$$\frac{u_0(0)u_1(1) - 1}{V_0(0,0)V_1(1,1)}. \tag{73}$$

Thus (73) has a unique solution if and only if

$$u_0(0)u_1(1) \neq 1. \tag{74}$$

In this case, the solution is given by

$$\begin{aligned} \phi(x) &= \frac{1}{1 - u_0(0)u_1(1)}\{[f_0 + u_1(1)f_1]V_0(0,0)w_0(x) \\ &\quad + [f_1 + u_0(0)f_0]V_1(1,1)w_1(x)\}. \end{aligned} \tag{75}$$

If $u_0(0)u_1(1) = 1$, then the problem is solvable if and only if (f_0, f_1) satisfies $u_0(0)f_0 + f_1 = 0$. In this case, there is a family of solutions given by

$$\phi(x) = V_1(1,1)f_1 w_1(x) + \lambda[w_0(x) + u_1(1)w_1(x)], \quad \lambda \in \mathbb{R}^1. \tag{76}$$

As a more interesting goal, we seek an expression for the solution given above by means of the reflecting Brownian motion $\{X_t^r, t \geq 0\}$ in $[0,1]$. By definition, the path of this process is reflected with respect to 0 or 1 whenever it reaches either point.

Let us set

$$\begin{aligned} e_q^r(t) &= \exp\left[\int_0^t q(X_s^r)\mathrm{d}s\right] \\ T_i &= \inf\{t > 0 : X_t^r = i\}, \quad i = 0, 1. \end{aligned} \tag{77}$$

Our extension of q implies that

$$u_0(0) = E^0[e_q^r(T_1)]; \quad u_1(1) = E^1[e_q^r(T_0)]. \tag{78}$$

Furthermore, for $x \in [0,1]$,

$$V_0(x,0) = \frac{1}{2}E^x\left\{\int_0^{T_1} e_q^r(t)L(\mathrm{d}t)\right\}$$

and (79)

$$V_1(x,1) = \frac{1}{2}E^x\left\{\int_0^{T_0} e_q^r(t)L(\mathrm{d}t)\right\},$$

where $L(\cdot)$ denotes the local time at the boundary $\{0,1\}$ for the reflecting Brownian motion, i.e.: $L(t) = 2L(t,0) + 2L(t,1)$, which explains the factor $\frac{1}{2}$ in (79).

Finally, we introduce a 'shuttle time' S as follows:

$$S = \inf\{t > T_0 \vee T_1 : X_t = 1 - X_{T_0 \vee T_1}\}.$$

Thus, S is the first time the path returns to the first boundary point which is hit after it has hit the other boundary point. Clearly, S is finite almost surely.

Theorem 9.31 *If $u_0(0)u_1(1) \neq 1$, then the solution of the Neumann's problem (71) is given by*

$$\phi(x) = \frac{1}{2[1 - u_0(0)u_1(1)]}E^x\left\{\int_0^S \exp\left[\int_0^t q(X_s^r)\mathrm{d}s\right] f(X_t^r)L(\mathrm{d}t)\right\}. \tag{80}$$

Proof We consider the case $\{T_0 < T_1\}$. On this set we have

$$\int_0^S = \int_0^{T_0} + \int_{T_0}^{T_1} + \int_{T_1}^{T_1 + T_0 \circ \theta_{T_1}}.$$

By repeated use of the strong Markov property and the fact that $L(\mathrm{d}t)$ is supported by the two points 0 and 1, we have

$$\begin{aligned} &E^x\left\{T_0 < T_1; \int_0^S e_q^r(t) f(X_t^r) L(\mathrm{d}t)\right\} \\ &= E^x[T_0 < T_1; e_q(T_0)]\left\{E^0\left[\int_0^{T_1} e_q^r(t) L(\mathrm{d}t)\right] f_0\right. \\ &\quad \left. + E^0[e_q^r(T_1)]E^1\left[\int_0^{T_0} e_q^r(t) L(\mathrm{d}t)\right] f_1\right\} \\ &= w_0(x)\{2V_0(0,0)f_0 + 2u_0(0)V_1(1,1)f_1\}. \end{aligned}$$

The last equation follows from (78) and (79).

Similarly, we have

$$E^x\left\{T_1 < T_0; \int_0^S e_q^r(t) f(X_t^r) L(\mathrm{d}t)\right\} = w_1(x)\{2V_1(1,1)f_1 + 2u_1(1)V_0(0,0)f_0\}.$$

Adding the two expectations above and using (70), we reduce (80) to (75). □

Notes on Chapter 9

This chapter is written largely independent of the rest of the book, except for a number of references.

The contents of Section 9.1 are based on Chung (1980), the first publication on the subject. There are some historical remarks there concerning the origin of the work.

Section 9.2 is an attempt to reconstruct the basic Sturm–Liouville theory from the probabilistic point of view. It may be regarded as a ramified gambler's ruin problem (with an exponentially accounted gain-and-loss). In elementary differential equations, the topic is called the 'two-point boundary value problem', in which the Green's function makes a somewhat mysterious appearance. These connections become clear in the probabilistic formulation; see the proof of Theorem 9.10 which is modeled on its high-dimensional analogue. For an exposition of the related material, see Chung (1982b), which contains another proof of Theorem 9.10 geared to the one-dimensional situation.

The notion of pseudo-gaugeability could have been introduced earlier in the book, but it was deemed dispensable in view of the Harnack inequality. In one dimension it arose from simple examples such as Example 4 in Section 9.3, communicated by K.B. Erickson who also alerted us to Theorem 9.17. It is possible to construct the complement ψ there by probabilistic considerations (Zhao, 1990b).

Theorem 9.21 was the answer to a problem posed by Pierre van Moerbeke concerning the 'ground state' of a problem in quantum physics. This was actually the first concrete result obtained by Chung (1980) which started his investigation of Schrödinger's equation by probabilistic methods. In retrospect, one may ask whether this result has to do with the time-reversibility of the process. This question was posed at the conference on time reversing in Santa Cruz, California, 1986. It has not yet been answered.

We mention that the condition (40) is the fundamental assumption made in Titchmarsh's classical volumes on the spectrum of the Schrödinger equation for $\mathbb{R}^1$, see Titchmarsh (1962). Any relevance of his purpose to the results here must be left to the appropriate experts to figure out.

The numerical example of α and β in Section 9.3 is due to Zhao.

For an entertaining anecdote concerning Proposition 9.25, see Chung (1985a). We are indebted to Philip Hartman for details of historical references to Proposition 9.25 by A. Kneser (1896), M. Bôcher (1900), H. Weyl (1909), and P. Hartman (1948). An analytic proof is contained in Hartman (1982, pages 350–358).

Proposition 9.28 is due to Chung (1983b); Theorem 9.29 is due to Zhao.

Sometimes, in the literature, the Dirichlet problem is referred to as the first boundary value problem, the C. Neumann problem is referred to as the second boundary value problem, while a mixture of both is referred to as the third boundary value problem. Once the normal derivative appears, a purely probabilistic formulation becomes a problem. Feller was conscious of this and called attention to it. This seems to have been forgotten apart from certain whisperings (rumours) about the Martin boundary. If we rely on the analytic equipment then the resulting development tends to merge with the usual operations of stochastic calculus, and its intuitive content begins to fade away. As it is not our intention to delve into the latter area we content ourselves with showing a little connection in the last section of the book, which is based on Chung and Zhao (1984). However, we also cite the existence of a gauge theorem for the Neumann problem (Chung and Hsu, 1986), and for the mixed (third) boundary value problem (Papanicolaou, 1990).

References

Aizenman, M., Simon, B. (1982): Brownian motion and Harnack's inequality for Schrödinger operators. Commun. Pure Appl. Math. **35**, 209–273.

Ancona, A. (1978): Principe de Harnack à la frontière et théorème de Fatou pour un opérateur elliptique dans un domaine Lipschitzien. Ann. Inst. Fourier **28**(4), 169–213.

Balkema, A.A., Chung, K.L. (1991): Paul Lévy's way to his local time. Seminar on Stochastic Processes (1990), Birkhäuser, Boston, 5–14.

Bañuelos, R. (1992): Lifetime and heat-kernel estimates in nonsmooth domains. Proc. Univ. Chicago Conf. on PDE with minimal smoothness, IMA Publications, vol. 42, 37–48.

Bass, R.F., Burdzy, K. (1990): A probabilistic proof of the boundary Harnack principle. Seminar on Stochastic Processes (1989), Birkhäuser, Boston, 1–16.

Boukricha, A., Hansen, W., Hueber, H. (1987): Continuous solutions of the generalized Schrödinger equation and perturbation of harmonic spaces. Expo. Math. **5**, 97–135.

Brossard, J. (1985): Noyeau de Poisson pour l'opérateur de Schrödinger. Duke Math. J. **52**, 199–210.

Chung, K.L. (1980): On stopped Feynman–Kac functional. Séminaire de Probabilités XIV (1978/79). (Lecture Notes in Math. 784) Springer, Berlin, 347–356.

Chung, K.L. (1981): Brownian motion on the line (I). J. Math. Res. Expo. **1**, 73–82.

Chung, K.L. (1982a): Lectures from Markov Processes to Brownian Motion. Springer, New York, 1982.

Chung, K.L. (1982b): Brownian motion on the line (II). J. Math. Res. Expo. **3**, 87–98.

Chung, K.L. (1983a): An inequality for boundary value problems. Seminar on Stochastic Processes (1982), Birkhäuser, Boston, 111–122.

Chung, K.L. (1983b): Properties of finite gauge with an application on local time. Probability and Math. Statistics, Uppsala, Sweden (1983), 16–23.

Chung, K.L. (1983/84): The gauge and conditional gauge theorem. Séminaire des probabilités XIX (1983/84), 496–503.

Chung, K.L. (1984a): Conditional gauges. Seminar on Stochastic Processes, (1983), Birkhäuser, Boston, 17–22.

Chung, K.L. (1984b): The lifetime of conditional Brownian motion in the plane. Ann. Inst. Henri Poincaré, Probab. Stat. **20**, 349–351.

Chung, K.L. (1984/85). Remark on the conditional gauge theorem. Séminaire des probabilités XX (1984/85), 423–425.

Chung, K.L. (1985a): A probability anecdote. J. Math. Res. Expo. **5**, 141–143.

Chung, K.L. (1985b): Probabilistic approach to boundary value problems for Schrödinger equation. Expo. Math. **3**, 175–178.

Chung, K.L. (1986a): Notes on the inhomogeneous Schrödinger equation. Seminar on Stochastic Processes (1984), Birkhäuser, Boston, 55–62.

Chung, K.L. (1986b): Doubly-Feller process with multiplicative functional. Seminar on Stochastic Processes (1985), Birkhäuser, Boston, 63–78.

Chung, K.L. (1987a): Green's function for a ball. Seminar on Stochastic Processes (1986), Birkhäuser, Boston, 1–13.

Chung, K.L. (1987b): Probability method in potential theory. Proc. Conf. Potential Theory (1987) (Lecture Notes in Math. 1344) Springer, Berlin, 42–54.
Chung, K.L. (1989): Gauge theorem for unbounded domains. Seminar on Stochastic Processes (1988), Birkhäuser, Boston, 87–98.
Chung, K.L. (1992): Greenian bounds for Markov processes. Potential Analysis **1**, 83–92.
Chung, K.L. (1995, 2001): Green, Brown, and Probability. World Scientific Publishing Co., Singapore.
Chung, K.L. (2000): A Course in Probability Theory, 3nd edn. Academic Press, New York.
Chung, K.L., Durrett, R., Zhao, Z. (1983): Extensions of domains with finite gauge. Math. Ann. **264**, 73–79.
Chung, K.L., Hsu, P. (1986): Gauge theorem for the Neumann problem. Seminar on Stochastic Processes (1984), Birkhäuser, Boston, 63–70.
Chung, K.L., Li, P. (1987): Comparison of probability and eigenvalue methods for the Schrödinger equation. Adv. Math., Suppl. Stud. **9**, 25–34.
Chung, K.L., Li, P., Williams, R.J. (1986): Comparison of probability and classical methods for the Schrödinger equation. Expo. Math. **4**, 271–278.
Chung, K.L., Rao, K.M. (1981): Feynman–Kac functional and the Schrodinger equation. Seminar on Stochastic Processes (1981), Birkhäuser, Boston, 1–29.
Chung, K.L., Rao, K.M. (1988): General gauge theorem for multiplicative functionals. Trans. Am. Math. Soc. **306**, 819–836.
Chung, K.L., Varadhan, S.R.S. (1980): Kac functional and Schrödinger equation. Studia Mathematica **LXVIII**, 249–260.
Chung, K.L., Williams, R.J. (1983): An Introduction to Stochastic Integration. Birkhäuser, Boston. Second edition 1990.
Chung, K.L., Zhao, Z. (1984): Some probabilistic results on Sturm–Liouville equation. J. Syst. Sci. Math. Sci. **4**(4), 320–324.
Courant, R., Hilbert, D. (1953): Methods of Mathematical Physics. Interscience, New York.
Cranston, M. (1985): Lifetime of conditional Brownian motion in Lipschitz domains. Z. Wahrsch. Verw. Gebiete **70**, 335–340.
Cranston, M., Fabes, E., Zhao, Z. (1988): Conditional gauge and potential theory for the Schrödinger operator. Trans. Am. Math. Soc. **307**, 174–194.
Cranston, R., McConnell, T.R. (1983): The lifetime of conditioned Brownian motion. Z. Wahrsch. Verw. Gebiete **65**, 1–11.
Cranston, M., Zhao, Z. (1987): Conditional transformation of drift formula and potential theory for $\frac{\Delta}{2} + v(\cdot)\nabla$. Commun. Math. Phys. **112**, 613–625.
Curtiss, J.H. (1978): Introduction to Functions of a Complex Variable. M. Dekker, New York.
Dahlberg, B. (1977): Estimates of harmonic measure, Arch. Ration. Mech. Anal. **65**, 275–288.
Doob, J.L. (1953): Stochastic Processes. Wiley and Sons, New York.
Doob, J.L. (1984): Classical Potential Theory and Its Probabilistic Counterpart. Springer, New York.
Dunford, N., Schwartz, J. (1958): Linear Operator, vol. I. Interscience, New York.
Dynkin, E.B. (1960). Theory of Markov Processes. Pergamon, Oxford, London.
Edmunds, D.E., Evans, W.D. (1987): Spectral Theory and Differential Operators. Clarendon Press, Oxford.
Falkner, N. (1983): Feynman–Kac functional and positive solutions of $\frac{1}{2}\Delta u + qu = 0$, Z. Wahrsch. Verw. Gebiete **65**, 19–34.
Falkner, N. (1987): Conditional Brownian motion in rapidly exhaustible domains. Ann. Probab. **15**, 1501–1504.
Freedman, D. (1971): Brownian Motion and Diffusion. Holden-Day, San Francisco.
Gesztesy, F., Zhao, Z. (1994): Domain perturbation, capacity, Brownian motion and ground states of Dirichlet Schrödinger operators. Math. Z. **215**, 143–150.

Gilbarg, D., Trudinger, N.S. (1977): Elliptic Partial Differential Equations of Second Order. Springer, New York.
Goelden, H.W. (1977). On non-degeneracy of the ground state of Schrödinger operators. Math. Z. **155**, 239–247.
Hartman, P. (1982): Ordinary Differential Equations, 2nd edn. Birkhäuser, Boston, 1982.
Herbst, I.W., Zhao, Z. (1989): Green's functions for the Schrödinger equation with short-range potentials. Duke Math. J. **59**, 475–519.
Hsu, P. (1985): Probabilistic approach to the Neumann problem. Commun. Pure Appl. Math. **38**, 445–472.
Hunt, G.A. (1956): Some theorems concerning Brownian motion. Trans. Am. Math. Soc. **81**, 294–319.
Hunt, R.R., Wheeden, R.L. (1968): On the boundary values of harmonic functions. Trans. Am. Math. Soc. **132**, 307–322.
Jerison, D.S., Kenig, C.E. (1982): Boundary behavior of harmonic functions in non-tangentially accessible domains. Adv. Math. **46**, 80–147.
Kac, M. (1951): On some connections between probability theory and differential and integral equations. Proc. Second Berkeley Symposium on Math. Stat. and Probability, University of California Press, Berkeley, 189–215.
Kac, M. (1959): Probability and Related Topics in Physical Sciences. Interscience, New York.
Kakutani, S. (1944): Two dimensional Brownian motion and harmonic functions. Proc. Imp. Acad. Tokyo **20**, 706–714.
Kakutani, S. (1945): Markov process and the Dirichlet problem. Proc. Japan. Acad. **21**, 227–233.
Kato, T. (1966): Perturbation Theory for Linear Operators. Springer, New York.
Kellogg, O.D. (1929): Foundations of Potential Theory. Springer, Berlin.
Khas'minskii, R. (1959): On positive solutions of the equation $Au+Vu = 0$. Theory Probab. Appl. **4**, 309–318.
Lévy, P. (1937): Théorie de l'Addition des Variables Aléatoires. Gauthier-Villars, Paris (2nd edn. 1954).
Lévy, P. (1948): Processus Stochastiques et Mouvement Brownien. Gauthier-Villars, Paris (2nd edn. 1965).
Lévy, P. (1970): Quelques Aspects de la Pensée d'un Mathématicien. Albert Blanchard, Paris.
Ma, Z., Zhao, Z. (1987): Truncated gauge and Schrödinger operator with both sign eigenvalues. Seminar on Stochastic Processes (1986), Birkhäuser, Boston, 149–153.
Mancino, M. (1989): Dal moto browniano all'equazione di Schrödinger. Tesi di laurea, Università degli Studi di Pisa.
Martin, R.S. (1941): Minimal positive harmonic functions. Trans. Am. Math. Soc. **49**, 137–172.
McConnell, T.R. (1990): A conformal inequality related to the conditional gauge theorem. Trans. Am. Math. Soc. **318**(2), 721–733.
Paley R., Wiener, N. (1934): Fourier Transforms in the Complex Domain. AMS Colloquium Publications, New York.
Papanicolaou, V.G. (1990): The probabilistic solution of the third boundary value problem for second order elliptic equations. Probab. Theory Relat. Fields **87**, 27–77.
Pop-Stojanovic, Z., Rao, M. (1990): Continuity of solutions of Schrödinger equation. Seminar on Stochastic Processes (1989), Birkhäuser, Boston, 193–195.
Port, S., Stone, C. (1978): Brownian Motion and Classical Potential Theory. Academic Press, New York.
Simon, B. (1979): Functional Integration and Quantum Physics. Academic Press, New York.
Simon, B. (1982): Schrödinger semigroup. Bull. Am. Math. Soc. **7**, 447–526.

Sturm, K.T. (1987): On the Dirichlet–Poisson problem for Schrödinger operators. C.R. Math. Acad. Sci., Soc. R. Can. **IX**, 149–154
Sturm, K.T. (1989): Störung von Hunt-Prozessen durch signierte additive Funktionale. Thesis, Erlangen.
Sturm, K.T. (1991): Gauge theorems for resolvents with application to Markov processes. Probab. Theory Relat. Fields **89**, 387–406.
Titchmarsh, E.C. (1962): Eigenfunction Expansions, Part 1, 2nd edn. Clarendon Press, Oxford.
Wermer, J. (1974): Potential Theory. (Lecture Notes in Math. 408) Springer, Berlin.
Widman, K.-O. (1967): Inequalities for the Green function and boundary continuity of the gradient of solutions of elliptic differential equations. Math. Scand. **21**, 13–67.
Wiener, N. (1923): Differential space. J. Math. Phys. MIT **2**, 131–174.
Wiener, N. (1924): The Dirichlet problem. J. Math. Phys. MIT **3**, 127–146.
Williams, R.J. (1985): A Feynman–Kac gauge for solvability of the Schrödinger equation. Adv. Appl. Math. **6**, 149–154.
Wu, J.-M.G. (1978): Comparisons of kernel functions, boundary Harnack principle and relative Fatou theorem on Lipschitz domains. Ann. Inst. Fourier **28**, 147–167.
Yosida, K. (1980): Functional Analysis, 6th edn. Springer, New York.
Zhao, Z. (1982): The local Feynman–Kac semigroup. J. Syst. Sci. Math. Sci. **2**(4), 314–326.
Zhao, Z. (1983): Conditional gauge with unbounded potential. Z. Wahrsch. Verw. Gebiete **65**, 13–18.
Zhao, Z. (1984): Uniform boundedness of conditional gauge and Schrödinger equations. Commun. Math. Phys. **93**, 19–31.
Zhao, Z. (1986): Green function for Schrödinger operator and conditioned Feynman–Kac gauge. J. Math. Anal. Appl. **116**, 309–334.
Zhao, Z. (1988): Green functions and conditioned gauge theorem for a 2-dimensional domain. Seminar on Stochastic Processes (1987), Birkhäuser, Boston, 283–294.
Zhao, Z. (1990a): Gaugeability for unbounded domains. Seminar on Stochastic Processes (1989), Birkhäuser, Boston, 207–214.
Zhao, Z. (1990b): An equivalence theorem for Schrödinger operators and its applications. In 'Diffusion Processes and Related Problems in Analysis, vol. 1'. Birkhäuser, Boston, 245–260.
Zhao, Z. (1990c): Subcriticality, positivity and gaugeability of the Schrödinger operator. Bull. Am. Math. Soc. **23**, 513–517.
Zhao, Z. (1991): A probabilistic principle and generalized Schrödinger perturbation. J. Funct. Anal. **101**, 162–176.

Index

Grundlehren der mathematischen Wissenschaften

A Series of Comprehensive Studies in Mathematics

A Selection

221. Mumford: Algebraic Geometry I: Complex Projective Varieties
222. Lang: Introduction to Modular Forms
223. Bergh/Löfström: Interpolation Spaces. An Introduction
224. Gilbarg/Trudinger: Elliptic Partial Differential Equations of Second Order
225. Schütte: Proof Theory
226. Karoubi: K-Theory. An Introduction
227. Grauert/Remmert: Theorie der Steinschen Räume
228. Segal/Kunze: Integrals and Operators
229. Hasse: Number Theory
230. Klingenberg: Lectures on Closed Geodesics
231. Lang: Elliptic Curves. Diophantine Analysis
232. Gihman/Skorohod: The Theory of Stochastic Processes III
233. Stroock/Varadhan: Multidimensional Diffusion Processes
234. Aigner: Combinatorial Theory
235. Dynkin/Yushkevich: Controlled Markov Processes
236. Grauert/Remmert: Theory of Stein Spaces
237. Köthe: Topological Vector Spaces II
238. Graham/McGehee: Essays in Commutative Harmonic Analysis
239. Elliott: Probabilistic Number Theory I
240. Elliott: Probabilistic Number Theory II
241. Rudin: Function Theory in the Unit Ball of C^n
242. Huppert/Blackburn: Finite Groups II
243. Huppert/Blackburn: Finite Groups III
244. Kubert/Lang: Modular Units
245. Cornfeld/Fomin/Sinai: Ergodic Theory
246. Naimark/Stern: Theory of Group Representations
247. Suzuki: Group Theory I
248. Suzuki: Group Theory II
249. Chung: Lectures from Markov Processes to Brownian Motion
250. Arnold: Geometrical Methods in the Theory of Ordinary Differential Equations
251. Chow/Hale: Methods of Bifurcation Theory
252. Aubin: Nonlinear Analysis on Manifolds. Monge-Ampère Equations
253. Dwork: Lectures on ρ-adic Differential Equations
254. Freitag: Siegelsche Modulfunktionen
255. Lang: Complex Multiplication
256. Hörmander: The Analysis of Linear Partial Differential Operators I
257. Hörmander: The Analysis of Linear Partial Differential Operators II
258. Smoller: Shock Waves and Reaction-Diffusion Equations
259. Duren: Univalent Functions
260. Freidlin/Wentzell: Random Perturbations of Dynamical Systems
261. Bosch/Güntzer/Remmert: Non Archimedian Analysis – A System Approach to Rigid Analytic Geometry
262. Doob: Classical Potential Theory and Its Probabilistic Counterpart
263. Krasnosel'skiĭ/Zabreĭko: Geometrical Methods of Nonlinear Analysis
264. Aubin/Cellina: Differential Inclusions
265. Grauert/Remmert: Coherent Analytic Sheaves
266. de Rham: Differentiable Manifolds
267. Arbarello/Cornalba/Griffiths/Harris: Geometry of Algebraic Curves, Vol. I
268. Arbarello/Cornalba/Griffiths/Harris: Geometry of Algebraic Curves, Vol. II
269. Schapira: Microdifferential Systems in the Complex Domain
270. Scharlau: Quadratic and Hermitian Forms

271. Ellis: Entropy, Large Deviations, and Statistical Mechanics
272. Elliott: Arithmetic Functions and Integer Products
273. Nikol'skiĭ: Treatise on the Shift Operator
274. Hörmander: The Analysis of Linear Partial Differential Operators III
275. Hörmander: The Analysis of Linear Partial Differential Operators IV
276. Liggett: Interacting Particle Systems
277. Fulton/Lang: Riemann-Roch Algebra
278. Barr/Wells: Toposes, Triples and Theories
279. Bishop/Bridges: Constructive Analysis
280. Neukirch: Class Field Theory
281. Chandrasekharan: Elliptic Functions
282. Lelong/Gruman: Entire Functions of Several Complex Variables
283. Kodaira: Complex Manifolds and Deformation of Complex Structures
284. Finn: Equilibrium Capillary Surfaces
285. Burago/Zalgaller: Geometric Inequalities
286. Andrianaov: Quadratic Forms and Hecke Operators
287. Maskit: Kleinian Groups
288. Jacod/Shiryaev: Limit Theorems for Stochastic Processes
289. Manin: Gauge Field Theory and Complex Geometry
290. Conway/Sloane: Sphere Packings, Lattices and Groups
291. Hahn/O'Meara: The Classical Groups and K-Theory
292. Kashiwara/Schapira: Sheaves on Manifolds
293. Revuz/Yor: Continuous Martingales and Brownian Motion
294. Knus: Quadratic and Hermitian Forms over Rings
295. Dierkes/Hildebrandt/Küster/Wohlrab: Minimal Surfaces I
296. Dierkes/Hildebrandt/Küster/Wohlrab: Minimal Surfaces II
297. Pastur/Figotin: Spectra of Random and Almost-Periodic Operators
298. Berline/Getzler/Vergne: Heat Kernels and Dirac Operators
299. Pommerenke: Boundary Behaviour of Conformal Maps
300. Orlik/Terao: Arrangements of Hyperplanes
301. Loday: Cyclic Homology
302. Lange/Birkenhake: Complex Abelian Varieties
303. DeVore/Lorentz: Constructive Approximation
304. Lorentz/v. Golitschek/Makovoz: Construcitve Approximation. Advanced Problems
305. Hiriart-Urruty/Lemaréchal: Convex Analysis and Minimization Algorithms I. Fundamentals
306. Hiriart-Urruty/Lemaréchal: Convex Analysis and Minimization Algorithms II. Advanced Theory and Bundle Methods
307. Schwarz: Quantum Field Theory and Topology
308. Schwarz: Topology for Physicists
309. Adem/Milgram: Cohomology of Finite Groups
310. Giaquinta/Hildebrandt: Calculus of Variations I: The Lagrangian Formalism
311. Giaquinta/Hildebrandt: Calculus of Variations II: The Hamiltonian Formalism
312. Chung/Zhao: From Brownian Motion to Schrödinger's Equation
313. Malliavin: Stochastic Analysis
314. Adams/Hedberg: Function Spaces and Potential Theory
315. Bürgisser/Clausen/Shokrollahi: Algebraic Complexity Theory
316. Saff/Totik: Logarithmic Potentials with External Fields
317. Rockafellar/Wets: Variational Analysis
318. Kobayashi: Hyperbolic Complex Spaces
319. Bridson/Haefliger: Metric Spaces of Non-Positive Curvature
320. Kipnis/Landim: Scaling Limits of Interacting Particle Systems
321. Grimmett: Percolation
322. Neukirch: Algebraic Number Theory
323. Neukirch/Schmidt/Wingberg: Cohomology of Number Fields
324. Liggett: Stochastic Interacting Systems: Contact, Voter and Exclusion Processes
325. Dafermos: Hyperbolic Conservation Laws in Continuum Physics
326. Waldschmidt: Diophantine Approximation on Linear Algebraic Groups

MIX
Papier aus verantwortungsvollen Quellen
Paper from responsible sources
FSC® C105338

If you have any concerns about our products,
you can contact us on
ProductSafety@springernature.com

In case Publisher is established outside the EU,
the EU authorized representative is:
Springer Nature Customer Service Center GmbH
Europaplatz 3, 69115 Heidelberg, Germany

Printed by Libri Plureos GmbH
in Hamburg, Germany